MODELS AND ALGORITHMS FOR BIOMOLECULES AND MOLECULAR NETWORKS

IEEE Press
445 Hoes Lane
Piscataway, NJ 08854

MODELS AND ALGORITHMS FOR BIOMOLECULES AND MOLECULAR NETWORKS

BHASKAR DASGUPTA

Department of Computer Science
University of Illinois at Chicago
Chicago, IL

JIE LIANG

Department of Bioengineering
University of Illinois at Chicago
Chicago, IL

IEEE Press Series in Biomedical Engineering
Metin Akay, *Series Editor*

IEEE
IEEE PRESS

WILEY

Library of Congress Cataloging-in-Publication Data is available.

ISBN: 978-0-470-60193-8

Printed in the United States of America

10 9 8 7 6 5 4 3 2 1

Dedicated to our spouses and students

CONTENTS

LIST OF FIGURES

LIST OF TABLES

FOREWORD

The subjects of this book are biomolecules and biomolecular networks. The first part of the book will cover surface and volume representation of the structures of biomolecules based on the idealized ball model. The underlying geometric constructs as well as their computation will then be discussed. This will be followed by the chapter on constructing effective scoring functions in different functional forms using either the statistical approach or the optimization approach, with the goal of identifying native-like protein structures or protein–protein interfaces, as well as constructing a general fitness landscape for protein design. The topic of sampling and estimation that can be used to generate biomolecular structures and to estimate their evolutionary patterns are then discussed, with equal emphasis on the Metropolis Monte Carlo (or Markov Chain Monte Carlo) approach and the chain growth (or sequential Monte Carlo) approach. This is followed by a chapter covering the topic of stochastic networks formed by interacting biomolecules and the framework of discrete chemical master equations, as well as computational methods for direct numerical computation and for sampling reaction trajectories of the probabilistic landscape of these networks.

The second part of the book will cover interaction networks of biomolecules. We will discuss stochastic models for networks with small copy numbers of molecular species, such as those arising in genetic circuits, protein synthesis, and transcription binding, and algorithms of computing the properties of stochastic molecular networks. We will then cover signal transduction networks that arise, for example, in complex interactions between the numerous constituents such as DNAs, RNAs, proteins, and small molecules in a complex biochemical system such as a cell. We will also discuss the experimental protocols and algorithmic methodologies necessary to synthesize these networks. Of special interest will be the synthesis of these networks from double-causal experimental evidences, as well as methods for reverse engineering of such networks based on suitable experimental protocols.

This book is written for graduate students, upper division undergraduate students, engineers, and scientists in academia and industries from a variety of disciplines, such as bioengineering, biophysics, electric engineering, chemical engineering, mathematics, biology, and computer science. It may also serve as a useful reference for researchers in these disciplines, including professional engineers and professional statisticians as well as practicing scientists in the pharmaceutical industry and the biotechnology industry. This book may be used as a monograph for learning important research topics and for finding algorithms and solutions to problems encountered in research and in practice.

ACKNOWLEDGMENTS

For Bhaskar DasGupta, this book could not have been written without collaboration with a large number of collaborators from different research areas, and he thanks all of them for their involvements. Special thanks go to his colleagues Réka Albert, Piotr Berman, and Eduardo Sontag for their enormous patience and contribution during collaborations. He would like to thank individually all the students and postdoctoral fellows involved in these projects (Anthony Gitter, Gamze Gürsoy, Rashmi Hegde, Gowri Sangeetha Sivanathan, Pradyut Pal, Paola Vera-Licona, Riccardo Dondi, Sema Kachalo, Ranran Zhang, Yi Zhang, Kelly Westbrooks, and German Enciso). Bhaskar DasGupta thankfully acknowledges generous financial support from the National Science Foundation through grants DBI-1062328, IIS-1064681, IIS-0346973, DBI-0543365, IIS-0610244, CCR-9800086, CNS-0206795, and CCF-0208749, along with generous support from the DIMACS Center of Rutgers University during his Sabbatical leave through their special focus on computational and mathematical epidemiology. Last, but not least, Bhaskar DasGupta thanks his wife Paramita Bandopadhyay for her help, understanding, and cooperation while the book was being written.

For Jie Liang, the material in this book draws on research collaborations with many colleagues, to whom he is grateful. Special thanks go to Rong Chen, Ken Dill, Linda Kenney, Herbert Edelsbrunner, and Shankar Subramaniam, with whom he has worked when embarking on new research directions. He also has the good fortunate of working with a group of talented students and postdoctoral researchers, who have contributed to research projects, some of which are reflected in material described in this book: Larisa Adamian, Andrew Binkowski, Youfang Cao, Joseph Dundas, Gamze Gürsoy, Changyu Hu, David Jiminez-Morales, Ronald Jauckups, Jr., Sema Kachalo, Xiang Li, Yingzi Li, Meishan Lin, Hsiao-Mei Lu, Chih-Hao Lu,

Hammad Naveed, Zheng Ouyang, Arun Setty, Nathan Stitziel, Ke Tang, Anna Terebus, Wei Tian, Jeffrey Tseng, Yaron Turpaz, Yun Xu, Jian Zhang, and Jinfeng Zhang. Jie Liang also thanks students at the University of Illinois at Chicago who have taken the bioinformatics courses he taught. Jie Liang thanks Xiang Li for co-writing the material for the chapter on scoring function, and he also thanks Ke Tang for help in preparing the figures. Jie Liang also acknowledges generous research support from the National Institutes of Health (GM079804, GM086145, GM68958, and GM081682), the National Science Foundation (DBI-0078270, DBI-0133856, DBI-0646035, DMS-0800257, and DBI-1062328), the Office of Naval Research (N000140310329 and N00014-06), the Whittaker Foundation, the Chicago Biomedical Consortium, and the Petroleum Research Fund (PRF#35616-G7). Finally, he wishes to thank his wife Re-Jin Guo for her understanding and patience during the period when the book was being written.

B. D. and J. L.

1

GEOMETRIC MODELS OF PROTEIN STRUCTURE AND FUNCTION PREDICTION

1.1 INTRODUCTION

Three-dimensional atomic structures of protein molecules provide rich information for understanding how these working molecules of a cell carry out their biological functions. With the amount of solved protein structures rapidly accumulating, computation of geometric properties of protein structure becomes an indispensable component in studies of modern biochemistry and molecular biology. Before we discuss methods for computing the geometry of protein molecules, we first briefly describe how protein structures are obtained experimentally.

There are primarily three experimental techniques for obtaining protein structures: X-ray crystallography, solution nuclear magnetic resonance (NMR), and recently freeze-sample electron microscopy (cryo-EM). In X-ray crystallography, the diffraction patterns of X-ray irradiation of a high-quality crystal of the protein molecule are measured. Since the diffraction is due to the scattering of X-rays by the electrons of the molecules in the crystal, the position, the intensity, and the phase of each recorded diffraction spot provide information for the reconstruction of an *electron density map* of atoms in the protein molecule. Based on independent information of the amino acid sequence, a model of the protein conformation is then derived by fitting model conformations of residues to the electron density map. An iterative process called *refinement* is then applied to improve the quality of the fit of the electron density map. The final model of the protein conformation consists of the coordinates of each of the non-hydrogen atoms [46].

Models and Algorithms for Biomolecules and Molecular Networks, First Edition. Bhaskar DasGupta and Jie Liang.

The solution NMR technique for solving protein structure is based on measuring the tumbling and vibrating motion of the molecule in solution. By assessing the chemical shifts of atomic nuclei with spins due to interactions with other atoms in the vicinity, a set of estimated distances between specific pairs of atoms can be derived from NOSEY spectra. When a large number of such distances are obtained, one can derive a set of conformations of the protein molecule, each being consistent with all of the distance constraints [10]. Although determining conformations from either X-ray diffraction patterns or NMR spectra is equivalent to solving an ill-posed inverse problem, a technique such as Bayesian Markov chain Monte Carlo with parallel tempering has been shown to be effective in obtaining protein structures from NMR spectra [52].

1.2 THEORY AND MODEL

1.2.1 Idealized Ball Model

The shape of a protein molecule is complex. The chemical properties of atoms in a molecule are determined by their electron charge distribution. It is this distribution that generates the scattering patterns of the X-ray diffraction. Chemical bonds between atoms lead to transfer of electronic charges from one atom to another, and the resulting isosurfaces of the electron density distribution depend not only on the location of individual nuclei but also on interactions between atoms. This results in an overall complicated isosurface of electron density [2].

The geometric model of macromolecule amenable to convenient computation is an idealized model, where the shapes of atoms are approximated by three-dimensional balls. The shape of a protein or a DNA molecule consisting of many atoms is then the space-filling shape taken by a set of atom balls. This model is often called the *interlocking hard-sphere model*, the *fused ball model*, the *space-filling model* [32,47,49,51], or the *union of ball* model [12]. In this model, details in the distribution of electron density (e.g., the differences between regions of covalent bonds and noncovalent bonds) are ignored. This idealization is quite reasonable, as it reflects the fact that the electron density reaches maximum at a nucleus and its magnitude decays almost spherically away from the point of the nucleus. Despite possible inaccuracy, this idealized model has found wide acceptance, because it enables quantitative measurement of important geometric properties (such as area and volume) of molecules. Insights gained from these measurements correlate well with experimental observations [9,21,32,48–50].

In this idealization, the shape of each atom is that of a ball, and its size parameter is the ball radius or atom radius. There are many possible choices for the parameter set of atomic radii [47,56]. Frequently, atomic radii are assigned the values of their van der Waals radii [7]. Among all these atoms, the hydrogen atom has the smallest mass and has a much smaller radius than those of other atoms.

For simplification, the model of *united atom* is often employed to approximate the union of a heavy atom and the hydrogen atoms connected by a covalent bond.

In this case, the radius of the heavy atom is increased to approximate the size of the union of the two atoms. This practice significantly reduces the total number of atom balls in the molecule. However, this approach has been questioned for possible inadequacy [60].

The mathematical model of this idealized model is that of the union of balls [12]. For a molecule M of n atoms, the ith atom is modeled as a ball b_i, whose center is located at $\boldsymbol{z}_i \in \mathbb{R}^3$, and the radius of this ball is $r_i \in \mathbb{R}$, namely, we have $b_i \equiv \{\boldsymbol{x} | \boldsymbol{x} \in \mathbb{R}^3, ||\boldsymbol{x} - \boldsymbol{z}_i|| \leq r_i\}$ parameterized by $(\boldsymbol{z}_i, r_i)$. The molecule M is formed by the union of a finite number n of such balls defining the set $\mathcal{B}$:

$$M = \bigcup \mathcal{B} = \bigcup_{i=1}^{n} \{b_i\}.$$

It creates a space-filling body corresponding to the union of the excluded volumes $\text{vol}\left(\bigcup_{i=1}^{n} \{b_i\}\right)$ [12]. When the atoms are assigned the van der Waals radii, the boundary surface $\partial \bigcup \mathcal{B}$ of the union of balls is called the *van der Waals* surface.

1.2.2 Surface Models of Proteins

Protein folds into native three-dimensional shape to carry out its biological functional roles. The interactions of a protein molecule with other molecules (such as ligand, substrate, or other protein) determine its functional roles. Such interactions occur physically on the surfaces of the protein molecule.

The importance of the protein surface was recognized very early on. Lee and Richards developed the widely used *solvent accessible surface* (SA) model, which is also often called the *Lee–Richards surface* model [32]. Intuitively, this surface is obtained by rolling a ball of radius r_s everywhere along the van der Waals surface of the molecule. The center of the solvent ball will then sweep out the solvent accessible surface. Equivalently, the solvent accessible surface can be viewed as the boundary surface $\partial \bigcup \mathcal{B}_{r_s}$ of the union of a set of inflated balls $\mathcal{B}_{r_s}$, where each ball takes the position of an atom, but with an inflated radius $r_i + r_s$ (Fig. 1.1a).

The solvent accessible surface in general has many sharp crevices and sharp corners. In hope of obtaining a smoother surface, one can take the surface swept out by the front instead of the center of the solvent ball. This surface is the *molecular surface* (MS model), which is also often called the *Connolly's surface* after Michael Connolly, who developed the first algorithm for computing molecular surface [9]. Both solvent accessible surface and molecular surface are formed by elementary pieces of simpler shape.

Elementary Pieces. For the solvent accessible surface model, the boundary surface of a molecule consists of three types of elements: the convex spherical surface pieces, arcs or curved line segments (possibly a full circle) formed by two intersecting spheres, and a vertex that is the intersection point of three atom spheres. The whole boundary surface of the molecules can be thought of as a surface formed by stitching these elements together.

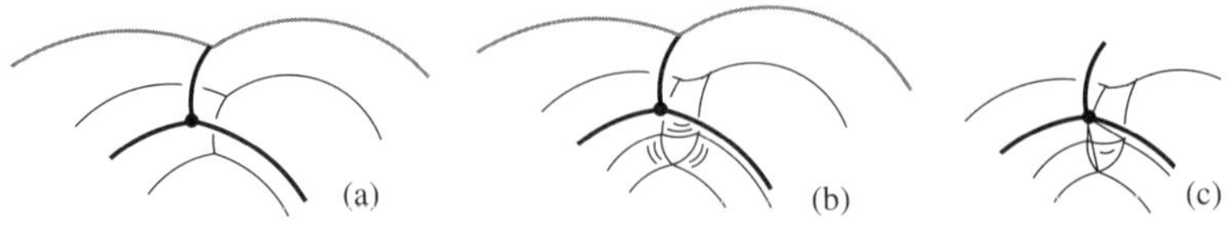

FIGURE 1.1 Geometric models of protein surfaces. (**a**) The solvent accessible surface (SA surface) is shown in the front. The van der Waals surface (beneath the SA surface) can be regarded as a shrunken version of the SA surface by reducing all atomic radii uniformly by the amount of the radius of the solvent probe $r_s = 1.4\text{Å}$. The elementary pieces of the solvent accessible surface are the three convex spherical surface pieces, the three arcs, and the vertex where the three arcs meet. (**b**) The molecular surface (MS, beneath the SA surface) also has three types of elementary pieces: the convex spherical pieces, which are shrunken version of the corresponding pieces in the solvent accessible surface, the concave toroidal pieces, and concave spherical surface. The latter two are also called the re-entrant surface. (**c**) The toroidal surface pieces in the molecular surface correspond to the arcs in the solvent accessible surface, and the concave spherical surface to the vertex. The set of elements in one surface can be continuously deformed to the set of elements in the other surface.

Similarly, the molecular surface swept out by the front of the solvent ball can also be thought of as being formed by elementary surface pieces. In this case, they are the convex spherical surface pieces, the toroidal surface pieces, and the concave or inverse spherical surface pieces (Fig. 1.1b). The latter two types of surface pieces are often called the ‘‘re-entrant surfaces’’ [9,49].

The surface elements of the solvent accessible surface and the molecular surface are closely related. Imagine a process where atom balls are shrunk or expanded. The vertices in solvent accessible surface becomes the concave spherical surface pieces, the arcs become the toroidal surfaces, and the convex surface pieces become smaller convex surface pieces (Fig. 1.1c). Because of this mapping, these two types of surfaces are combinatorially equivalent and have similar topological properties; that is, they are homotopy equivalent.

However, the SA surface and the MS surface differ in their metric measurement. In concave regions of a molecule, often the front of the solvent ball can sweep out a larger volume than the center of the solvent ball. A void of size close to zero in the solvent accessible surface model will correspond to a void of the size of a solvent ball ($4\pi r_s^3/3$). It is therefore important to distinguish these two types of measurement when interpreting the results of volume calculations of protein molecules. The intrinsic structures of these fundamental elementary pieces are closely related to several geometric constructs we describe below.

1.2.3 Geometric Constructs

Voronoi Diagram. Voronoi diagram (Fig. 1.2a), also known as Voronoi tessellation, is a geometric construct that has been used for analyzing protein packing in the early days of protein crystallography [18,20,47]. For two-dimensional Voronoi diagram, we consider the following analogy. Imagine a vast forest containing a number of

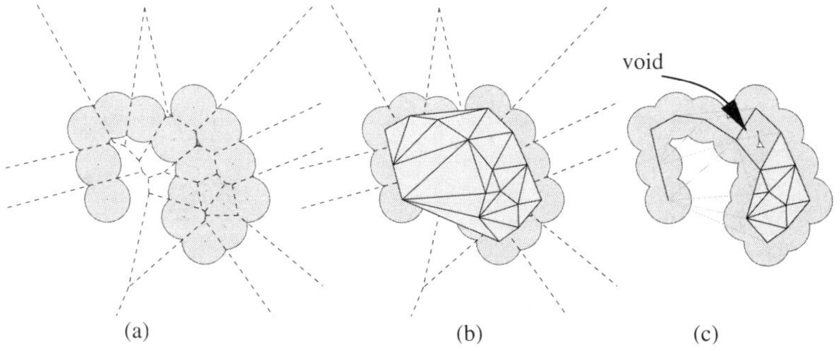

FIGURE 1.2 Geometry of a simplified two-dimensional model molecule to illustrate the geometric constructs and the procedure mapping the Voronoi diagram to the Delaunay triangulation. (**a**) The molecule is formed by the union of atom disks of uniform size. Voronoi diagram is in dashed lines. (**b**) The shape enclosed by the boundary polygon is the *convex hull*. It is tessellated by the *Delaunay triangulation*. (**c**) The alpha shape of the molecule is formed by removing those Delaunay edges and triangles whose corresponding Voronoi edges and Voronoi vertices do not intersect with the body of the molecule. A molecular void is represented in the alpha shape by two empty triangles.

fire observation towers. Each fire ranger is responsible for putting out any fire closer to his/her tower than to any other tower. The set of all trees for which a ranger is responsible constitutes the Voronoi cell associated with his/her tower, and the map of ranger responsibilities, with towers and boundaries marked, constitutes the Voronoi diagram.

We formalize this for three-dimensional space. Consider the point set S of atom centers in three-dimensional space $\mathbb{R}^3$. The *Voronoi region* or *Voronoi cell* V_i of an atom b_i with atom center $\boldsymbol{z}_i \in \mathbb{R}^3$ is the set of all points that are at least as close to $\boldsymbol{z}_i$ as to any other atom centers in $\mathcal{S}$:

$$V_i = \{\boldsymbol{x} \in \mathbb{R}^3 | ||\boldsymbol{x} - \boldsymbol{z}_i|| \leq ||\boldsymbol{x} - \boldsymbol{z}_j||, \boldsymbol{z}_j \in \mathcal{S}\}.$$

We can have an alternative view of the Voronoi cell of an atom b_i. Consider the distance relationship between atom center $\boldsymbol{z}_i$ and the atom center $\boldsymbol{z}_k$ of another atom b_k. The plane bisecting the line segment connecting points $\boldsymbol{z}_i$ and $\boldsymbol{z}_k$ divides the full $\mathbb{R}^3$ space into two half-spaces, where points in one half-space is closer to $\boldsymbol{z}_i$ than to $\boldsymbol{z}_k$, and points in the other allspice is closer to $\boldsymbol{z}_k$ than to $\boldsymbol{z}_i$. If we repeat this process and take $\boldsymbol{z}_k$ in turn from the set of all atom centers other than $\boldsymbol{z}_i$, we will have a number of half-spaces where points are closer to $\boldsymbol{z}_i$ than to each of the atom center $\boldsymbol{z}_k$. The Voronoi region V_i is then the common intersections of these half-spaces, which is convex (see exercises). When we consider atoms of different radii, we replace the Euclidean distance $||\boldsymbol{x} - \boldsymbol{z}_i||$ with the power distance defined as $\pi_i(\boldsymbol{x}) \equiv ||\boldsymbol{x} - \boldsymbol{z}_i||^2 - r_i^2$.

Delaunay Tetrahedrization. Delaunay triangulation in $\mathbb{R}^2$ or Delaunay tetrahedrization in $\mathbb{R}^3$ is a geometric construct that is closely related to the Voronoi diagram

(Fig. 1.2b). In general, it uniquely tessellates or tile up the space of the *convex hull* of the atom centers in $\mathbb{R}^3$ with tetrahedra. Convex hull for a point set is the smallest convex body that contains the point set[1]. The Delaunay tetrahedrization of a molecule can be obtained from the Voronoi diagram. Consider that the Delaunay tetrahedrization is formed by gluing four types of primitive elements together: vertices, edges, triangles, and tetrahedra. Here vertices are just the atom centers. We obtain a Delaunay edge by connecting atom centers z_i and z_j if and only if the Voronoi regions V_i and V_j have a common intersection, which is a planar piece that may be either bounded or extended to infinity. We obtain a Delaunay triangle connecting atom centers z_i, z_j, and z_k if the common intersection of Voronoi regions V_i, V_j and V_k exists, which is either a line segment, a half-line, or a line in the Voronoi diagram. We obtain a Delaunay tetrahedra connecting atom centers z_i, z_j, z_k, and z_l if and only if the Voronoi regions V_i, V_j, V_k, and V_l intersect at a point.

1.2.4 Topological Structures

Delaunay Complex. The structures in both Voronoi diagram and Delaunay tetrahedrization are better described with concepts from algebraic topology. We focus on the intersection relationship in the Voronoi diagram and introduce concepts formalizing the primitive elements. In $\mathbb{R}^3$, between two to four Voronoi regions may have common intersections. We use *simplices* of various dimensions to record these intersection or overlap relationships. We have vertices σ_0 as 0-simplices, edges σ_1 as 1-simplices, triangles σ_2 as 2-simplices, and tetrahedra σ_3 as 3-simplices. Each of the Voronoi plane, Voronoi edge, and Voronoi vertices corresponds to a 1-simplex (Delaunay edge), 2-simplex (Delaunay triangle), and 3-simplex (Delaunay tetrahedron), respectively. If we use 0-simplices to represent the Voronoi cells and add them to the simplices induced by the intersection relationship, we can think of the Delaunay tetrahedrization as the structure obtained by "gluing" these simplices properly together. Formally, these simplices form a *simplicial complex* $\mathcal{K}$:

$$\mathcal{K} = \left\{ \sigma_{|I|-1} \Big| \bigcap_{i \in I} V_i \neq \emptyset \right\},$$

where I is an index set for the vertices representing atoms whose Voronoi cells overlap, and $|I| - 1$ is the dimension of the simplex.

Alpha Shape and Protein Surfaces. Imagine we can turn a knob to increase or decrease the size of all atoms simultaneously. We can then have a model of growing

[1]For a two-dimensional toy molecule, we can imagine that we put nails at the locations of the atom centers and tightly wrap a rubber band around these nails. The rubber band will trace out a polygon. This polygon and the region enclosed within is the convex hull of the set of points corresponding to the atom centers. Similarly, imagine if we can tightly wrap a tin-foil around a set of points in three-dimensional space, the resulting convex body formed by the tin-foil and space enclosed within is the convex hull of this set of points in $\mathbb{R}^3$.

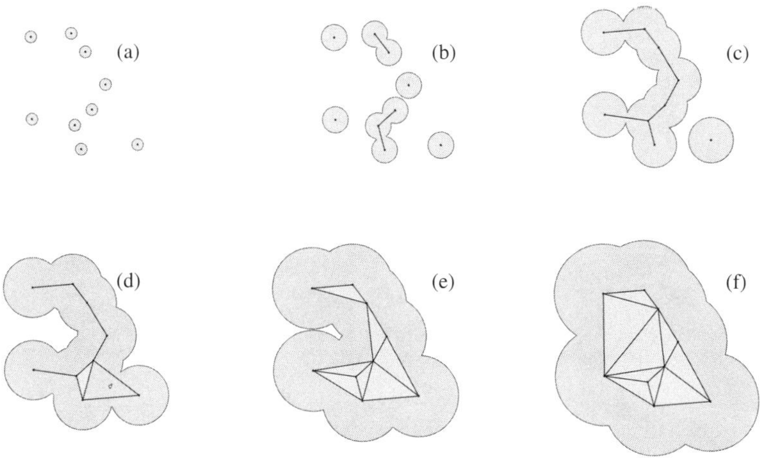

FIGURE 1.3 The family of alpha shapes or dual simplicial complexes for a two-dimensional toy molecule. (**a**) We collect simplices from the Delaunay triangulation as atoms grow by increasing the α value. At the beginning as α grows from $-\infty$, atoms are in isolation and we only have vertices in the alpha shape. (**b** and **c**) When α is increased such that some atom pairs start to intersect, we collect the corresponding Delaunay edges. (**d**) When three atoms intersect as α increases, we collect the corresponding Delaunay triangles. When $\alpha = 0$, the collection of vertices, edges, and triangles form the dual simplicial complex $\mathcal{K}_0$, which reflects the topological structure of the protein molecule. (**e**) More edges and triangles from the Delaunay triangulation are now collected as atoms continue to grow. (**f**) Finally, all vertices, edges, and triangles are now collected as atoms are grown to large enough size. We get back the full original Delaunay complex.

balls and obtain further information from the Delaunay complex about the shape of a protein structure. Formally, we use a parameter $\alpha \in \mathbb{R}$ to control the size of the atom balls. For an atom ball b_i of radius r_i, we modified its radius r_i at a particular α value to $r_i(\alpha) = (r_i^2 + \alpha)^{1/2}$. When $-r_i < \alpha < 0$, the size of an atom is shrunk. The atom could even disappear if $\alpha < 0$ and $|\alpha| > r_i$. With this construction of α, the weighted Voronoi diagram is invariant with regard to α (see exercises). We start to collect the simplices at different α values as we increase α from $-\infty$ to $+\infty$ (see Fig. 1.3 for a two-dimensional example). At the beginning, we only have vertices. When α is increased such that two atoms are close enough to intersect, we collect the corresponding Delaunay edge that connects these two atom centers. When three atoms intersect, we collect the corresponding Delaunay triangle spanning these three atom centers. When four atoms intersect, we collect the corresponding Delaunay tetrahedron.

At any specific α value, we have a *dual simplicial complex* or *alpha complex* $\mathcal{K}_\alpha$ formed by the collected simplices. If all atoms take the incremented radius of $r_i + r_s$ and $\alpha = 0$, we have the dual simplicial complex $\mathcal{K}_0$ of the protein molecule. When α is sufficiently large, we have collected all simplices and we get the full Delaunay complex. This series of simplicial complexes at different α values form a family of

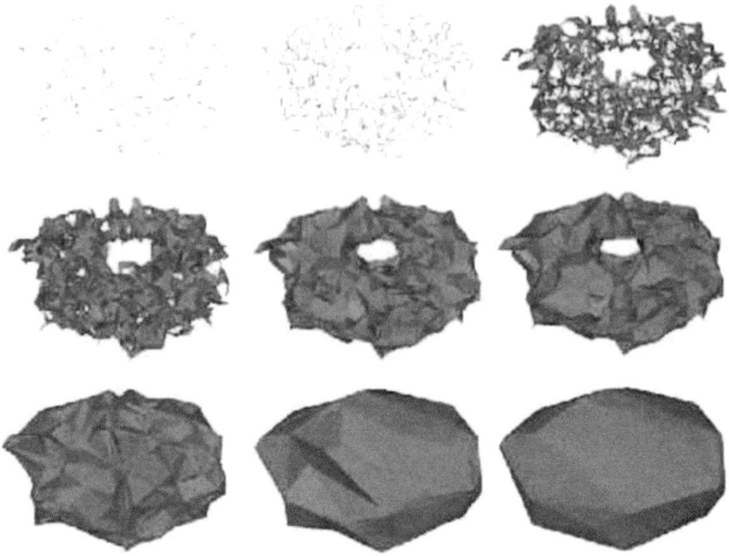

FIGURE 1.4 An illustration of a family of alpha shapes of HIV-1 protease as α value increases from left to right and top to bottom. As α increases, more edges, triangles, and tetrahedra enter the collection of simplices. At each α value, the collected simplices form a simplicial complex. When α is sufficiently large, we obtain the full Delaunay tetrahedrization.

shapes (Fig. 1.3), called *alpha shapes*, each faithfully represents the geometric and topological property of the protein molecule at a particular resolution parameterized by the α value. Figure 1.4 illustrates an example of the alpha shapes of the HIV-1 protease at different α values.

An equivalent way to obtain the alpha shape at $\alpha = 0$ is to take a subset of the simplices, with the requirement that the corresponding intersections of Voronoi cells must overlap with the body of the union of the balls. We obtain the dual complex or alpha shape $\mathcal{K}_0$ of the molecule at $\alpha = 0$ (Fig. 1.2c):

$$\mathcal{K}_0 = \left\{ \sigma_{|I|-1} \,\middle|\, \bigcap_{i \in I} V_i \cap \bigcup \mathcal{B} \neq \emptyset \right\}.$$

Alpha shapes provides a guide map for computing geometric properties of the structures of biomolecules. Take the molecular surface as an example: The re-entrant surfaces are formed by the concave spherical patch and the toroidal surface. These can be mapped from the boundary triangles and boundary edges of the alpha shape, respectively [14]. Recall that a triangle in the Delaunay tetrahedrization corresponds to the intersection of three Voronoi regions, that is, a Voronoi edge. For a triangle on the boundary of the alpha shape, the corresponding Voronoi edge intersects with the body of the union of balls by definition. In this case, it intersects with the solvent accessible surface at the common intersecting vertex when the three atoms overlap.

This vertex corresponds to a concave spherical surface patch in the molecular surface. For an edge on the boundary of the alpha shape, the corresponding Voronoi plane coincides with the intersecting plane when two atoms meet, which intersect with the surface of the union of balls on an arc. This line segment corresponds to a toroidal surface patch. The remaining part of the surface are convex pieces, which correspond to the vertices, namely, the atoms on the boundary of the alpha shape.

The numbers of toroidal pieces and concave spherical pieces are exactly the numbers of boundary edges and boundary triangles in the alpha shape, respectively. Because of the restriction of bond length and the excluded volume effects, the number of edges and triangles in molecules are roughly on the order of $\mathcal{O}(n)$ [38].

1.2.5 Metric Measurements

We have described the relationship between the simplices and the surface elements of the molecule. Based on this type of relationship, we can compute efficiently size properties of the molecule. We take the problem of volume computation as an example.

Consider a grossly incorrect way to compute the volume of a protein molecule using the solvent accessible surface model. We could define that the volume of the molecule is the summation of the volumes of individual atoms, whose radii are inflated to account for solvent probe. By doing so, we would have significantly inflated the value of the true volume, because we neglected to consider volume overlaps. We can explicitly correct this by following the inclusion–exclusion formula: When two atoms overlap, we subtract the overlap; when three atoms overlap, we first subtract the pair overlaps, and (we then) add back the triple overlap, and so on. This continues when there are four, five, or more atoms intersecting. At the combinatorial level, the principle of inclusion–exclusion is related to the Gauss–Bonnet theorem used by Connolly [9]. The corrected volume $V(\mathcal{B})$ for a set of atom balls $\mathcal{B}$ can then be written as

$$V(\mathcal{B}) = \sum_{\substack{\mathrm{vol}(\bigcap T)>0 \\ T\subset\mathcal{B}}} (-1)^{\dim(T)-1}\,\mathrm{vol}\left(\bigcap T\right), \tag{1.1}$$

where vol $\left(\bigcap T\right)$ represents volume overlap of various degree, $T \subset \mathcal{B}$ is a subset of the balls with nonzero volume overlap: vol $\left(\bigcap T\right) > 0$.

However, the straightforward application of this inclusion–exclusion formula does not work. The degree of overlap can be very high: Theoretical and simulation studies showed that the volume overlap can be up to 7–8 degrees [29,45]. It is difficult to keep track of these high degree of volume overlaps correctly during computation. In addition, it is also difficult to compute the volume of these overlaps, because of the proliferation of different types of combinations of intersecting balls. That is, we need to quantify the k-volume overlap of each of the $\binom{7}{k}$ possible overlap patterns when k of the 7 balls overlap. Similarly, the volumes of $\binom{8}{k}$ overlapping atoms for all of $k = 2, \ldots, 7$ need to be quantified [45]. It turns out that for three-dimensional molecules, overlaps of five or more atoms at a time can always be reduced to a "+" or a "−" signed combination of overlaps of four or fewer atom balls [12]. This requires that the 2-body, 3-body, and 4-body terms in Eq. (1.1) enter the formula if and only if the corresponding edge σ_{ij} connecting the two balls (1-simplex), triangles σ_{ijk} spanning the three balls (2-simplex), and tetrahedron σ_{ijkl} cornered on the four balls (3-simplex) all exist in the dual simplicial complex $\mathcal{K}_0$ of the molecule [12,38]. Atoms corresponding to these simplices will all have volume overlaps. In this case, we have the simplified exact expansion:

$$V(\mathcal{B}) = \sum_{\sigma_i \in \mathcal{K}} \text{vol}(b_i) - \sum_{\sigma_{ij} \in \mathcal{K}} \text{vol}(b_i \cap b_j) + \sum_{\sigma_{ijk} \in \mathcal{K}} \text{vol}(b_i \cap b_j \cap b_k) - \sum_{\sigma_{ijkl} \in \mathcal{K}} \text{vol}(b_i \cap b_j \cap b_k \cap b_l).$$

The same idea is applicable for the calculation of surface area of molecules.

An Example. An example of area computation by the alpha shape is shown in Fig. 1.5. Let b_1, b_2, b_3, b_4 be the four disks. To simplify the notation, we write A_i for the area of b_i, A_{ij} for the area of $b_i \cap b_j$, and A_{ijk} for the area of $b_i \cap b_j \cap b_k$. The total area of the union, $b_1 \cup b_2 \cup b_3 \cup b_4$, is

$$\begin{aligned} A_{\text{total}} &= (A_1 + A_2 + A_3 + A_4) \\ &\quad - (A_{12} + A_{23} + A_{24} + A_{34}) \\ &\quad + A_{234}. \end{aligned}$$

We add the area of b_i if the corresponding vertex belongs to the alpha complex (Fig. 1.5), we subtract the area of $b_i \cap b_j$ if the corresponding edge belongs to the alpha complex, and we add the area of $b_i \cap b_j \cap b_k$ if the corresponding triangle belongs to the alpha complex. Note that without the guidance of the alpha complex,

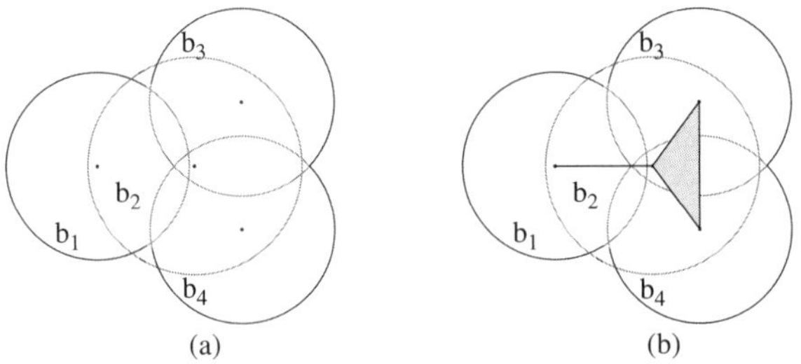

FIGURE 1.5 An example of analytical area calculation. (A) Area can be computed using the direct inclusion–exclusion. (B) The formula is simplified without any redundant terms when using the alpha shape.

the inclusion-exclusion formula may be written as

$$\begin{aligned} A_{\text{total}} = {} & (A_1 + A_2 + A_3 + A_4) \\ & - (A_{12} + A_{13} + A_{14} + A_{23} + A_{24} + A_{34}) \\ & + (A_{123} + A_{124} + A_{134} + A_{234}) \\ & - A_{1234}. \end{aligned}$$

This contains 6 canceling redundant terms: $A_{13} = A_{123}$, $A_{14} = A_{124}$, and $A_{134} = A_{1234}$. Computing these terms would be wasteful. Such redundancy does not occur when we use the alpha complex: The part of the Voronoi regions contained in the respective atom balls for the redundant terms do not intersect. Therefore, the corresponding edges and triangles do not enter the alpha complex. In two dimensions, we have terms of at most three disk intersections, corresponding to triangles in the alpha complex. Similarly, in three dimensions the most complicated terms are intersections of four spherical balls, and they correspond to tetrahedra in the alpha complex.

Voids and Pockets. Voids and pockets represent the concave regions of a protein surface. Because shape-complementarity is the basis of many molecular recognition processes, binding and other activities frequently occur in pocket or void regions of protein structures. For example, the majority of enzyme reactions take place in surface pockets or interior voids.

The topological structure of the alpha shape also offers an effective method for computing voids and pockets in proteins. Consider the Delaunay tetrahedra that are not included in the alpha shape. If we repeatedly merge any two such tetrahedra on the condition that they share a 2-simplex triangle, we will end up with discrete sets of tetrahedra. Some of them will be completely isolated from the outside, and some of them are connected to the outside by triangle(s) on the boundary of the alpha shape. The former corresponds to voids (or cavities) in proteins, whereas the latter corresponds to *pockets* and *depressions* in proteins.

A pocket differs from a depression in that it must have an opening that is at least narrower than one interior cross section. Formally, the *discrete flow* [17] explains the distinction between a depression and a pocket. In a two-dimensional Delaunay triangulation, the empty triangles that are not part of the alpha shape can be classified into obtuse triangles and acute triangles. The largest angle of an obtuse triangle is more than 90 degrees, and the largest angle of an acute triangle is less than 90 degrees. An empty obtuse triangle can be regarded as a ''source'' of empty space that ''flows'' to its neighbor, and an empty acute triangle can be regarded to be a ''sink'' that collects flow from its obtuse empty neighboring triangle(s). In Fig. 1.6a, obtuse triangles 1, 3, 4, and 5 flow to the acute triangle 2, which is a sink. Each of the discrete empty spaces on the surface of protein can be organized by the flow systems of the corresponding empty triangles: Those that flow together belong to the same discrete empty space. For a pocket, there is at least one sink among the empty triangles. For a depression, all triangles are obtuse, and the discrete flow goes from

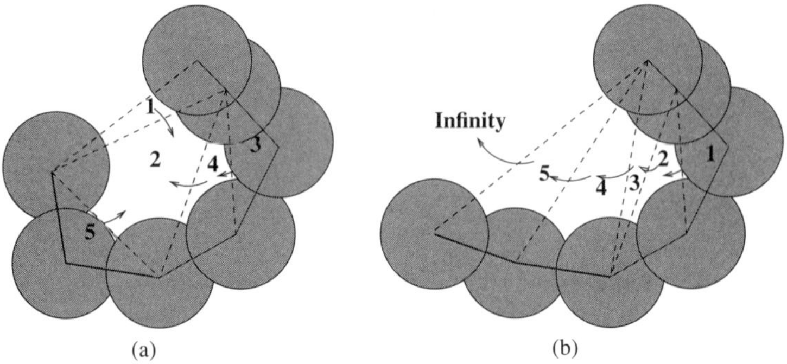

FIGURE 1.6 Discrete flow of empty space illustrated for two-dimensional disks.

one obtuse triangle to another, from the innermost region to outside the convex hull. The discrete flow of a depression therefore goes to infinity. Figure 1.6b gives an example of a depression formed by a set of obtuse triangles.

Once voids and pockets are identified, we can apply the inclusion–exclusion principle based on the simplices to compute the exact size measurement (e.g., volume and area) of each void and pocket [17,39]. Figure 1.7 shows the computed binding surface pockets on Ras21 protein and FtsZ protein.

The distinction between voids and pockets depends on the specific set of atomic radii and the solvent radius. When a larger solvent ball is used, the radii of all atoms will be inflated by a larger amount. This could lead to two different outcomes. A void or pocket may become completely filled and disappear. On the other hand, the inflated atoms may not fill the space of a pocket, but may close off the opening of the pocket. In this case, a pocket becomes a void. A widely used practice in the

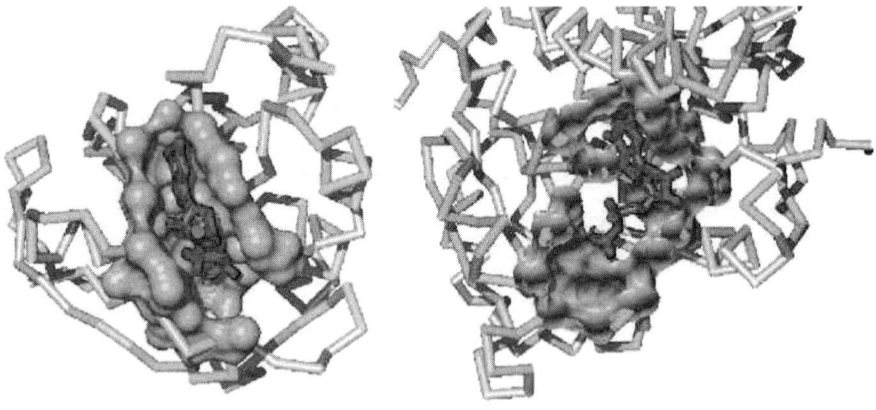

FIGURE 1.7 The computed surface pockets of binding sites on Ras21 protein and FtsZ protein.

past was to adjust the solvent ball and repeatedly compute voids, in the hope that some pockets will become voids and hence be identified by methods designed for cavity/void computation. The pocket algorithm [17] and tools such as CastP [11,40] often makes this unnecessary.

1.3 ALGORITHM AND COMPUTATION

Computing Delaunay Tetrahedrization and Voronoi Diagram. It is easier to discuss the computation of tetrahedrization first. The incremental algorithm developed in [16] can be used to compute the weighted tetrahedrization for a set of atoms of different radii. For simplicity, we sketch the outline of the algorithm below for two-dimensional unweighted Delaunay triangulation.

The intuitive idea of the algorithm can be traced back to the original observation of Delaunay. For the Delaunay triangulation of a point set, the circumcircle of an edge and a third point forming a Delaunay triangle must not contain a fourth point. Delaunay showed that if all edges in a particular triangulation satisfy this condition, the triangulation is a Delaunay triangulation. It is easy to come up with an arbitrary triangulation for a point set. A simple algorithm to convert this triangulation to the Delaunay triangulation is therefore to go through each of the triangles and then make corrections using "flips" discussed below if a specific triangle contains an edge violating the above condition. The basic ingredients for computing Delaunay tetrahedrization are generalizations of these observations. We discuss the concept of *locally Delaunay* edge and the *edge-flip* primitive operation below.

Locally Delaunay Edge. We say an edge *ab* is locally Delaunay if either it is on the boundary of the convex hull of the point set or it belongs to two triangles *abc* and *abd*, and the circumcircle of *abc* does not contain *d* (e.g., edge *cd* in Fig. 1.8a).

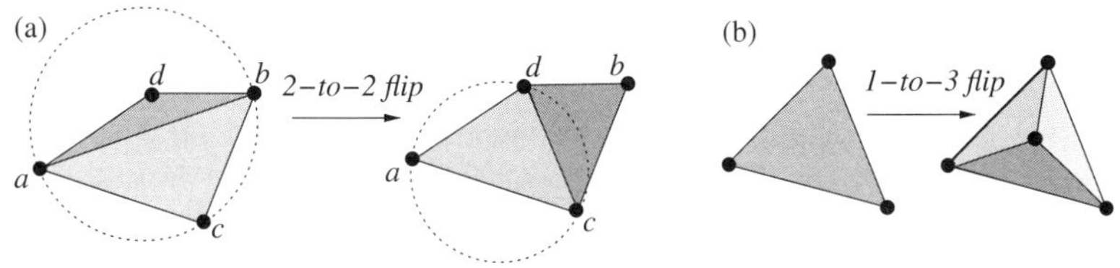

FIGURE 1.8 An illustration of *locally Delaunay edge* and *flips*. (**a**) For the quadrilateral *abcd*, edge *ab* is not locally Delaunay, as the circumcircle passing through edge *ab* and a third point *c* contains a fourth point *d*. Edge *cd* is locally Delaunay, as *b* is outside the circumcircle *adc*. An *edge-flip* or *2-to-2 flip* replaces edge *ab* by edge *cd*, and replace the original two triangles *abc* and *adb* with two new triangles *acd* and *bcd*. (**b**) When a new vertex is inserted, we replace the old triangle containing this new vertex with three new triangles. This is called *1-to-3* flip.

Edge-Flip. If ab is not locally Delaunay (edge ab in Fig. 1.8a), then the union of the two triangles $abc \cup abd$ is a convex quadrangle $acbd$, and edge cd is locally Delaunay. We can replace edge ab by edge cd. We call this an *edge-flip* or *2-to-2 flip*, as two old triangles are replaced by two new triangles.

We recursively check each boundary edge of the quadrangle $abcd$ to see if it is also locally Delaunay after replacing ab by cd. If not, we recursively edge-flip it.

Incremental Algorithm for Delaunay Triangulation. Assume that we have a finite set of points (namely, atom centers) $\mathcal{S} = \{z_1, z_2, \ldots, z_i, \ldots, z_n\}$. We start with a large auxiliary triangle that contains all these points. We insert the points one by one. At all times, we maintain a Delaunay triangulation $\mathcal{D}_i$ up to insertion of point z_i.

After inserting point z_i, we search for the triangle τ_{i-1} that contains this new point. We then add z_i to the triangulation and split the original triangle τ_{i-1} into three smaller triangles. This split is called *1-to-3 flip*, as it replaces one old triangle with three new triangles. We then check if each of the three edges in τ_{i-1} still satisfies the locally Delaunay requirement. If not, we perform a recursive edge-flip. This algorithm is summarized in Algorithm I.

Algorithm I Delaunay Triangulation

```
Obtain random ordering of points {z_1, ···, z_n};
for i = 1 to n do
    find τ_{i-1} such z_i ∈ τ_{i-1};
    add z_i, and split τ_{i-1} into three triangles (1-to-3 flip);
    while any edge ab not locally Delaunay do
        flip ab to other diagonal cd (2-to-2 edge flip);
    end while
end for
```

In $\mathbb{R}^3$, the algorithm of tetrahedrization becomes more complex, but the same basic ideas apply. In this case, we need to locate a tetrahedron instead of a triangle that contains the newly inserted point. The concept of locally Delaunay is replaced by the concept of locally convex, and there are flips different than the 2-to-2 flip in $\mathbb{R}^3$ [16]. Although an incremental approach (i.e., sequentially adding points) is not necessary for Delaunay triangulation in $\mathbb{R}^2$, it is necessary in $\mathbb{R}^3$ to avoid non-flippable cases and to guarantee that the algorithm will terminate. This incremental algorithm has excellent expected performance [16].

The computation of Voronoi diagram is conceptually easy once the Delaunay triangulation is available. We can take advantage of the mathematical duality and compute all of the Voronoi vertices, edges, and planar faces from the Delaunay tetrahedra, triangles, and edges (see exercises). Because one point z_i may be an vertex of many Delaunay tetrahedra, the Voronoi region of z_i therefore may contain many Voronoi vertices, edges, and planar faces. The efficient quad-edge data structure can be used for software implementation [24].

Volume and Area Computation. Let V and A denote the volume and area of the molecule, respectively, $\mathcal{K}_0$ for the alpha complex, σ for a simplex in $\mathcal{K}$, i for a vertex, ij for an edge, ijk for a triangle, and $ijkl$ for a tetrahedron. The algorithm for volume and area computation can be written as Algorithm II.

Algorithm II Volume and Area Measurement

$V := A := 0.0$;
for all $\sigma \in \mathcal{K}$ **do**
  if σ is a vertex i **then**
    $V := V + \text{vol}(b_i)$; $A := A + \text{area}(b_i)$;
  end if
  if σ is an edge ij **then**
    $V := V - \text{vol}(b_i \cap b_j)$; $A := A - \text{area}(b_i \cap b_j)$;
  end if
  if σ is a triangle ijk **then**
    $V := V + \text{vol}(b_i \cap b_j \cap b_k)$; $A := A + \text{area}(b_i \cap b_j \cap b_k)$;
  end if
  if σ is a tetrahedron $ijkl$ **then**
    $V := V - \text{vol}(b_i \cap b_j \cap b_k \cap b_l)$; $A := A - \text{area}(b_i \cap b_j \cap b_k \cap b_l)$;
  end if
end for

Additional details of volume and area computation can be found in references 14 and 38.

Software. The CASTP webserver for pocket computation can be found at cast.engr.uic.edu. There are other studies that compute or use Voronoi diagrams of protein structures [8,23,25], although not all computes the weighted version which allows atoms to have different radii.

In this short description of algorithm, we have neglected many details important for geometric computation. For example, the problem of how to handle geometric degeneracy, namely, when three points are collinear, or when four points are coplanar. Interested readers should consult the excellent monograph by Edelsbrunner for a detailed treatise of these and other important topics in computational geometry [13].

1.4 APPLICATIONS

1.4.1 Protein Packing

An important application of the Voronoi diagram and volume calculation is the measurement of protein packing. Tight packing is an important feature of protein structure [47,48] and is thought to play important roles in protein stability and folding dynamics [33]. The packing density of a protein is measured by the ratio of its van der Waals volume and the volume of the space it occupies. One approach is to calculate

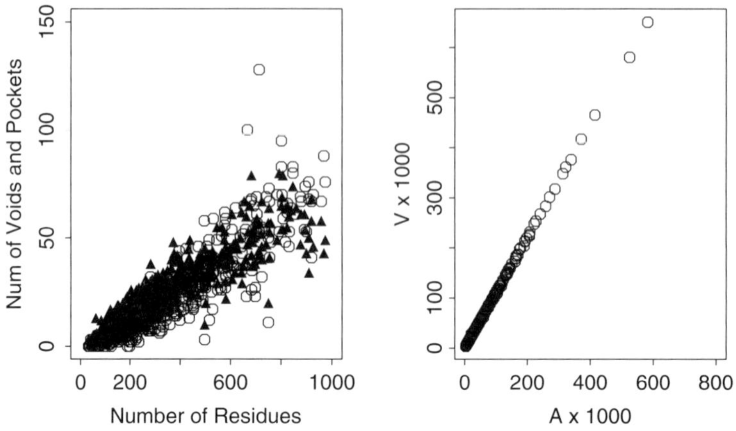

FIGURE 1.9 Voids and pockets for a set of 636 proteins representing most of the known protein folds, and the scaling behavior of the geometric properties of proteins. (left) The number of voids and pockets detected with a 1.4 Å probe is linearly correlated with the number of residues in a protein. Only proteins with less than 1000 residues are shown. Solid triangles and empty circles represent the pockets and the voids, respectively. (right) The van der Waals (vdw) volume and van der Waals area of proteins scale linearly with each other. Similarly, molecular surface (ms) volume also scales linearly with molecular surface area using a probe radius of 1.4 Å. (Data not shown. Figure adapted from reference 37.)

the packing density of buried residues and atoms using the Voronoi diagram [47,48]. This approach was also used to derive radii parameters of atoms [56].

Based on the computation of voids and pockets in proteins, a detailed study surveying major representatives of all known protein structural folds showed that there is a substantial amount of voids and pockets in proteins [37]. On average, every 15 residues introduces a void or a pocket (Fig. 1.9 (left side)). For a perfectly solid three-dimensional sphere of radius r, the relationship between volume $V = 4\pi r^3/3$ and surface area $A = 4\pi r^2$ is $V \propto A^{3/2}$. In contrast, Fig. 1.9 (right side) shows that the van der Waals volume scales linearly with the van der Waals surface areas of proteins. The same linear relationship holds irrespective of whether we relate molecular surface volume and molecular surface area, or solvent accessible volume and solvent accessible surface area. This and other scaling behavior point out that protein interior is not packed as tight as solid [37]. Rather, packing defects in the form of voids and pockets are common in proteins.

If voids and pockets are prevalent in proteins, an interesting question is what is then the origin of the existence of these voids and pockets. This question was studied by examining the scaling behavior of packing density and coordination number of residues through the computation of voids, pockets, and edge simplices in the alpha shapes of random compact chain polymers [62]. For this purpose, a 32-state discrete state model was used to generate a large ensemble of compact selfavoiding walks. This is a difficult task, as it is very challenging to generate a large number of independent conformations of very compact chains that are self-avoiding. The results in reference 63 showed that it is easy for compact random chain polymers to have similar scaling behavior of packing density and coordination number with chain length. This suggests that proteins are not optimized by evolution to eliminate voids and pockets, and the existence of many pockets and voids is random in nature and is due to the generic requirement of compact chain polymers. The frequent occurrence and the origin of voids and pockets in protein structures raise a challenging question: How can we distinguish voids and pockets that perform biological functions such as binding from those formed by random chance? This question is related to the general problem of protein function prediction.

1.4.2 Predicting Protein Functions from Structures

Conservation of protein structures often reveals a very distant evolutionary relationship, which are otherwise difficult to detect by sequence analysis [55]. Comparing protein structures can provide insightful ideas about the biochemical functions of proteins (*e.g.*, active sites, catalytic residues, and substrate interactions) [26,42,44].

A fundamental challenge in inferring protein function from structure is that the functional surface of a protein often involves only a small number of key residues. These interacting residues are dispersed in diverse regions of the primary sequences and are difficult to detect if the only information available is the primary sequence. Discovery of local spatial motifs from structures that are functionally relevant has been the focus of many studies.

Graph-Based Methods for Spatial Patterns in Proteins. To analyze local spatial patterns in proteins, Artymiuk *et al.* developed an algorithm based on subgraph isomorphism detection [1]. By representing residue side-chains as simplified pseudo-atoms, a molecular graph is constructed to represent the patterns of side-chain pseudo-atoms and their interatomic distances. A user-defined query pattern can then be searched rapidly against the Protein Data Bank for similarity relationship. Another widely used approach is the method of geometric hashing. By examining spatial patterns of atoms, Fischer *et al.* developed an algorithm that can detect surface similarity of proteins [19,43]. This method has also been applied by Wallace *et al.* for the derivation and matching of spatial templates [59]. Russell developed a different algorithm that detects side-chain geometric patterns common to two protein structures [53]. With the evaluation of statistical significance of measured root mean square distance, several new examples of convergent evolution were discovered, where common patterns of side-chains were found to reside on different tertiary folds.

These methods have a number of limitations. Most require a user-defined template motif, restricting their utility for automated database-wide search. In addition, the size of the spatial pattern related to protein function is also often restricted.

Predicting Protein Functions by Matching Pocket Surfaces. Protein functional surfaces are frequently associated with surface regions of prominent concavity [30,40]. These include pockets and voids, which can be accurately computed as we have discussed. Computationally, one wishes to automatically identify voids and pockets on protein structures where interactions exist with other molecules such as substrate, ions, ligands, or other proteins.

Binkowski et al. developed a method for predicting protein function by matching a surface pocket or void on a protein of unknown or undetermined function to the pocket or void of a protein of known function [4,6]. Initially, the Delaunay tetrahedrization and alpha shapes for almost all of the structures in the PDB databank are computed [11]. All surface pockets and interior voids for each of the protein structures are then exhaustively computed [17,39]. For each pocket and void, the residues forming the wall are then concatenated to form a short sequence fragment of amino acid residues while ignoring all intervening residues that do not participate in the formation of the wall of the pocket or void. Two sequence fragments, one from the query protein and another from one of the proteins in the database, both derived from pocket or void surface residues, are then compared using dynamic programming. The similarity score for any observed match is assessed for statistical significance using an empirical randomization model constructed for short sequence patterns.

For promising matches of pocket/void surfaces showing significant sequence similarity, we can further evaluate their similarity in shape and in relative orientation. The former can be obtained by measuring the coordinate root mean square distance (RMSD) between the two surfaces. The latter is measured by first placing a unit sphere at the geometric center $\boldsymbol{z}_0 \in \mathbb{R}^3$ of a pocket/void. The location of each residue $\boldsymbol{z} = (x, y, z)^T$ is then projected onto the unit sphere along the direction of the vector from the geometric center: $\boldsymbol{u} = (\boldsymbol{z} - \boldsymbol{z}_0)/||\boldsymbol{z} - \boldsymbol{z}_0||$. The projected pocket is represented by a collection of unit vectors located on the unit sphere, and the original orientation of residues in the pocket is preserved. The RMSD distance of the two sets of unit vectors derived from the two pockets are then measured, which is called the ORMSD for *orientation RMSD* [4]. This allows similar pockets with only minor conformational changes to be detected [4].

The advantage of the method of Binkowski *et al.* is that it does not assume prior knowledge of functional site residues and does not require *a priori* any similarity in either the full primary sequence or the backbone fold structures. It has no limitation in the size of the spatially derived motif and can successfully detect patterns small and large. This method has been successfully applied to detect similar functional surfaces among proteins of the same fold but low sequence identities and among proteins of different fold [4,5].

Function Prediction through Models of Protein Surface Evolution. To match local surfaces such as pockets and voids and to assess their sequence similarity, an effective scoring matrix is critically important. In the original study of Binkowski et al., the BLOSUM matrix was used. However, this is problematic, as BLOSUM matrices were derived from analysis of precomputed large quantities of sequences, while the information of the particular protein of interest has limited or no influence. In addition, these precomputed sequences include buried residues in protein core, whose conservation reflects the need to maintain protein stability rather than to maintain protein function. In references 57 and 58, a continuous time Markov process was

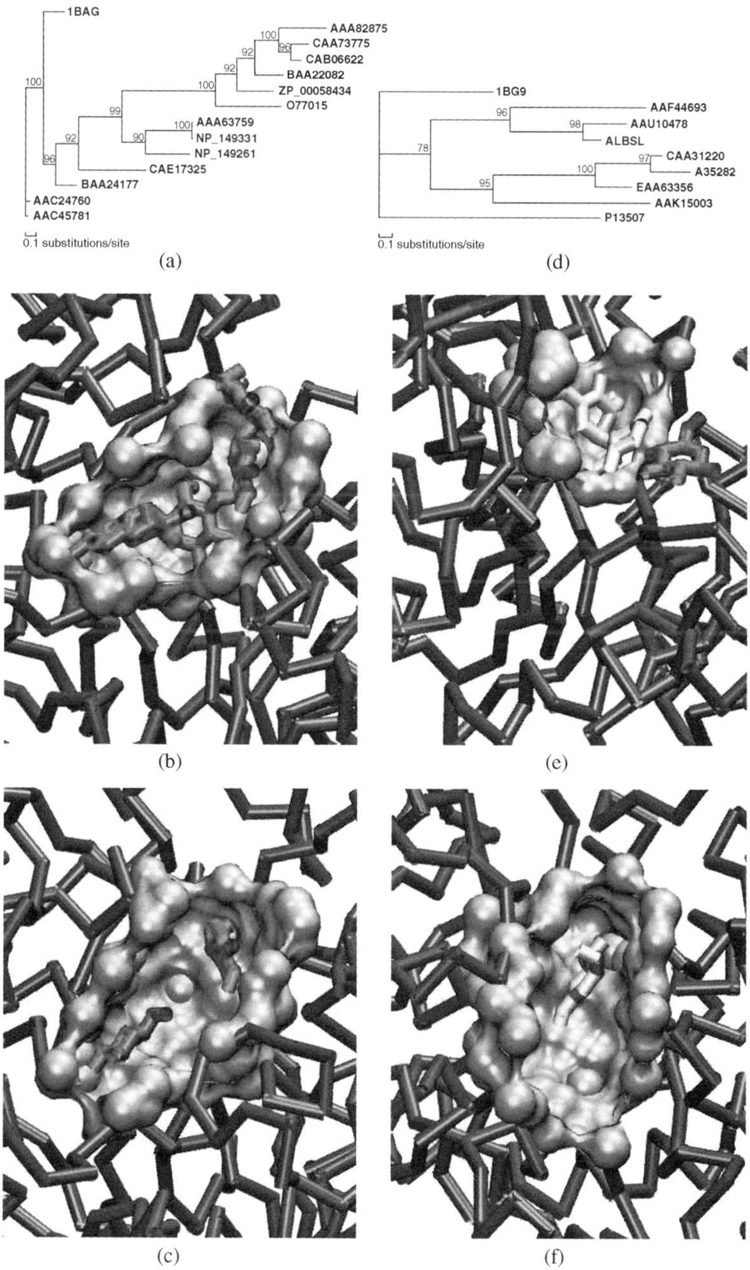

FIGURE 1.10 Protein function prediction as illustrated by the example of alpha amylases. Two template binding surfaces are used to search the database of protein surfaces to identify protein structures that are of similar functions. **(a)** The phylogenetic tree for the template PDB structure `1bag` from *B. subtilis*. **(b)** The template binding pocket of alpha amylase on `1bag`. **(c)** A matched binding surface on a different protein structure (`1b2y` from human, full sequence identity 22%) obtained by querying with `1bag`. **(d)** The phylogenetic tree for the template structure `1bg9` from *H. vulgare*. **(e)** The template binding pocket on `1bg9`. **(f)** A matched binding surface on a different protein structure (`1u2y` from human, full sequence identity 23%) obtained by querying with `1bg9` (Adapted from reference 58).

developed to explicitly model the substitution rates of residues in binding pockets. Using a Bayesian Markov chain Monte Carlo method, the residue substitution rates at functional pocket are estimated. The substitution rates are found to be very different for residues in the binding site and residues on the remaining surface of proteins. In addition, substitution rates are also very different for residues in the buried core and residues on the solvent exposed surfaces. These rates are then used to generate a set of scoring matrices of different time intervals for residues located in the functional pocket. Application of protein-specific and region-specific scoring matrices in matching protein surfaces result in significantly improved sensitivity and specificity in protein function prediction [57,58].

In a large-scale study of predicting protein functions from structures, a subset of 100 enzyme families are collected from a total of 286 enzyme families containing between 10 and 50 member protein structures with known Enzyme Classification (E.C.) labels. By estimating the substitution rate matrix for residues on the active site pocket of a query protein, a series of scoring matrices of different evolutionary time is derived. By searching for similar pocket surfaces from a database of 770,466 pockets derived from the CASTP database (with the criterion that each must contain at least 8 residues), this method can recover active site surfaces on enzymes similar to that on the query structure at an accuracy of 92% or higher. An example of identifying human amylase using template surfaces from *B. subtilis* and from barley is shown in Fig. 1.10.

The method of surface matching based on evolutionary model is also especially effective in solving the challenging problems of protein function prediction of orphan structures of unknown function (such as those obtained in structural genomics projects), which have only sequence homologs that are themselves hypothetical proteins with unknown functions.

1.5 DISCUSSION AND SUMMARY

A major challenge in studying protein geometry is to understand our intuitive notions of various geometric aspects of molecular shapes and to quantify these notions with mathematical models that are amenable to fast computation. The advent of the union of ball model of protein structures enabled rigorous definition of important geometric concepts such as solvent accessible surface and molecular surface. It also led to the development of algorithms for area and volume calculations of proteins. Deep understanding of the topological structure of molecular shapes is also based on the idealized union of ball model [12]. A success in approaching these problems is exemplified in the development of the pocket algorithm [17]. Another example is the recent development of a rigorous definition of protein–protein binding or interaction interface and algorithm for its computation [3].

Perhaps a more fundamental problem we face is to identify important structural and chemical features that are the determinants of biological problems of interest. For example, we would like to know the shape features that have significant influence on protein solvation, protein stability, ligand specific binding, and protein conformational changes. It is not clear whether our current geometric intuitions are sufficient or are the correct or the most relevant ones. There may still be important unknown shape properties of molecules that elude us at the moment.

An important application of geometric computation of protein structures is to detect patterns important for protein function. The shape of local surface regions on a protein structure and their chemical texture are the basis of its binding interactions with other molecules. Proteins fold into specific native structure to form these local regions for carrying out various biochemical functions. The geometric shape and chemical pattern of the local surface regions and how they change dynamically are therefore of fundamental importance in computational studies of proteins.

Another important application is the development of geometric potential functions. Potential functions are important for generating conformations, for distinguishing native and near native conformations from other decoy conformations in protein structure predictions [34,36,54,63] and in protein–protein docking [35]. They are also important for peptide and protein design [27,35].

We have not described in detail the approach of studying protein geometry using graph theory. In addition to side-chain pattern analysis briefly discussed earlier, the graph-based protein geometric model also has led to a number of important insights, including the optimal design of model proteins formed by hydrophobic and polar residues [28] and methods for optimal design of side-chain packing [31,61].

Further development of descriptions of geometric shape and topological structure, as well as algorithms for their computation, will provide a solid foundation for studying many important biological problems. The other important tasks are then to show how these descriptors may be effectively used to deepen our biological insights and to develop accurate predictive models of biological phenomena. For example, in computing protein–protein interfaces, a challenging task is to discriminate surfaces that are involved in protein binding from other nonbinding surface regions and to understand in what fashion this depends on the properties of the binding partner protein.

Undoubtedly, evolution plays central roles in shaping up the function and stability of protein molecules. The method of analyzing residue substitution rates using continuous-time Markov models [57,58] and the method of surface mapping of conservation entropy and phylogeny [22,41] only scratches the surface of this important issue. Much remains to be done in incorporating evolutionary information in protein shape analysis for understanding biological functions.

Remark. The original work of Lee and Richards can be found in reference 32, where they also formulated the molecular surface model [49]. Michael Connolly developed the first method for the computation of the molecular surface [9]. Tsai *et al.* described a method for obtaining atomic radii parameter [56]. The mathematical theory of the union of balls and alpha shape was developed by Herbert Edelsbrunner and colleagues [12,15]. Algorithm for computing weighted Delaunay tetrahedrization can be found in reference 16 or in a concise monograph with in-depth discussion of geometric computing [13]. Details of area and volume calculations can be found in references 14, 38 and 39. The theory of pocket computation and applications can be found in references 17 and 40. Richards and Lim offered a comprehensive review on protein packing and protein folding [50]. A detailed packing analysis of proteins can be found in reference 37. The study on inferring protein function by matching surfaces is described in references 4,58. The study of the evolutionary model of protein binding pocket and its application in protein function prediction can be found in reference 8.

Summary. The accumulation of experimentally solved molecular structures of proteins provides a wealth of information for studying many important biological problems. With the development of a rigorous model of the structure of protein molecules, various shape properties, including surfaces, voids, and pockets, and measurements of their metric properties can be computed. Geometric algorithms have found important applications in protein packing analysis, in developing potential functions, in docking, and in protein function prediction. It is likely the further development of geometric models and algorithms will find important applications in answering additional biological questions.

REFERENCES

1. P. J. Artymiuk, A. R. Poirrette, H. M. Grindley, D. W. Rice, and P. Willett. A graph-theoretic approach to the identification of three-dimensional patterns of amino acid side-chains in protein structure. *J. Mol. Biol.*, **243**:327–344, 1994.
2. R. F. W. Bader. *Atoms in Molecules: A Quantum Theory*. The International Series of Monographs on Chemistry, no. 22. Oxford University Press, New York, 1994.
3. Y. Ban, H. Edelsbrunner, and J. Rudolph. Interface surfaces for protein–protein complexes. In *RECOMB*, pages 205–212, 2004.
4. T. A. Binkowski, L. Adamian, and J. Liang. Inferring functional relationship of proteins from local sequence and spatial surface patterns. *J. Mol. Biol.*, **332**:505–526, 2003.
5. T. A. Binkowski, P. Freeman, and J. Liang. pvSOAR: Detecting similar surface patterns of pocket and void surfaces of amino acid residues on proteins. *Nucleic Acid Res.*, **32**:W555–W558, 2004.
6. T. A. Binkowski, A. Joachimiak, and J. Liang. Protein surface analysis for function annotation in high-throughput structural genomics pipeline. *Protein Sci.*, **14**(12):2972–81, 2005.
7. A. Bondi. VDW volumes and radii. *J. Phys. Chem.*, **68**:441–451, 1964.
8. S. Chakravarty, A. Bhinge, and R. Varadarajan. A procedure for detection and quantitation of cavity volumes proteins. Application to measure the strength of the hydrophobic driving force in protein folding. *J Biol Chem.*, **277**(35):31345–31353, 2002.
9. M. L. Connolly. Analytical molecular surface calculation. *J. Appl. Cryst.*, **16**:548–558, 1983.
10. G. M. Crippen and T. F. Havel. *Distance Geometry and Molecular Conformation*. John Wiley & Sons, New York, 1988.
11. J. Dundas, Z. Ouyang, J. Tseng, A. Binkowski, Y. Turpaz, and J. Liang. CASTp: Computed atlas of surface topography of proteins with structural and topographical mapping of

functionally annotated residues. *Nucleic Acids Res.*, **34**(Web Server issue):W116–W118, 2006.

12. H. Edelsbrunner. The union of balls and its dual shape. *Discrete Comput. Geom.*, **13**:415–440, 1995.
13. H. Edelsbrunner. *Geometry and Topology for Mesh Generation.* Cambridge University Press, New York, 2001.
14. H. Edelsbrunner, M. Facello, P. Fu, and J. Liang. Measuring proteins and voids in proteins. In *Proc. 28th Ann. Hawaii Int'l Conf. System Sciences*, Vol. 5, pp. 256–264, IEEE Computer Society Press, Los Alamitos, CA, 1995.
15. H. Edelsbrunner and E. P. Mücke. Three-dimensional alpha shapes. *ACM Trans. Graphics*, **13**:43–72, 1994.
16. H. Edelsbrunner and N. R. Shah. Incremental topological flipping works for regular triangulations. *Algorithmica*, **15**:223–241, 1996.
17. H. Edeslbrunner, M. Facello, and J. Liang. On the definition and the construction of pockets in macromolecules. *Disc. Appl. Math.*, **88**:18–29, 1998.
18. J. L. Finney. Volume occupation, environment and accessibility in proteins. The problem of the protein surface. *J. Mol. Biol.*, **96**:721–732, 1975.
19. D. Fischer, R. Norel, H. Wolfson, and R. Nussinov. Surface motifs by a computer vision technique: Searches, detection, and implications for protein-ligand recognition. *Proteins: Struct., Funct. Genet.*, **16**:278–292, 1993.
20. B. J. Gellatly and J. L. Finney. Calculation of protein volumes: An alternative to the Voronoi procedure. *J. Mol. Biol.*, 161:305–322, 1982.
21. M. Gerstein, F. M. Richards, M. S. Chapman, and M. L. Connolly. *Protein surfaces and volumes: measurement and use, International Tables for Crystallography, Volume F: Crystallography of biological macromolecules,* pp 531–545. Springer, Netherlands, 2001.
22. F. Glaser, T. Pupko, I. Paz, R.E. Bell, D. Shental, E. Martz, and N. Ben-Tal. Consurf: Identification of functional regions in proteins by surface-mapping of phylogenetic information. *Bioinformatics*, 19(1):163–164, 2003.
23. A Goede, R Preissner, and C Frömmel. Voronoi cell: New method for allocation of space among atoms: Elimination of avoidable errors in calculation of atomic volume and density. *J. Comput. Chem.*, **18**(9):1113–1123, 1997.
24. L. Guibas and J. Stolfi. Primitives for the manipulation of general subdivisions and the computation of Voronoi diagrams. *ACM Trans. Graphiques*, **4**:74–123, 1985.
25. Y. Harpaz, M. Gerstein, and C Chothia. Volume changes on protein folding. *Structure (London, England: 1993)*, **2**(7):641–649, 1994.
26. L. Holm and C. Sander. New structure: Novel fold? *Structure*, **5**:165–171, 1997.
27. C. Hu, X. Li, and J. Liang. Developing optimal nonlinear scoring function for protein design. *Bioinformatics*, **20**:3080–3098, 2004.
28. J. Kleinberg. Efficient algorithms for protein sequence design and the analysis of certain evolutionary fitness landscapes. In *RECOMB*, pp. 205–212, 2004.
29. K. W. Kratky. Intersecting disks (and spheres) and statistical mechanics. I. Mathematical basis. *J. Stat. Phys.*, **25**:619–634, 1981.
30. R. A. Laskowski, N. M. Luscombe, M. B. Swindells, and J. M. Thornton. Protein clefts in molecular recognition and function. *Protein Sci.*, **5**:2438–2452, 1996.

31. A. Leaver-Fay, B. Kuhlman, and J. Snoeyink. An adaptive dynamic programming algorithm for the side chain placement problem. In *Pacific Symposium on Biocomputing*, pp. 17–28, 2005.
32. B. Lee and F. M. Richards. The interpretation of protein structures: Estimation of static accessibility. *J. Mol. Biol.*, **55**:379–400, 1971.
33. M. Levitt, M. Gerstein, E. Huang, S. Subbiah, and J. Tsai. Protein folding: The endgame. *Annu. Rev. Biochem.*, **66**:549–579, 1997.
34. X. Li, C. Hu, and J. Liang. Simplicial edge representation of protein structures and alpha contact potential with confidence measure. *Proteins*, **53**:792–805, 2003.
35. X. Li and J. Liang. Computational design of combinatorial peptide library for modulating protein–protein interactions. *Pac. Symp. Biocomput.*, pages 28–39, 2005.
36. X. Li and J. Liang. Geometric cooperativity and anticooperativity of three-body interactions in native proteins. *Proteins*, **60**(1):46–65, 2005.
37. J. Liang and K. A. Dill. Are proteins well-packed? *Biophys. J.*, **81**:751–766, 2001.
38. J. Liang, H. Edelsbrunner, P. Fu, P. V. Sudhakar, and S. Subramaniam. Analytical shape computing of macromolecules I: Molecular area and volume through alpha-shape. *Proteins*, **33**:1–17, 1998.
39. J. Liang, H. Edelsbrunner, P. Fu, P. V. Sudhakar, and S. Subramaniam. Analytical shape computing of macromolecules II: Identification and computation of inaccessible cavities inside proteins. *Proteins*, **33**:18–29, 1998.
40. J. Liang, H. Edelsbrunner, and C. Woodward. Anatomy of protein pockets and cavities: Measurement of binding site geometry and implications for ligand design. *Protein Sci.*, **7**:1884–1897, 1998.
41. O. Lichtarge, H. R. Bourne, and F. E. Cohen. An evolutionary trace method defines binding surfaces common to protein families. *J. Mol. Biol.*, **257**(2):342–358, 1996.
42. A. C. R. Martin, C. A. Orengo, E. G. Hutchinson, A. D. Michie, A. C. Wallace, M. L. Jones, and J. M. Thornton. Protein folds and functions. *Structure*, **6**:875–884, 1998.
43. R. Norel, D. Fischer, H. J. Wolfson, and R. Nussinov. Molecular surface recognition by a computer vision-based technique. *Protein Engineering*, **7**(1):39–46, 1994.
44. C. A. Orengo, A. E. Todd, and J. M. Thornton. From protein structure to function. *Curr. Opini. Struct. Biol.*, **9**(4):374–382, 1999.
45. M. Petitjean. On the analytical calculation of van der waals surfaces and volumes: Some numerical aspects. *J. Comput. Chem.*, **15**:507–523, 1994.
46. G. Rhodes. *Crystallography Made Crystal Clear: A Guide for Users of Macromolecular Models*. Academic Press, Walthan, MA, 1999.
47. F. M. Richards. The interpretation of protein structures: Total volume, group volume distributions and packing density. *J. Mol. Biol.*, **82**:1–14, 1974.
48. F. M. Richards. Areas, volumes, packing, and protein structures. *Annu. Rev. Biophys. Bioeng.*, **6**:151–176, 1977.
49. F. M. Richards. Calculation of molecular volumes and areas for structures of known geometries. *Methods Enzymol.*, **115**:440–464, 1985.
50. F. M. Richards and W. A. Lim. An analysis of packing in the protein folding problem. *Q. Rev. Biophys.*, **26**:423–498, 1994.

51. T. J. Richmond. Solvent accessible surface area and excluded volume in proteins: analytical equations for overlapping spheres and implications for the hydrophobic effect. *J. Mol. Biol.*, **178**:63–89, 1984.
52. W. Rieping, M. Habeck, and M. Nilges. Inferential structure determination. *Science*, **309**(5732):303–306, 2005.
53. R. Russell. Detection of protein three-dimensional side-chain patterns: New examples of convergent evolution. *J. Mol. Biol.*, **279**:1211–1227, 1998.
54. R. K. Singh, A. Tropsha, and I. I. Vaisman. Delaunay tessellation of proteins: four body nearest-neighbor propensities of amino-acid residues. *J. Comp. Biol.*, **3**:213–221, 1996.
55. A. E. Todd, C. A. Orengo, and J. M. Thornton. Evolution of function in protein superfamilies, from a structural perspective. *J. Mol. Biol.*, **307**:1113–1143, 2001.
56. J. Tsai, R. Taylor, C. Chothia, and M. Gerstein. The packing density in proteins: Standard radii and volumes. *J. Mol. Biol.*, **290**(1):253–66, 1999.
57. Y.Y. Tseng and J. Liang. Estimating evolutionary rate of local protein binding surfaces: A bayesian monte carlo approach. *Proceedings of 2005 IEEE-EMBC Conference*, 2005.
58. Y. Y. Tseng and J. Liang. Estimation of amino acid residue substitution rates at local spatial regions and application in protein function inference: A Bayesian Monte Carlo approach. *Mol. Biol. Evol.*, **23**(2):421–436, 2006.
59. A. C. Wallace, N. Borkakoti, and J. M. Thornton. TESS: a geometric hashing algorithm for deriving 3d coordinate templates for searching structural databases. Application to enzyme active sites. *Protein Sci.*, **6**:2308–2323, 1997.
60. J.M. Word, S.C. Lovell, J.S. Richardson, and D.C. Richardson. Asparagine and glutamine: using hydrogen atom contacts in the choice of side-chain amide orientation. *J. Mol. Biol.*, **285**(4):1735–1747, 1999.
61. J. Xu. Rapid protein side-chain packing via tree decomposition. In *RECOMB*, pp. 423–439, 2005.
62. J. Zhang, R. Chen, C. Tang, and J. Liang. Origin of scaling behavior of protein packing density: A sequential monte carlo study of compact long chain polymers. *J. Chem. Phys.*, **118**:6102–6109, 2003.
63. W. Zheng, S. J. Cho, I. I. Vaisman, and A. Tropsha. A new approach to protein fold recognition based on Delaunay tessellation of protein structure. In R. B. Altman, A. K. Dunker, L. Hunter, and T. E. Klein, editors, *Pacific Symposium on Biocomputing'97*, pp. 486–497. World Scientific, Singapore, 1997.

EXERCISES

1.1 For two points $\boldsymbol{x}_1, \boldsymbol{x}_2 \in \mathbb{R}^d$, the line through $\boldsymbol{x}_1$ and $\boldsymbol{x}_2$ can be written as $\{\boldsymbol{x} | \boldsymbol{x} = \boldsymbol{x}_1 + \lambda(\boldsymbol{x}_2 - \boldsymbol{x}_1),\ \lambda \in \mathbb{R}\}$. Equivalently, we can define the line as

$$\{\boldsymbol{x} | \boldsymbol{x} = (1 - \lambda)\boldsymbol{x}_1 + \lambda \boldsymbol{x}_2, \quad \lambda \in \mathbb{R}\},$$

or

$$\{\boldsymbol{x} | \boldsymbol{x} = p_1 \boldsymbol{x}_1 + p_2 \boldsymbol{x}_2, \quad p_1, p_2 \in \mathbb{R}, \quad p_1 + p_2 = 1\}.$$

A closed line segment joining $\boldsymbol{x}_1$ and $\boldsymbol{x}_2$ is

$$[\boldsymbol{x}_1, \boldsymbol{x}_2] = \{\boldsymbol{x} | \boldsymbol{x} = (1 - \lambda)\boldsymbol{x}_1 + \lambda \boldsymbol{x}_2, \quad 0 \leq \lambda \leq 1\}.$$

Similarly, an open line segment joining $\boldsymbol{x}_1$ and $\boldsymbol{x}_2$ is

$$(\boldsymbol{x}_1, \boldsymbol{x}_2) = \{\boldsymbol{x} | \boldsymbol{x} = (1 - \lambda)\boldsymbol{x}_1 + \lambda \boldsymbol{x}_2, \quad 0 < \lambda < 1\}.$$

A set $S \subseteq \mathbb{R}^d$ is convex if the closed line segment joining every two points of S is in S. Equivalently, S is convex if for $\boldsymbol{x}_1, \boldsymbol{x}_2 \in S, \quad \lambda \in \mathbb{R}, \quad 0 \leq \lambda \leq 1$ we obtain

$$(1 - \lambda)\boldsymbol{x}_1 + \lambda \boldsymbol{x}_2 \in S.$$

For a nonzero vector $\boldsymbol{w} \in \mathbb{R}^d, \boldsymbol{w} \neq \boldsymbol{0}$, and $b \in \mathbb{R}$, the point set $\{\boldsymbol{x} | \boldsymbol{x} \in \mathbb{R}^n, \ \boldsymbol{w} \cdot \boldsymbol{x} < b\}$ is an *open half-space* in $\mathbb{R}^d$, and the set $\{\boldsymbol{x} | \boldsymbol{x} \in \mathbb{R}^n, \ w\boldsymbol{x} \leq b\}$ is a *closed half-space* in $\mathbb{R}^d$. Show with proof that:

(a) Both an open half-space and a closed half-space are convex.

(b) If $A_1, \ldots, A_n$ is a family of convex sets in $\mathbb{R}^d$, then their intersection $\bigcap_{i=1}^n A_i$ is a convex set. Specifically, the intersection of a set of half-spaces—for example, a Voronoi cell—is convex.

1.2 We can follow the dual relationship to compute the Voronoi diagram from the constructed Delaunay triangulation. In three-dimensional space, a Delaunay vertex corresponds to an atom ball, a Delaunay edge corresponds to a Voronoi plane, a Delaunay triangle corresponds to a Voronoi edge, and a Delaunay tetrahedron corresponds to a Voronoi vertex. To obtain the coordinates of a Voronoi vertex $\boldsymbol{v} = (v_1, v_2, v_3) \in \mathbb{R}^3$ from a Delaunay tetrahedron formed by four atoms $b_i(\boldsymbol{z}_i, r_i)$, $b_j(\boldsymbol{z}_j, r_j)$, $b_k(\boldsymbol{z}_k, r_k)$, and $b_l(\boldsymbol{z}_l, r_l)$, which are located at $\boldsymbol{z}_i$, $\boldsymbol{z}_j$, $\boldsymbol{z}_k$, and $\boldsymbol{z}_l$, with radii r_i, r_j, r_k and r_l, respectively, we use the fact that the power distance

$$\pi_i(\boldsymbol{v}) \equiv ||\boldsymbol{v} - \boldsymbol{z}_i||^2 - r_i^2$$

from $\boldsymbol{v}$ to $b_i(\boldsymbol{z}_i, r_i)$ is the same as $\pi_j(\boldsymbol{v})$, $\pi_k(\boldsymbol{v})$, and $\pi_l(\boldsymbol{v})$. Denote this power distance as R^2.

(a) Write down the set of quadratic equations whose solution will provide $\boldsymbol{r} = (r_1, r_2, r_3)$ and R^2.

(b) Define functions $\lambda(\boldsymbol{v}) \equiv \boldsymbol{v} \cdot \boldsymbol{v} - R^2$, and $\lambda(\boldsymbol{z}_i) \equiv \boldsymbol{z}_i \cdot \boldsymbol{z}_i - r_i^2$, and define $\lambda(\boldsymbol{z}_j)$, $\lambda(\boldsymbol{z}_k)$, $\lambda(\boldsymbol{z}_k)$, and $\lambda(\boldsymbol{z}_l)$ similarly. Use $\frac{\lambda(\boldsymbol{v})}{2}$, $\frac{\lambda(\boldsymbol{z}_i)}{2}$, $\frac{\lambda(\boldsymbol{z}_j)}{2}$, $\frac{\lambda(\boldsymbol{z}_k)}{2}$, and $\frac{\lambda(\boldsymbol{z}_l)}{2}$ to simplify the system of quadratic equations into a system of linear equations, whose solution will give $\boldsymbol{r}$ and R^2.

(c) Write down the set of linear equations that determine the Voronoi line dual to a Delaunay triangle.

(d) Write down the linear equation that determines the Voronoi plane dual to a Delaunay edge.

1.3 By growing atom balls using a parameter α, we can generate a family of unions of balls, in which the size of each atom is inflated from r_i to $r_i(\alpha) = (r_i^2 + \alpha)^{1/2}$ [12,15]. We now examine the corresponding Voronoi diagrams.

(**a**) In the Voronoi diagram, every point $\boldsymbol{x}$ on the separator surface for the two original atoms $(\boldsymbol{z}_i, r_i)$ and $(\boldsymbol{z}_j, r_j)$ has equal power distances $\pi_i(\boldsymbol{x})$ and $\pi_j(\boldsymbol{x})$ to the two atoms. Write down the equation for the separator surface. Is the separator surface elliptic, parabolic, or planar?

(**b**) Now we inflate both atoms by α such that we have two new balls with different radii $(\boldsymbol{z}_i, r_i(\alpha))$ and $(\boldsymbol{z}_j, r_j(\alpha))$. Write down the equation for the separator surface.

(**c**) What is the relationship between these two separator surfaces? What is the relationship between the two corresponding Voronoi diagrams?

1.4 The Voronoi diagrams can be generalized using different distance functions. When considering atoms of different radii, instead of replacing the Euclidean distance $||\boldsymbol{x} - \boldsymbol{z}_i||$ with the power distance $\pi_i(\boldsymbol{x})$, we can use the additive distance:

$$d_i(\boldsymbol{x}) \equiv ||\boldsymbol{x} - \boldsymbol{z}_i|| - r_i.$$

The resulting Voronoi diagram is called the additively weighted Voronoi diagram.

(**a**) Write down the equation for the separator surface formed by the set of points with equal additive distances to the two atoms $(\boldsymbol{z}_i, r_i)$ and $(\boldsymbol{z}_j, r_j)$. Is the separator surface elliptic, parabolic, or planar?

(**b**) Now we inflate both atoms by α such that we have two new balls with different radii $(\boldsymbol{z}_i, r_i + \alpha)$ and $(\boldsymbol{z}_j, r_j + \alpha)$. Write down the equation for the separator surface. Is the separator surface elliptic, parabolic, or planar?

(**c**) Is there a simple relationship between these two separator surfaces or between the two corresponding Voronoi diagrams?

2

SCORING FUNCTIONS FOR PREDICTING STRUCTURE AND BINDING OF PROTEINS

2.1 INTRODUCTION

In the experimental work that led to the recognition of the 1972 Nobel prize, Christian Anfinsen showed that a completely unfolded protein ribonuclease could refold spontaneously to its biologically active conformation. This observation indicated that the sequence of amino acids of a protein contains all of the information needed to specify its three-dimensional structure [5,6]. The automatic *in vitro* refolding of denatured proteins was further confirmed in many other protein systems [50]. Anfinsen's experiments led to the thermodynamic hypothesis of protein folding, which postulates that a native protein folds into a three-dimensional structure in equilibrium, in which the state of the whole protein–solvent system corresponds to the global minimum of free energy under physiological conditions.

Based on this thermodynamic hypothesis, computational studies of proteins, including structure prediction, folding simulation, and protein design, all depend on the use of a potential function for calculating the effective energy of the molecule. In protein structure prediction, a potential function is used either to guide the conformational search process or to select a structure from a set of possible sampled candidate structures. Potential function has been developed through an inductive approach [113], where the parameters are derived by matching the results from quantum-mechanical calculations on small molecules to experimentally measured thermodynamic properties of simple molecular systems. These potential functions are then generalized to the macromolecular level based on the assumption that the

Models and Algorithms for Biomolecules and Molecular Networks, First Edition. Bhaskar DasGupta and Jie Liang.

complex phenomena of macromolecular systems result from the combination of a large number of interactions as found in the most basic molecular systems. For example, a version of the potential function in the CHARMM force field takes the form of

$$\begin{aligned} U = &\sum_{\text{bonds}} k_b(b-b_0)^2 + \sum_{\text{angles}} k_\theta(\theta-\theta_0)^2 + \sum_{\text{dihedrals}} k_\phi[1+\cos(n\phi-\delta)] \\ &+ \sum_{\text{impropers}} k_\omega(\omega-\omega_0)^2 + \sum_{\text{Urey-Bradley}} k_u(u-u_0)^2 \\ &+ \sum_{\text{nonbonded}} \epsilon\left[\left(\frac{R_{\min,ij}}{r_{ij}}\right)^{12} - \left(\frac{R_{\min,ij}}{r_{ij}}\right)^{6}\right] + \frac{q_i q_j}{\epsilon \cdot r_{ij}}. \end{aligned} \tag{2.1}$$

Here the first term accounts for bond stretching, with k_b the bond force constant and b_0 the resting bond distance. The second term accounts for the bond angles, with k_θ the angel force constant and θ_0 the stationary angle between three atoms. The third term accounts for the dihedral angles, with k_ϕ the dihedral force constant, n the multiplicity of the function, ϕ the dihedral angle, and δ the phase shift. The fourth term accounts for improper out-of-plane bending, with k_ω the bending force constant and $\omega - \omega_0$ the out-of-plane bending angle. The fifth term accounts for Urey–Bradley cross-term angle bending, with k_u the force constant and $u - u_0$ the distance in the harmonic potential. The sixth term takes the form of the Lennard-Jones potential and accounts for the van der Waals interactions between the (i, j) pair of atoms, which are separated by at least three bonds. The last term is to account for the electrostatic energy and takes the form of a Coulombic potential [15].

Such potential functions are often referred to as "physics-based," "semiempirical" effective potential function, or a force field [15,55,70,90,127]. The physics-based potential functions have been extensively studied and have found wide use in protein folding studies [15,29,67]. Nevertheless, it is difficult to use physics-based potential functions for protein structure prediction, because they are based on the full atomic model and therefore require high computational costs. In addition, such potential function may not fully capture all of the important physical interactions.

Another type of potential function is developed following a deductive approach by extracting parameters of the potential functions from a database of known protein structures [113]. Because this approach incorporates physical interactions (electrostatic, van der Walls, cation–π interactions) only implicitly and the extracted potentials do not necessarily reflect true energies, it is often referred to as "knowledge-based effective energy function," "empirical potential function," or "scoring function." This approach became attractive partly due to the rapidly growing database of experimentally determined three-dimensional protein structures. Successes in protein folding, protein–protein docking, and protein design have been achieved using knowledge-based scoring functions [48,51,100,102,123]. An example of the empirical potential function used in the ROSETTA software for protein structure prediction and design is described in details in reference 100. In this chapter, we focus our discussion on this type of potential functions.

We first discuss the theoretical frameworks and methods for developing knowledge-based potential functions. We then discuss in some detail the Miyazawa–Jernigan contact statistical potential, distance-dependent statistical potentials, and geometric statistical potentials. We also describe a geometric model for developing both linear and nonlinear potential functions by optimization. Applications of knowledge-based potential functions in protein–decoy discrimination, in protein–protein interactions, and in protein design are then described. Several issues of knowledge-based potential functions are further discussed.

2.2 GENERAL FRAMEWORK OF SCORING FUNCTION AND POTENTIAL FUNCTION

Different approaches have been developed to extract knowledge-based scoring functions or potential functions from protein structures. They can be categorized into two groups. One prominent group of knowledge-based potentials are those derived from statistical analysis of database of protein structures [80,87,104,115]. In this class of potentials, the interacting potential between a pair of residues are estimated from its relative frequency in database when compared with that of a reference state or a null model [52,69,80,88,104,110,114,126]. A different class of knowledge-based potentials are based on the principle of optimization. In this case, the set of parameters for the potential functions are optimized by some criterion — for example, by maximizing the energy gap between known native conformation and a set of alternative (or decoy) conformations [8,27,28,42,48,81,84,116,118,124,125].

There are three main ingredients for developing a knowledge-based potential function. We first need *protein descriptors* to describe the sequence and the shape of the native protein structure in a format that is suitable for computation. We then need to decide on a *functional form* of the potential function. Finally, we need a *method to derive the values of the parameters* for the potential function.

2.2.1 Protein Representation and Descriptors

To describe the geometric shape of a protein and its sequence of amino acid residues, a protein can be represented by a d-dimensional descriptor $\boldsymbol{c} \in \mathbb{R}^d$. For example, a widely used method is to count nonbonded contacts of 210 types of amino acid residue pairs in a protein structure. In this case, the count vector $\boldsymbol{c} \in \mathbb{R}^d, d = 210$, is used as the protein descriptor. Once the structural conformation of a protein $\boldsymbol{s}$ and its amino acid sequence $\boldsymbol{a}$ is given, the protein descriptions $f : (\boldsymbol{s}, \boldsymbol{a}) \mapsto \mathbb{R}^d$ will fully determine the d-dimensional vector $\boldsymbol{c}$. In the case of contact descriptor, f corresponds to the mapping provided by specific contact definition; for example, two residues are in contact if their distance is below a cutoff threshold distance. At the residue level, the coordinates of C_α, C_β, or side-chain center can be used to represent the location of a residue. At the atomic level, the coordinates of atoms are directly used, and contact may be defined by the spatial proximity of atoms. In addition, other features of protein structures can be used as protein descriptors, including distances between residue or atom pairs, solvent accessible surface areas, dihedral angles of backbones and side-chains, and packing densities.

2.2.2 Functional Form

The form of the potential function $H : \mathbb{R}^d \mapsto \mathbb{R}$ determines the mapping of a d-dimensional descriptor $\boldsymbol{c}$ to a real energy value. A widely used functional form for potential function H is the weighted linear sum of pairwise contacts [80,87,104,115,118,124]:

$$H(f(s, \boldsymbol{a})) = H(\boldsymbol{c}) = \boldsymbol{w} \cdot \boldsymbol{c} = \sum_i w_i c_i, \tag{2.2}$$

where "·" denotes inner product of vectors and c_i is the number of occurrence of the ith type of descriptor. Once the weight vector $\boldsymbol{w}$ is specified, the potential function is fully defined. In Subsection 2.4.3, we will discuss a nonlinear form potential function.

2.2.3 Deriving Parameters of Potential Functions

For statistical knowledge-based potential functions, the weight vector $\boldsymbol{w}$ for linear potential is derived by characterization of the frequency distributions of structural descriptors from a database of experimentally determined protein structures. For the optimized knowledge-based linear potential function, $\boldsymbol{w}$ is obtained through optimization. We describe the details of these two approaches below.

2.3 STATISTICAL METHOD

2.3.1 Background

In statistical methods, the observed frequencies of protein structural features are converted into effective free energies, based on the assumption that frequently observed structural features correspond to low-energy states [87,112,115]. This idea was first proposed by Tanaka and Scheraga in their work to estimate potentials for pairwise interaction between amino acids [115]. Miyazawa and Jernigan (1985) significantly extended this idea and derived a widely used statistical potentials, where solvent terms are explicitly considered and the interactions between amino acids are modeled by contact potentials. Sippl (1990) and others [80,104,137] derived distance-dependent energy functions to incorporate both short-range and long-range pairwise interactions. The pairwise terms were further augmented by incorporating dihedral angles [57,95], solvent accessibility, and hydrogen-bonding [95]. Singh and Tropsha (1996) derived potentials for higher-order interactions [111]. More recently, Ben-Naim (1997) presented three theoretical examples to demonstrate the nonadditivity of three-body interactions [10]. Li and Liang (2005) identified three-body interactions in native proteins based on an accurate geometric model, and they quantified systematically the nonadditivities of three-body interactions [75].

2.3.2 Theorctical Model

At the equilibrium state, an individual molecule may adopt many different conformations or microscopic states with different probabilities. It is assumed that the distribution of protein molecules among the microscopic states follows the Boltzmann distribution, which connects the potential function $H(\boldsymbol{c})$ for a microstate $\boldsymbol{c}$ to its *probability of occupancy* $\pi(\boldsymbol{c})$. This probability $\pi(\boldsymbol{c})$ or the Boltzmann factor is

$$\pi(\boldsymbol{c}) = \exp[-H(\boldsymbol{c})/kT]/Z(\boldsymbol{a}), \tag{2.3}$$

where k and T are the Boltzmann constant and the absolute temperature measured in Kelvin, respectively. The partition function $Z(\boldsymbol{a})$ is defined as

$$Z(\boldsymbol{a}) \equiv \sum_{\boldsymbol{c}} \exp[-H(\boldsymbol{c})/kT]. \tag{2.4}$$

It is a constant under the true energy function once the sequence $\boldsymbol{a}$ of a protein is specified, and it is independent of the representation $f(\boldsymbol{s}, \boldsymbol{a})$ and descriptor $\boldsymbol{c}$ of the protein. If we are able to measure the probability distribution $\pi(\boldsymbol{c})$ accurately, we can obtain the knowledge-based potential function $H(\boldsymbol{c})$ from the Boltzmann distribution:

$$H(\boldsymbol{c}) = -kT \ln \pi(\boldsymbol{c}) - kT \ln Z(\boldsymbol{a}). \tag{2.5}$$

The partition function $Z(\boldsymbol{a})$ cannot be obtained directly from experimental measurements. However, at a fixed temperature, $Z(\boldsymbol{a})$ is a constant and has no effect on the different probability of occupancy for different conformations.

In order to obtain an knowledge-based potential function that encodes the sequence–structure relationship of proteins, we have to remove background interactions $H'(\boldsymbol{c})$ that are independent of the protein sequence and the protein structure. These generic energetic contributions are referred collectively as that of the *reference state* [112]. An *effective potential energy* $\Delta H(\boldsymbol{c})$ is then obtained as

$$\Delta H(\boldsymbol{c}) = H(\boldsymbol{c}) - H'(\boldsymbol{c}) = -kT \ln \left[\frac{\pi(\boldsymbol{c})}{\pi'(\boldsymbol{c})}\right] - kT \ln \left[\frac{Z(\boldsymbol{a})}{Z'(\boldsymbol{a})}\right], \tag{2.6}$$

where $\pi'(\boldsymbol{c})$ is the probability of a sequence adopting a conformation specified by the vector $\boldsymbol{c}$ in the reference state. Since $Z(\boldsymbol{a})$ and $Z'(\boldsymbol{a})$ are both constants, $-kT \ln(Z(\boldsymbol{a})/Z'(\boldsymbol{a}))$ is also a constant that does not depend on the descriptor vector $\boldsymbol{c}$. If we assume that $Z(\boldsymbol{a}) \approx Z'(\boldsymbol{a})$ as in reference 112, the effective potential energy can be calculated as:

$$\Delta H(\boldsymbol{c}) = -kT \ln \left[\frac{\pi(\boldsymbol{c})}{\pi'(\boldsymbol{c})}\right]. \tag{2.7}$$

To calculate $\pi(\boldsymbol{c})/\pi'(\boldsymbol{c})$, one can further assume that the probability distribution of each descriptor is independent, and we have $\pi(\boldsymbol{c})/\pi'(\boldsymbol{c}) = \prod_i \left[\frac{\pi(c_i)}{\pi'(c_i)}\right]$. Furthermore, by assuming that each occurrence of the ith descriptor is independent, we have $\prod_i \left[\frac{\pi(c_i)}{\pi'(c_i)}\right] = \prod_i \prod_{c_i} \left[\frac{\pi_i}{\pi'_i}\right]$, where π_i and π'_i are the probability of and ith-type structural feature in native proteins and the reference state, respectively. In a linear potential function, the right-hand side of Eq. (2.7) can be calculated as

$$-kT \ln \left[\frac{\pi(\boldsymbol{c})}{\pi'(\boldsymbol{c})}\right] = -kT \sum_i c_i \ln \left[\frac{\pi_i}{\pi'_i}\right]. \tag{2.8}$$

Correspondingly, to calculate the effective potential energy $\Delta H(\boldsymbol{c})$ of the system, one often assumes that $\Delta H(\boldsymbol{c})$ can be decomposed into various basic energetic terms. For a linear potential function, $\Delta H(\boldsymbol{c})$ can be calculated as

$$\Delta H(\boldsymbol{c}) = \sum_i \Delta H(c_i) = \sum_i c_i w_i. \tag{2.9}$$

If the distribution of each c_i is assumed to be linearly independent to the others in the native protein structures, we have

$$w_i = -kT \ln \left[\frac{\pi_i}{\pi'_i}\right]. \tag{2.10}$$

In another words, the probability of each structural feature in native protein structures follows the Boltzmann distribution. This is the *Boltzmann assumption* made in nearly all statistical potential functions. Finkelstein (1995) summarized protein structural features which are observed to correlate with the Boltzmann distribution. These include the distribution of residues between the surface and interior of globules, the occurrence of various ϕ, ψ, χ angles, *cis* and *trans* prolines, ion pairs, and empty cavities in protein globules [33].

The probability π_i can be estimated by counting frequency of the ith structural feature after combining all structures in the database. The probability π_i is determined once a database of crystal structures is given. The probability π'_i is calculated as the probability of the ith structural feature in the reference state. The choice of the reference state has important effects and is critical for developing knowledge-based statistical potential function.

2.3.3 Miyazawa–Jernigan Contact Potential

Because of the importance of the Miyazawa–Jernigan model in developing statistical knowledge-based potential and its wide use, we discuss the Miyazawa–Jernigan contact potential in details. This also gives an exposure of different technical aspects of developing statistical knowledge-based potential functions.

Residue Representation and Contact Definition. In the Miyazawa–Jernigan model, the lth residue is represented as single ball located at its side-chain center $\boldsymbol{z}_l$. If the lth residue is a Gly residue, which lacks a side chain, the positions of the C^α atom is taken as $\boldsymbol{z}_l$. A pair of residues (l, m) are defined to be in contact if the distance between their side-chain centers is less than a threshold $\theta = 6.5$ Å. Neighboring residues l and m along amino acid sequences ($|l - m| = 1$) are excluded from statistical counting because they are likely to be in spatial contact that does not reflect the intrinsic preference for inter-residue interactions. Thus, a contact between the lth and mth residues is defined as $\Delta_{(l,m)}$:

$$\Delta_{(l,m)} = \begin{cases} 1, & \text{if } |\boldsymbol{z}_l - \boldsymbol{z}_m| \leq \theta \text{ and } |l - m| > 1 \\ 0, & \text{otherwise,} \end{cases},$$

where $|\boldsymbol{z}_l - \boldsymbol{z}_m|$ is the Euclidean distance between the lth and mth residues. Hence, the total number count of (i, j) contacts of residue type i with residue type j in protein p is

$$n_{(i,j);\,p} = \sum_{\substack{l,m,\\ l<m}} \Delta_{(l,m)} \qquad \text{if } (\square(l), \square(m)) = (i, j) \text{ or } (j, i), \tag{2.11}$$

where $\square(l)$ is the residue type of the lth amino acid residue. The total number count of (i, j) contacts in all proteins are then

$$n_{(i,j)} = \sum_{p} n_{(i,j);\,p}, \qquad i, j = 1, 2, \ldots, 20. \tag{2.12}$$

Coordination and Solvent Assumption. The number of different types of pairwise residue–residue contacts $n_{(i,j)}$ can be counted directly from the structure of proteins following Eq. (2.12). We also need to count the number of residue–solvent contacts. Since solvent molecules are not consistently present in X-ray crystal structures, and therefore cannot be counted exactly, Miyazawa and Jernigan made an assumption based on the model of an effective solvent molecule, which has the volume of the average volume of the 20 types of residues. Physically, one effective solvent molecule may represent several real water molecules or other solvent molecules. The number of residue–solvent contacts $n_{(i,0)}$ can be estimated as

$$n_{(i,0)} = q_i n_{(i)} - \left(\sum_{\substack{j=1;\\ j \neq i}}^{20} n_{(i,j)} + 2n_{(i,i)} \right), \tag{2.13}$$

where the subscript 0 represents the effective solvent molecule; the other indices i and j represent the types of amino acids; $n_{(i)}$ is the number of residue type i in the set of proteins; q_i is the mean coordination number of buried residue i, calculated as the number of contacts formed by a buried residue of type i averaged over a structure database. Here the assumption is that residues make the same number of contacts on average, with either effective solvent molecules (first term in Eq. (2.13), or other residues (second term in Eq. (2.13)).

For convenience, we calculate the total number of residues $n_{(r)}$, of residue–residue contacts $n_{(r,r)}$, of residue–solvent contacts $n_{(r,0)}$, and of pairwise contacts of any type $n_{(\cdot,\cdot)}$ as follows:

$$n_{(r)} = \sum_{i=1}^{20} n_{(i)}, \qquad n_{(i,r)} = n_{(r,i)} = \sum_{j=1}^{20} n_{(i,j)}, \qquad n_{(r,r)} = \sum_{i=1}^{20} n_{(i,r)}$$

$$n_{(r,0)} = n_{(0,r)} = \sum_{i=1}^{20} n_{(i,0)}, \qquad n_{(\cdot,\cdot)} = n_{(r,r)} + n_{(r,0)} + n_{(0,0)}.$$

Chemical Reaction Model. Miyazawa and Jernigan (1985) developed a physical model based on hypothetical chemical reactions. In this model, residues of type i and j in solution need to be desolvated before they can form a contact. The overall reaction is the formation of (i, j) contacts, depicted in Fig. 2.1. The total free energy change to form one pair of (i, j) contact from fully solvated residues of i and j is (Fig. 2.1a):

$$e_{(i,j)} = (E_{(i,j)} + E_{(0,0)}) - (E_{(i,0)} + E_{(j,0)}), \tag{2.14}$$

where $E_{(i,j)}$ is the absolute contact energy between the ith and jth types of residues, and $E_{(i,j)} = E_{(j,i)}$, $E_{(i,0)}$ are the absolute contact energy between the ith residue and effective solvent, and $E_{(i,0)} = E_{(0,i)}$; likewise for $E_{(j,0)}$; $E_{(0,0)}$ are the absolute contact energies of solvent–solvent contacts $(0, 0)$.

The overall reaction can be decomposed into two steps (Fig. 2.1b). In the first step, residues of type i and type j, initially fully solvated, are desolvated or "demixed from solvent" to form self-pairs (i, i) and (j, j). The free energy changes $e_{(i,i)}$ and

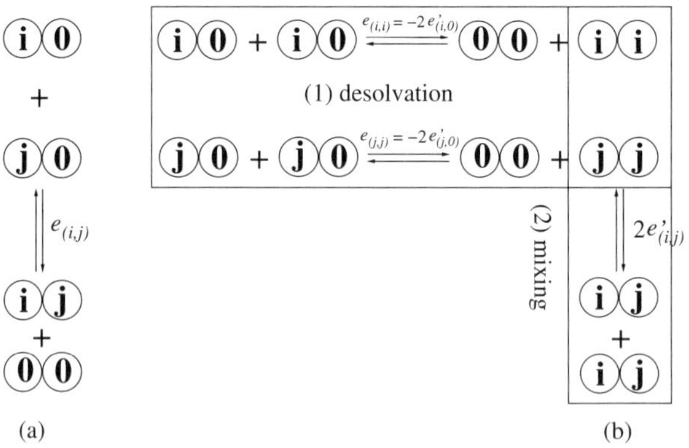

FIGURE 2.1 The Miyazawa–Jernigan model of chemical reaction. Amino acid residues first go through the desolvation process and then mix together to form pair contact interactions. The associated free energies of desolvation $e_{(i,i)}$ and mixing $e'_{(i,j)}$ can be obtained from the equilibrium constants of these two processes.

$e_{(j,j)}$ upon this desolvation step can be seen from the desolvation process (horizontal box) in Fig. 2.1 as

$$\begin{aligned} e_{(i,i)} &= E_{(i,i)} + E_{(0,0)} - 2E_{(i,0)}, \\ e_{(j,j)} &= E_{(j,j)} + E_{(0,0)} - 2E_{(j,0)}, \end{aligned} \tag{2.15}$$

where $E_{(i,i)}$, $E_{(j,j)}$ are the absolute contact energies of self-pair (i,i) and (j,j), respectively. In the second step, the contacts in (i,i) and (j,j) pairs are broken and residues of type i and residues of type j are mixed together to form two (i,j) pairs. The free energy change upon this mixing step $2e'_{(i,j)}$ is (vertical box in Fig. 2.1):

$$2e'_{(i,j)} = 2E_{(i,j)} - (E_{(i,i)} + E_{(j,j)}). \tag{2.16}$$

Denote the free energy changes upon the mixing of residue of type i and solvent as $e'_{(i,0)}$, We have

$$-2e'_{(i,0)} = e_{(i,i)} \quad \text{and} \quad -2e'_{(j,0)} = e_{(j,j)}, \tag{2.17}$$

which can be obtained from Eq. (2.15) and Eq. (2.16) after substituting "j" with "0". Following the reaction model of Fig. 2.1b, the total free energy change to form one pair of (i,j) can be written as

$$2e_{(i,j)} = 2e'_{(i,j)} + e_{(i,i)} + e_{(j,j)} \tag{2.18a}$$

$$= 2e'_{(i,j)} - 2e'_{(i,0)} - 2e'_{(j,0)}. \tag{2.18b}$$

Contact Energy Model. The total energy of the system is due to the contacts between residue–residue, residue–solvent, solvent–solvent:

$$\begin{aligned} E_c &= \sum_{i=0}^{20} \sum_{\substack{j=0; \\ j\geq i}}^{20} E_{(i,j)}\, n_{(i,j)} \\ &= \sum_{i=1}^{20} \sum_{\substack{j=1; \\ j\geq i}}^{20} E_{(i,j)}\, n_{(i,j)} + \sum_{i=1}^{20} E_{(i,0)}\, n_{(i,0)} + E_{(0,0)}\, n_{(0,0)}. \end{aligned} \tag{2.19}$$

Because the absolute contact energies $E_{(i,j)}$ are difficult to measure and knowledge of this value is unnecessary for studying the dependence of energy on protein conformation, we can simplify Eq. (2.19) further. Our goal is to separate out terms that do not depend on contact interactions and hence do not depend on the conformation of the molecule. Equation (2.19) can be rewritten as

$$E_c = \sum_{i=0}^{20} (2E_{(i,0)} - E_{(0,0)})\, q_i n_{(i)}/2 + \sum_{i=1}^{20} \sum_{\substack{j=1;\\ j \geq i}}^{20} e_{(i,j)} n_{(i,j)} \tag{2.20a}$$

$$= \sum_{i=0}^{20} E_{(i,i)} q_i n_{(i)}/2 + \sum_{i=0}^{20} \sum_{\substack{j=0;\\ j \geq i}}^{20} e'_{(i,j)} n_{(i,j)} \tag{2.20b}$$

by using Eq. (2.13) and Eq. (2.14). Here only the second terms in Eqs. (2.20a) and (2.20b) depend on protein conformations. Therefore, one needs only to estimate either $e_{(i,j)}$ or $e'_{(i,j)}$. Since the number of residue–residue contacts can be counted directly while the number of residue–solvent contacts is more difficult to obtain, Eq. (2.20a) is more convenient for calculating the total contact energy of protein conformations. Both $e_{(i,j)}$ and $e'_{(i,j)}$ are termed as *effective contact energies*, and their values were reported in reference 90.

Estimating Effective Contact Energies: quasi-chemical Approximation. The effective contact energies $e_{(i,j)}$ in Eq. (2.20a) can be estimated in kT unit by assuming that the solvent and solute molecules are in quasi-chemical equilibrium for the reaction depicted in Fig. 2.1a:

$$e_{(i,j)} = -\ln \frac{[m_{(i,j)}/m_{(\cdot,\cdot)}][m_{(0,0)}/m_{(\cdot,\cdot)}]}{[m_{(i,0)}/m_{(\cdot,\cdot)}][m_{(j,0)}/m_{(\cdot,\cdot)}]} = -\ln \frac{m_{(i,j)} m_{(0,0)}}{m_{(i,0)} m_{(j,0)}}, \tag{2.21}$$

where $m_{(i,j)}$, $m_{(i,0)}$, and $m_{(0,0)}$ are the contact numbers of pairs between residue type i and j, residue type i and solvent, and solvent and solvent, respectively. $m_{(\cdot,\cdot)}$ is the total number of contacts in the system and is canceled out. Similarly, $e'_{(i,j)}$ and $e'_{(i,0)}$ can be estimated from the model depicted in Fig. 2.1b:

$$2e'_{(i,j)} = -\ln \frac{[m_{(i,j)}]^2}{m_{(i,i)} m_{(j,j)}}, \tag{2.22a}$$

$$2e'_{(i,0)} = -\ln \frac{[m_{(i,0)}]^2}{m_{(i,i)} m_{(0,0)}}. \tag{2.22b}$$

Based on these models, two different techniques have been developed to obtain effective contact energy parameters. Following the hypothetical reaction in Fig. 2.1a, $e_{(i,j)}$ can be directly estimated from Eq. (2.21), as was done by Zhang and Kim [131]. Alternatively, one can follow the hypothetical two-step reaction in Fig. 2.1b and estimate each term in Eq. (2.18b) for $e_{(i,j)}$ by using Eq. (2.22). Because the second approach leads to additional insight about the desolvation effects ($e'_{(i,0)}$) and the mixing effects ($e'_{(i,j)}$) in contact interactions, we follow this approach in subsequent discussions. The first approach will become self-evident after our discussion.

Models of Reference State. In reality, the true fraction $\frac{m_{(i,j)}}{m_{(\cdot,\cdot)}}$ of contacts of (i, j) type among all pairwise contacts $(\cdot, \cdot)$ is unknown. One can approximate this by calculating its mean value from sampled structures in the database. We have

$$\frac{m_{(i,j)}}{m_{(\cdot,\cdot)}} \approx \frac{\sum_p n_{(i,j);p}}{\sum_p n_{(\cdot,\cdot);p}}, \qquad \frac{m_{(i,0)}}{m_{(\cdot,\cdot)}} \approx \frac{\sum_p n_{(i,0);p}}{\sum_p n_{(\cdot,\cdot);p}}, \qquad \frac{m_{(0,0)}}{m_{(\cdot,\cdot)}} \approx \frac{\sum_p n_{(0,0);p}}{\sum_p n_{(\cdot,\cdot);p}},$$

where i and $j \neq 0$. However, this yields a biased estimation of $e'_{(i,j)}$ and $e_{(i,j)}$. When effective solvent molecules, residues of ith and jth types are randomly mixed, $e'_{(i,j)}$ will not equal to 0 as should be because of differences in amino acid composition among proteins in the database. Therefore, a reference state must be used to remove this bias.

In the work of Miyazawa and Jernigan, the effective contact energies for mixing two types of residues $e'_{(i,j)}$ and for solvating a residue $e'_{(i,0)}$ are estimated based on two different random mixture reference states [87]. In both cases, the contacting pairs in a structure are randomly permuted, but the global conformation is retained. Hence, the total number of residue–residue, residue–solvent, and solvent–solvent contacts remain unchanged.

The first random mixture reference state for desolvation contains the same set of residues of the protein p and a set of effective solvent molecules. We denote the overall number of $(i, i), (i, 0), (0, 0)$ contacts in this random mixture state after summing over all proteins as $c'_{(i,i)}$, $c'_{(i,0)}$, and $c'_{(0,0)}$, respectively. $c'_{(i,i)}$ can be computed as

$$c'_{(i,i)} = \sum_p \left[\frac{q_i\, n_{i;p}}{\sum_k q_k\, n_{k;p}} \right]^2 \cdot n_{(\cdot,\cdot);p}, \tag{2.23}$$

where Miyazawa and Jernigan assumed that the average coordination number of residue i in all proteins is q_i. Therefore, a residue of type i makes $q_i n_{i;p}$ number of contacts in protein p. Similarly, the number of $(i, 0)$ contacts $c'_{(i,0)}$ can be computed as

$$c'_{(i,0)} = \sum_p \left[\frac{q_i\, n_{i;p}}{\sum_k q_k\, n_{k;p}} \right] n_{(\cdot,0);p}. \tag{2.24}$$

From the horizontal box in Fig. 2.1, the effective contact energy $e'_{(i,0)}$ can now be computed as

$$2e'_{(i,0)} = -\ln \left[\frac{n^2_{(i,0)}}{n_{(i,i)} n_{(0,0)}} \Big/ \frac{c'^{\,2}_{(i,0)}}{c'_{(i,i}c'_{(0,0)}} \right] \qquad (i \neq 0). \tag{2.25}$$

The second random mixture reference state for mixing contains the exact same set of residues as the protein p, but have all residues randomly mixed. We denote the number of (i, j) contacts in this random mixture as $c_{(i,j);p}$. The overall number of (i, j) contacts in the full protein set $c_{(i,j)}$ is the sum of $c_{(i,j);p}$ over all proteins:

$$c_{(i,j)} = \sum_p \left[\frac{n_{(i,\cdot);p}}{n_{(\cdot,\cdot);p}}\right] \left[\frac{n_{(j,\cdot);p}}{n_{(\cdot,\cdot);p}}\right] \cdot n_{(\cdot,\cdot);p}. \tag{2.26}$$

From the vertical box in Fig. 2.1, the effective contact energy $e'_{(i,j)}$ can now be computed as

$$2e'_{(i,j)} = -\ln\left[\frac{n^2_{(i,j)}}{n_{(i,i)}n_{(j,j)}} \Big/ \frac{c^2_{(i,j)}}{c_{(i,i)}c_{(j,j)}}\right], \qquad i \text{ or } j \neq 0. \tag{2.27}$$

The compositional bias is removed by the denominator in Eq. (2.27), and $e'_{(i,j)}$ now equals to 0.

Although $c'_{(0,0)}$ can be estimated from Eq. (2.22b) by assuming that $e'_{(i,0)} = 0$ in a reference state, Zhang and DeLisi simplified the Miyazawa–Jernigan process by further assuming that the numbers of solvent–solvent contacts in both reference states to be the same as in the native state [135]:

$$c'_{(0,0)} = n_{(0,0)}. \tag{2.28}$$

Therefore, $c'_{(0,0)}$ and $n_{(0,0)}$ are canceled out in Eq. (2.25) and not needed for calculating $e'_{(i,0)}$. This treatment systematically subtracts a constant scaling energy from all effective energies $e_{(i,j)}$ and should produce exactly the same relative energy values for protein conformations as Miyazawa–Jernigan's original work, with the difference of a constant offset value. In fact, Miyazawa and Jernigan (1996) showed that this constant scaling energy is the effective contact energy $e_{\hat{r}\hat{r}}$ between the average residue $\hat{r}$ of the 20 residue types and suggested that $e_{(i,j)} - e_{\hat{r}\hat{r}}$ be used to measure the stability of a protein structure [88].

Hydrophobic nature of Miyazawa–Jernigan contact potential. In the relation of Eq. (2.18b), $e_{(i,j)} = e'_{(i,j)} - (e'_{(i,0)} + e'_{(j,0)})$, the Miyazawa–Jernigan effective contact energy $e_{(i,j)}$ is composed of two types of terms: the desolvation terms $e'_{(i,0)}$ and $e'_{(j,0)}$ and the mixing term $e'_{(i,j)}$. The desolvation term of residue type i, that is, $-e'_{(i,0)}$ or $e_{(i,i)}/2$ (Fig. 2.1), is the energy change due to the desolvation of residue i, the formation of the i–i self-pair, and the solvent–solvent pair. The value of this term $e_{(i,i)}/2$ should correlate well with the hydrophobicity of residue type i [72,87], although for charged amino acids this term also incorporates unfavorable electrostatic potentials of self-pairing. The mixing term $e'_{(i,j)}$ is the energy change accompanying the mixing of two different types of amino acids of i and j to form a contact pair

i–j after breaking self-pairs i–i and j–j. Its value measures the tendency of different residues to mix together. For example, the mixing between two residues with opposite charges are more favorable than mixing between other types of residues, because of the favorable electrostatic interactions.

Important insights into the nature of residue–residue contact interactions can also be obtained by a quantitative analysis of the desolvation terms and the mixing terms. Among different types of contacts, the average difference of the desolvation terms is 9 times larger than that of the mixing terms (see Table 2.1 taken from reference 88). Thus, a comparison of the values of $(e_{(i,i)} + e_{jj})/2$ and $e'_{(i,j)}$ shows that the desolvation term plays the dominant role in determining the energy difference among different conformations. The importance of hydrophobicity in the Miyazawa–Jernigan contact energies reveals that the hydrophobic effect is the dominant driving force for protein folding. This conclusion justifies the HP model proposed by Chan and Dill (1990), where only hydrophobic interactions are included in studies of simple models of protein folding [20].

2.3.4 Distance-Dependent Potential Function

In the Miyazawa–Jernigan potential function, interactions between amino acids are assumed to be short-ranged, and a distance cutoff is used to define the occurrence of a contact. This type of statistical potential is referred to as the "contact potential." Another class of statistical potential allows modeling of residue interactions that are distance-dependent. The distance of interactions are usually divided into a number of small intervals or bins, and the potential functions are derived by applying Eq. (2.10) for individual distance intervals.

Formulation of Distance-Dependent Potential Functions. In distance-dependent statistical potential functions, Eq. (2.10) can be written in several forms. To follow the conventional notations, we use (i, j) to represent the kth protein descriptor c_k for pairwise interactions between residue type i and residue type j. From Eq. (2.10), we have

$$\begin{aligned}\Delta H(i,j;\,d) &= -\ln \frac{\pi(i,j;\,d)}{\pi'(i,j;\,d)} = -\ln \frac{n_{(i,j;\,d)}/n}{\pi'(i,j;\,d)}\\ &= -\ln \frac{n_{(i,j;\,d)}}{n'_{(i,j;\,d)}},\end{aligned} \tag{2.29a}$$

where $(i, j;\, d)$ represents an interaction between a specific residue pair (i, j) at distance d, $\Delta H(i, j;\, d)$ is the the contribution from the (i, j) type of residue pairs at distance d, $\pi(\beta, j;\, d)$ and $\pi'(i, j;\, d)$ are the observed and expected probabilities of this distance-dependent interaction, respectively, $n_{(i,j;\,d)}$ is the observed number of $(i, j;\, d)$ interactions, n is the observed total number of all pairwise interactions in a database, $n'_{(i,j;\,d)}$ and is the expected number of $(\beta, j;\, d)$ interactions when the total number of all pairwise interactions in the reference state is set to be n.

TABLE 2.1 Miyazawa–Jernigan Contact Energies in kT Units; $e_{(i,j)}$ for Upper Half and Diagonal and $e'_{(i,j)}$ for Lower Half

	Cys	Met	Phe	Ile	Leu	Val	Trp	Tyr	Ala	Gly	Thr	Ser	Asn	Gln	Asp	Glu	His	Arg	Lys	Pro
Cys	-5.44	-4.99	-5.80	-5.50	-5.83	-4.96	-4.95	-4.16	-3.57	-3.16	-3.11	-2.86	-2.59	-2.85	-2.41	-2.27	-3.60	-2.57	-1.95	-3.07
Met	0.46	-5.46	-6.56	-6.02	-6.41	-5.32	-5.55	-4.91	-3.94	-3.39	-3.51	-3.03	-2.95	-3.30	-2.57	-2.89	-3.98	-3.12	-2.48	-3.45
Phe	0.54	-0.20	-7.26	-6.84	-7.28	-6.29	-6.16	-5.66	-4.81	-4.13	-4.28	-4.02	-3.75	-4.10	-3.48	-3.56	-4.77	-3.98	-3.36	-4.25
Ile	0.49	-0.01	0.06	-6.54	-7.04	-6.05	-5.78	-5.25	-4.58	-3.78	-4.03	-3.52	-3.24	-3.67	-3.17	-3.27	-4.14	-3.63	-3.01	-3.76
Leu	0.57	0.01	0.03	-0.08	-7.37	-6.48	-6.14	-5.67	-4.91	-4.16	-4.34	-3.92	-3.74	-4.04	-3.40	-3.59	-4.54	-4.03	-3.37	-4.20
Val	0.52	0.18	0.10	-0.01	-0.04	-5.52	-5.18	-4.62	-4.04	-3.38	-3.46	-3.05	-2.83	-3.07	-2.48	-2.67	-3.58	-3.07	-2.49	-3.32
Trp	0.30	-0.29	0.00	0.02	0.08	0.11	-5.06	-4.66	-3.82	-3.42	-3.22	-2.99	-3.07	-3.11	-2.84	-2.99	-3.98	-3.41	-2.69	-3.73
Tyr	0.64	-0.10	0.05	0.11	0.10	0.23	-0.04	-4.17	-3.36	-3.01	-3.01	-2.78	-2.76	-2.97	-2.76	-2.79	-3.52	-3.16	-2.60	-3.19
Ala	0.51	0.15	0.17	0.05	0.13	0.08	0.07	0.09	-2.72	-2.31	-2.32	-2.01	-1.84	-1.89	-1.70	-1.51	-2.41	-1.83	-1.31	-2.03
Gly	0.68	0.46	0.62	0.62	0.65	0.51	0.24	0.20	0.18	-2.24	-2.08	-1.82	-1.74	-1.66	-1.59	-1.22	-2.15	-1.72	-1.15	-1.87
Thr	0.67	0.28	0.41	0.30	0.40	0.36	0.37	0.13	0.10	0.10	-2.12	-1.96	-1.88	-1.90	-1.80	-1.74	-2.42	-1.90	-1.31	-1.90
Ser	0.69	0.53	0.44	0.59	0.60	0.55	0.38	0.14	0.18	0.14	-0.06	-1.67	-1.58	-1.49	-1.63	-1.48	-2.11	-1.62	-1.05	-1.57
Asn	0.97	0.62	0.72	0.87	0.79	0.77	0.30	0.17	0.36	0.22	0.02	0.10	-1.68	-1.71	-1.68	-1.51	-2.08	-1.64	-1.21	-1.53
Gln	0.64	0.20	0.30	0.37	0.42	0.46	0.19	-0.12	0.24	0.24	-0.08	0.11	-0.10	-1.54	-1.46	-1.42	-1.98	-1.80	-1.29	-1.73
Asp	0.91	0.77	0.75	0.71	0.89	0.89	0.30	-0.07	0.26	0.13	-0.14	-0.19	-0.24	-0.09	-1.21	-1.02	-2.32	-2.29	-1.68	-1.33
Glu	0.91	0.30	0.52	0.46	0.55	0.55	0.00	-0.25	0.30	0.36	-0.22	-0.19	-0.21	-0.19	0.05	-0.91	-2.15	-2.27	-1.80	-1.26
His	0.65	0.28	0.39	0.66	0.67	0.70	0.08	0.09	0.47	0.50	0.16	0.26	0.29	0.31	-0.19	-0.16	-3.05	-2.16	-1.35	-2.25
Arg	0.93	0.38	0.42	0.41	0.43	0.47	-0.11	-0.30	0.30	0.18	-0.07	-0.01	-0.02	-0.26	-0.91	-1.04	0.14	-1.55	-0.59	-1.70
Lys	0.83	0.31	0.33	0.32	0.37	0.33	-0.10	-0.46	0.11	0.03	-0.19	-0.15	-0.30	-0.46	-1.01	-1.28	0.23	0.24	-0.12	-0.97
Pro	0.53	0.16	0.25	0.39	0.35	0.31	-0.33	-0.23	0.20	0.13	0.04	0.14	0.18	-0.08	0.14	0.07	0.15	-0.05	-0.04	-1.75

Source: Miyazawa and Jernigan [88].

Since the expected joint probability $\pi'(i, j;\, d)$ for the reference is not easy to estimate, Sippl (1990) replaces Eq. (2.10) with

$$\Delta H(i, j;\, d) = -\ln \frac{\pi(i, j \mid d)}{\pi'(i, j \mid d)} = -\ln \frac{n_{(i,j;\,d)}/n_{(d)}}{\pi'(i, j \mid d)} = -\ln \frac{n_{(i,j;\,d)}}{n'_{(i,j;\,d)}}, \tag{2.29b}$$

where $\pi(i, j \mid d)$ and $\pi'(i, j \mid d)$ are the observed and expected probability of interaction of residue pairs (i, j) given the distance interval d, respectively; $n_{(d)}$ is the observed total number of all pairwise interactions at the distance d; $n'_{(i,j;\,d)} = \pi'(i, j \mid d) \cdot n(d)$ is the expected number of (i, j) interactions at d when the total number of all pairwise interactions at this distance d in the reference state is set to $n_{(d)}$. There are several variations of potential function of this form, including the *''Knowledge-Based Potential function''* (KBP) by Lu and Skolnick (2001) [80].

In the work of developing the *''Residue-specific All-atom Probability Discriminatory Function''* (RAPDF) [104], Samudrala and Moult (1998) alternatively replaced Eq. (2.10) with

$$\Delta H(i, j;\, d) = -\ln \frac{\pi(d \mid i, j)}{\pi'(d \mid i, j)} = -\ln \frac{n_{(i,j;\,d)}/n_{(i,j)}}{\pi'(d \mid i, j)} = -\ln \frac{n_{(i,j;\,d)}}{n'_{(i,j;\,d)}}, \tag{2.29c}$$

where $\pi(d \mid i, j)$ and $\pi'(d \mid i, j)$ are the observed and expected probability of interaction at the distance d for a given pair of residues (i, j), respectively and $n_{(i,j)}$ is the observed total number of interactions for (i, j) pairs regardless of the distance. $n'_{(i,j;\,d)} = \pi'(d \mid i, j) \cdot n_{(i,j)}$ is the expected number of (i, j) interactions at distance d when the total number of (i, j) interactions in the reference state is set to $n_{(d)}$.

The knowledge-based potential functions of Eqs. (2.29a), (2.29b), and (2.29c) can all be written using the unifying formula based on the number counts of interactions:

$$\Delta H(i, j;\, d) = -\ln \left[\frac{n_{(i,j;\,d)}}{n'_{(i,j;\,d)}} \right]. \tag{2.30}$$

Clearly, the different ways of assigning $n'_{(i,j;\,d)}$ make the potential functions differ from each other significantly, since the method to calculate $n_{(i,j;\,d)}$ is essentially the same for many potential functions. In other words, the model of reference state used to compute $n'_{(i,j;\,d)}$ is critical for distance-dependent energy functions.

Different Models of Reference States. Sippl (1990) first proposed the ''uniform density'' model of reference state, where the probability density function for a pair of contacting residues (i, j) is uniformly distributed along the distance vector connecting them: $\pi'(i, j \mid d) = \pi'(i, j)$ [112]. Lu and Skolnick made use of this type of reference state to calculate the expected number of (i, j) interactions at distance d as [80]

$$n'_{(i,j;\,d)} = \pi'(i, j \mid d) \cdot n_{(d)} = \pi'(i, j) \cdot n_{(d)}.$$

The expected probability $\pi'(i, j)$ is estimated using the random mixture approximation as

$$\pi'(i, j) = \chi_i \chi_j,$$

where χ_i and χ_j are the mole fractions of residue type i and j, respectively.

Samudrala and Moult (1998) made use of another type of reference state, where the probability of the distance between a pair of residues (i, j) being d is independent of the contact types (i, j) [104]:

$$\pi'(d \mid i, j) = \pi'(d).$$

The expected number of (i, j) interactions at distance d in Eq. (2.29c) becomes

$$n'_{(i,j;\,d)} = \pi'(d \mid i, j) \cdot n_{(i,j)} = \pi'(d) \cdot n_{(i,j)},$$

where $\pi'(r)$ is estimated from $\pi(r)$:

$$\pi'(d) = \pi(d) = n_{(d)}/n.$$

Ideal Gas Reference State. In the uniform density model of Sippl, the same density of a particular residue pair (i, j) along a line could result from very different volume distribution of (i, j) pairs in specific regions of the protein. For example, one spherical shell proximal to the molecular center could be sparsely populated with residues, and another distant shell could be densely populated, but all may have the same density of (i, j) pairs along the same radial vector. Zhou and Zhou (2002) developed a new reference state (called DFIRE for ''Distance-scaled, Finite Ideal-gas REference state'') where residues follow uniform distribution everywhere in the protein [137]. Assuming that residues can be modeled as noninteracting points (i.e., as ideal gas molecules), the distribution of noninteracting pairs should follow the uniform distribution not only along any vector lines, but also in the whole volume of the protein.

When the distance between a pair of residues (i, j) is at a threshold distance $d_\theta = 14.5$ Å, the interaction energy between them can be considered to be 0. Therefore, residue type i and type j form pairs at the distance d_θ purely by random chance, and the observed number of (i, j) pairs at the distance d_θ can be considered the same as the expected number of (i, j) pairs at the distance d_θ in the reference state. Denote v_d as the volume of a spherical shell of width Δd at a distance d from the center. The expected number of interactions (i, j) at the distance d after volume correction is

$$n'_{(i,j;\,d)} = n_{(i,j;\,d_\theta)} \cdot \frac{v_d}{d_\theta} = n_{(i,j,d_\theta)} \cdot \left(\frac{d}{d_\theta}\right)^{\alpha} \frac{\Delta d}{\Delta d_\theta}.$$

For a protein molecule, $n'_{(i,j;\,d)}$ will not increase as r^2 because of its finite size. In addition, it is well known that the volume of protein molecule cannot be treated as a solid body, as there are numerous voids and pockets in the interior. This implies that the number density for a very large molecule will also not scale as d^2 [76]. Zhou and Zhou (2002) assumed that $n'_{(i,j;\,d)}$ increase in d^α rather than d^2, where the exponent α needs to be determined. To estimate the α value, each protein p in the database is reshaped into a ball of radius $c_p R_{g;\,p}$, where $R_{g;\,p}$ is the radius of gyration of the protein p, and residues are distributed uniformly in this reshaped ball. Here c_p takes the value so that in the reshaped molecule, the number of total interacting pairs at d_θ distance is about the same as that observed in the native protein p, namely,

$$\sum_{(i,j)} n'_{(i,j;\,d_\theta)} = \sum_{(i,j)} n_{(i,j;\,d_\theta)}$$

for protein p. Once the value of c_p is determined and hence the effective radius $c_p R_{g;\,p}$ for each native protein is known, the number of interacting pairs $n_{(d)}$ at distance d can be counted directly from the reshaped ball. Zhou and Zhou further defined a reduced distance-dependent function $f(d) = n_{(d)}/d^\alpha$ and the relative fluctuation δ of $f(d)$:

$$\delta = \left[\frac{1}{n_b} \sum_d (f(d) - \bar{f})^2/(\bar{f}) \right]^{1/2},$$

where $\bar{f} = \sum_d f(d)/n_b$, and n_b is the total number of distance shells, all of which has the same thickness. α is then estimated by minimizing the relative fluctuation δ. The rationale is that since idealized residues are points and are uniformly distributed in the reshaped ball, δ should be 0. In their study, α was found to be 1.61 [137].

2.3.5 Geometric Potential Functions

The effectiveness of potential function also depends on the representation of protein structures. Another class of knowledge-based statistical potentials is based on the computation of various geometric constructs that reflect the shape of the protein molecules more accurately. These geometric constructs include the Voronoi diagram [82], the Delaunay triangulation [19,65,111,136], and the alpha shape [73–75] of the protein molecules. Geometric potential functions have achieved significant successes in many fields. For example, the potential function developed by McConkey et al. is based on the Voronoi diagram of the atomic structures of proteins and is among one of the best performing atom-level potential functions in decoy discrimination [82]. Because the alpha shape of the molecule contains rich topological, combinatorial, and metric information and has a strong theoretical foundation, we discuss the alpha potential functions in more detail below as an example of this class of potential function.

Geometric Model. In Miyazawa–Jernigan and other contact potential functions, pairwise contact interactions are declared if two residues or atoms are within a

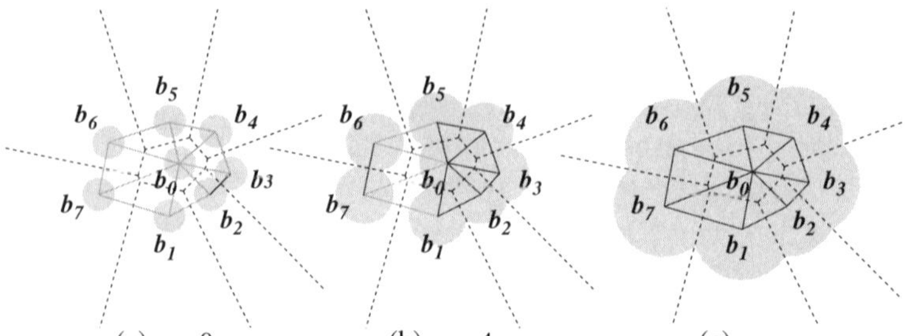

FIGURE 2.2 Schematic drawing of the Delaunay complex and the alpha shape of a two-dimensional molecule. The Voronoi region of a ball is the set of points closest to it when measured in power distance. If two Voronoi regions share a boundary – that is, if there is a Voronoi edge (dashed line) – we draw a Delaunay edge (solid line in grey or black) between these two Voronoi vertices. A Delaunay edge is therefore the *dual* of a Voronoi edge. All Delaunay edges incident to ball residue b_i form the *1-star* for b_i, denoted as $St_1(b_i)$. When the balls are inflated by increasing the α value, more balls overlap, and more Voronoi edges intersect with the balls. Therefore, more *dual* Delaunay edges are included in the alpha shape (shown as black solid line segments). **(a)** When $\alpha = 0.0$, the balls are not inflated and there is only one alpha edge $\sigma_{2,3}$ between ball b_2 and ball b_3. **(b)** When $\alpha = 4.0$, the balls are inflated and their radii are $\sqrt{r^2 + 4.0}$. There are ten alpha edges. For a ball b_i, the set of residue balls connected to it by alpha edges are called the near neighbors of the ball. The number of this set of residue balls is defined as the *degree of near neighbors* of the residue ball b_i, denoted as ρ_i. For example, $\rho_0 = 5$ and $\rho_7 = 1$. **(c)** When $\alpha = \infty$, all the Delaunay edges become alpha edges ($\alpha = 16.0$ is used for drawing). Hence, all long-range interactions not intervened by a third residue are included.

specific cutoff distance. Contacts by distance cutoff can potentially include many implausible noncontacting neighbors, which have no significant physical interaction [13]. Whether or not a pair of residues can make physical contact depends not only on the distance between their center positions (such as C_α or C_β, or geometric centers of side chain), but also on the size and the orientations of side-chains [13]. Furthermore, two atoms close to each other may in fact be shielded from contact by other atoms. By occupying the intervening space, other residues can block a pair of residues from direct interacting with each other. Inclusion of these fictitious contact interactions would be undesirable.

The alpha potential solves this problem by identifying interacting residue pairs following the edges computed in the alpha shape. When the parameter α is set to be 0, residue contact occurs if residues or atoms from nonbonded residues share a Voronoi edge, and this edge is at least partially contained in the body of the molecule. Figure 2.2 illustrates the basic ideas.

Distance and Packing-Dependent Alpha Potential. For two nonbonded residue balls b_i of radius r_i with its center located at $\mathbf{z}_i$ and b_j of radius r_j at $\mathbf{z}_j$, they form an alpha contact $(i, j \mid \alpha)$ if their Voronoi regions intersect and these residue balls also intersect after their radii are inflated to $r_i(\alpha) = (r_i^2 + \alpha)^{1/2}$ and $r_j(\alpha) = (r_j^2 + \alpha)^{1/2}$, respectively. That is, the alpha contact $(i, j \mid \alpha)$ exists when

$$|z_i - z_j| < (r_i{}^2 + \alpha)^{1/2} + (r_j{}^2 + \alpha)^{1/2}, \qquad \sigma_{i,j} \in \mathcal{K}_\alpha \text{ and } |i - j| > 1.$$

We further define the *1-star* for each residue ball b_i as $St_1(b_i) = \{(b_i, b_j) \in \mathcal{K}_\alpha$, namely, the set of 1-simplices with b_i as a vertex. The *near neighbors* of b_i are derived from $St_1(b_i)$ and are defined as

$$\mathcal{N}_\alpha(b_i) \equiv \{b_j | \sigma_{i,j} \in \mathcal{K}_\alpha\}, \qquad \alpha = 4.0.$$

and the *degree of near neighbors* ρ_i of residue b_i is defined as the size of this set of residues:

$$\rho_i \equiv |\mathcal{N}_\alpha(b_i)|, \qquad \alpha = 4.0.$$

The degree of near neighbors ρ_i is a parameter related to the local packing density and hence indirectly related to the solvent accessibility around the residue ball b_i (Fig. 2.2b). A large ρ_i value indicates high local packing density and less solvent accessibility, and a small ρ_i value indicates low local packing density and high solvent accessibility. Similarly, the *degree of near neighbors* for a pair of residues is defined as

$$\rho_{(i,j)} \equiv |\mathcal{N}_\alpha(b_i, b_j)| = |\mathcal{N}_\alpha(b_i)| + |\mathcal{N}_\alpha(b_j)|, \quad \alpha = 4.0.$$

Reference State and Collection of Noninteracting Pairs. We denote the shortest path length between residue b_i and residue b_j as $L_{(i,j)}$, which is the fewest number of alpha edges ($\alpha = 4$) that connects b_i and b_j. The reference state of the alpha potential is based on the collection of all noninteracting residue pairs (i, j):

$$\{(i, j) | L_{(i,j)} = 3\}.$$

Any (i, j) pair in this reference state is intercepted by two residues (Fig. 2.3). We assume that there are no attractive or repulsive interactions between them, because of the shielding effect by the two intervening residues. Namely, residue i and residue j form a pair only by random chance, and any properties associated with b_i, such as packing density, side-chain orientation, are independent of the same properties associated with b_j.

Statistical Model: Pairwise Potential and Desolvation Potential. According to Eq. (2.10), the packing and distance-dependent statistical potential of residue pair (k, l) at the packing environment $\rho_{(k,l)}$ and the distance specified by α is given by

$$H(k, l, \rho_{(k,l)} \mid \alpha) = -k_b T \ln \left(\frac{\pi_{(k,l,\rho_{(k,l)} \mid \alpha)}}{\pi'_{(k,l,\rho_{(k,l)})}} \right). \tag{2.31}$$

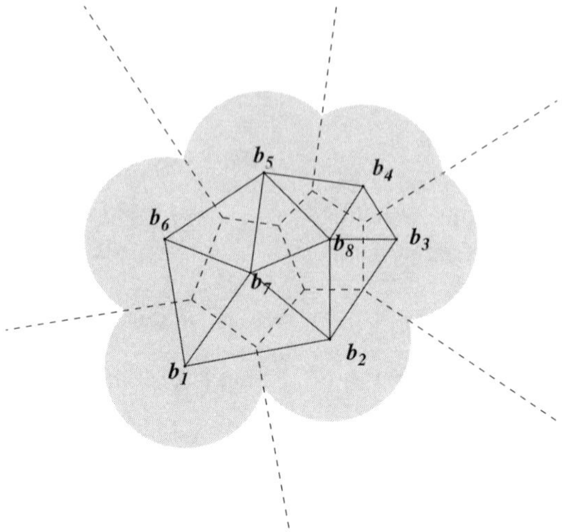

FIGURE 2.3 Schematic illustration of noninteracting pairs of residues. (b_1, b_4) is considered as a noninteracting pair because the shortest length $L_{(1,4)}$ is equal to three; that is, the interaction between b_1 and b_4 is blocked by two residues b_7 and b_8. Likewise, (b_3, b_6) is considered as a noninteracting pair as well.

Here, $\pi_{(k,l,\rho_{(k,l)} \mid \alpha)}$ is the observed probability:

$$\pi_{(k,l,\rho_{(k,l)} \mid \alpha)} = \frac{n_{(k,l,\rho_{(k,l)},\alpha)}}{n_{(\alpha)}}, \tag{2.32}$$

where $n_{(k,l,\rho_{(k,l)},\alpha)}$ is the number of residue pair (k, l) at the packing environment $\rho_{(k,l)}$ and the distance specified by α, and $n_{(\alpha)}$ is the total number of residue pairs at the distance specified by α. $\pi'_{(k,l,\rho_{(k,l)})}$ is the expected probability:

$$\pi'_{(k,l,\rho_{(k,l)})} = \frac{n'_{(k,l,\rho_{(k,l)})}}{n'}, \tag{2.33}$$

where $n'_{(k,l,\rho_{(k,l)})}$ is the number of residue pairs (k, l) at the packing environment $\rho_{(k,l)}$ in reference state, and n' is the total number of noninteracting residue pairs at the reference state.

The desolvation potential of residue type k to have ρ near neighbors $H(\rho \mid k)$ is estimated simply by using Eq. (2.10):

$$H(\rho \mid k) = \frac{\pi_{(\rho \mid k)}}{\pi'_{(\rho \mid k)}} = \frac{[n_{(k,\rho)}/n_{(k)}]}{[n_{(r,\rho)}/n_{(r)}]}, \tag{2.34}$$

where r represents all 20 residue types.

For a protein structure, the total internal energy is estimated by the summation of the desolvation energy and pairwise interaction energy in the particular desolvated environment:

$$H(s,a) = \sum_{k,\rho} H(\rho \mid k) \cdot n_{(k,\rho)} + \frac{1}{2} \sum_{k,l,\rho_{k,l},\alpha} H(k,l,\rho_{(k,l)} \mid \alpha) \cdot n_{(k,l,\rho_{(k,l)},\alpha)}. \tag{2.35}$$

2.4 OPTIMIZATION METHOD

There are several drawbacks of knowledge-based potential function derived from statistical analysis of database. These include the neglect of chain connectivity in the reference state, as well as the problematic implicit assumption of Boltzmann distribution [10,116,117]. We defer a detailed discussion to Section 2.6.1.

An alternative method to develop potential functions for proteins is by optimization. For example, in protein design, we can use the thermodynamic hypothesis of Anfinsen to require that the native amino acid sequence $\boldsymbol{a}_N$ mounted on the native structure $\boldsymbol{s}_N$ has the best (lowest) fitness score compared to a set of alternative sequences (sequence decoys) taken from unrelated proteins known to fold into a different fold $\mathcal{D} = \{\boldsymbol{s}_N, \boldsymbol{a}_D\}$ when mounted on the same native protein structure $\boldsymbol{s}_N$:

$$H(f(\boldsymbol{s}_N, \boldsymbol{a}_N)) < H(f(\boldsymbol{s}_N, \boldsymbol{a}_D)) \qquad \text{for all } (\boldsymbol{s}_N, \boldsymbol{a}_D) \in \mathcal{D}.$$

Equivalently, the native sequence will have the highest probability to fit into the specified native structure. This is the same principle described in references 25, 71 and 107. Sometimes we can further require that the difference in score must be greater than a constant $b > 0$ [106]:

$$H(f(\boldsymbol{s}_N, \boldsymbol{a}_N)) + b < H(f(\boldsymbol{s}_N, \boldsymbol{a}_D)) \qquad \text{for all } (\boldsymbol{s}_N, \boldsymbol{a}_D) \in \mathcal{D}.$$

Similarly, for protein structure prediction and protein folding, we require that the native amino acid sequence $\boldsymbol{a}_N$ mounted on the native structure $\boldsymbol{s}_N$ has the lowest energy compared to a set of alternative conformations (structural decoys) $\mathcal{D} = \{\boldsymbol{s}_D, \boldsymbol{a}_N\}$:

$$H(f(\boldsymbol{s}_N, \boldsymbol{a}_N)) < H(f(\boldsymbol{s}_D, \boldsymbol{a}_N)) \qquad \text{for all } \boldsymbol{s}_D \in \mathcal{D}$$

and

$$H(f(\boldsymbol{s}_N, \boldsymbol{a}_N)) + b < H(f(\boldsymbol{s}_D, \boldsymbol{a}_S)) \qquad \text{for all } (\boldsymbol{s}_D, \boldsymbol{a}_N) \in \mathcal{D}.$$

when we insist to maintain an energy gap between the native structure and decoy conformations. For linear potential function, we have

$$\boldsymbol{w} \cdot \boldsymbol{c}_N + b < \boldsymbol{w} \cdot \boldsymbol{c}_D \qquad \text{for all } \boldsymbol{c}_D = f(\boldsymbol{s}_D, \boldsymbol{a}_N). \tag{2.36}$$

Our goal is to find a set of parameters through optimization for the potential function such that all these inequalities are satisfied.

There are three key steps in developing effective knowledge-based scoring function using optimization: (1) the functional form, (2) the generation of a large set of decoys for discrimination, and (3) the optimization techniques. The initial step of choosing an appropriate functional form is important. Knowledge-based pairwise potential functions are usually all in the form of weighted linear sum of interacting residue pairs. In this form, the weight coefficients are the parameters of the potential function to be optimized for discrimination. This is the same functional form used in statistical potential, where the weight coefficients are derived from database statistics. The objectives of optimization are often maximization of energy gap between native protein and the average of decoys, or energy gap between native and decoys with lowest score, or the z-score of the native protein [8,27,42,44,60,61,81,84,85,116,118,124,125].

2.4.1 Geometric Nature of Discrimination

There is a natural geometric view of the inequality requirement for weighted linear sum scoring functions. A useful observation is that each of the inequalities divides the space of $\mathbb{R}^d$ into two halves separated by a hyperplane (Fig. 2.4a). The hyperplane for Eq. (2.36) is defined by the normal vector $(\boldsymbol{c}_N - \boldsymbol{c}_D)$ and its distance $b/||\boldsymbol{c}_N - \boldsymbol{c}_D||$ from the origin. The weight vector $\boldsymbol{w}$ must be located in the half-space opposite to the direction of the normal vector $(\boldsymbol{c}_N - \boldsymbol{c}_D)$. This half-space can be written as $\boldsymbol{w} \cdot (\boldsymbol{c}_N - \boldsymbol{c}_D) + b < 0$. When there are many inequalities to be satisfied simultaneously, the intersection of the half-spaces forms a convex polyhedron [31]. If the weight vector is located in the polyhedron, all the inequalities are satisfied. Scoring functions with such weight vector $\boldsymbol{w}$ can discriminate the native protein sequence from the set of all decoys. This is illustrated in Fig. 2.4a for a two-dimensional toy example, where each straight line represents an inequality $\boldsymbol{w} \cdot (\boldsymbol{c}_N - \boldsymbol{c}_D) + b < 0$ that the scoring function must satisfy.

For each native protein i, there is one convex polyhedron $\mathcal{P}_i$ formed by the set of inequalities associated with its decoys. If a scoring function can discriminate simultaneously n native proteins from a union of sets of sequence decoys, the weight vector $\boldsymbol{w}$ must be located in a smaller convex polyhedron $\mathcal{P}$ that is the intersection of the n convex polyhedra:

$$\boldsymbol{w} \in \mathcal{P} = \bigcap_{i=1}^{n} \mathcal{P}_i.$$

There is yet another geometric view of the same inequality requirements. If we now regard $(\boldsymbol{c}_N - \boldsymbol{c}_D)$ as a point in $\mathbb{R}^d$, the relationship $\boldsymbol{w} \cdot (\boldsymbol{c}_N - \boldsymbol{c}_D) + b < 0$ for all sequence decoys and native proteins requires that all points $\{\boldsymbol{c}_N - \boldsymbol{c}_D\}$ be located on one side of a different hyperplane, which is defined by its normal vector $\boldsymbol{w}$ and its distance $b/||\boldsymbol{w}||$ to the origin (Fig. 2.4b). We can show that such a hyperplane exists if the origin is not contained within the convex hull of the set of points $\{\boldsymbol{c}_N - \boldsymbol{c}_D\}$ [48].

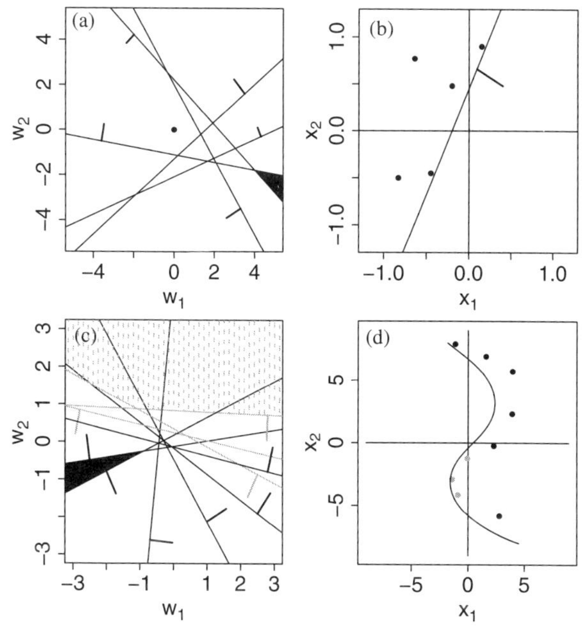

FIGURE 2.4 Geometric views of the inequality requirement for the protein scoring function. Here we use a two-dimensional toy example for illustration. (**a**) In the first geometric view, the space $\mathbb{R}^2$ of $\boldsymbol{w} = (w_1, w_2)$ is divided into two half-spaces by an inequality requirement, represented as a hyperplane $\boldsymbol{w} \cdot (\boldsymbol{c}_N - \boldsymbol{c}_D) + b < 0$. The hyperplane, which is a line in $\mathbb{R}^2$, is defined by the normal vector $(\boldsymbol{c}_N - \boldsymbol{c}_D)$, and its distance $b/||\boldsymbol{c}_N - \boldsymbol{c}_D||$ from the origin. Here this distance is set to 1.0. The normal vector is represented by a short line segment whose direction points away from the straight line. A feasible weight vector $\boldsymbol{w}$ is located in the half-space opposite to the direction of the normal vector $(\boldsymbol{c}_N - \boldsymbol{c}_D)$. With the given set of inequalities represented by the lines, any weight vector $\boldsymbol{w}$ located in the shaped polygon can satisfy all inequality requirement and provides a linear scoring function that has perfect discrimination. (**b**) A second geometric view of the inequality requirement for linear protein scoring function. The space $\mathbb{R}^2$ of $\boldsymbol{x} = (x_1, x_2)$, where $\boldsymbol{x} \equiv (\boldsymbol{c}_N - \boldsymbol{c}_D)$, is divided into two half-spaces by the hyperplane $\boldsymbol{w} \cdot (\boldsymbol{c}_N - \boldsymbol{c}_D) + b < 0$. Here the hyperplane is defined by the normal vector $\boldsymbol{w}$ and its distance $b/||\boldsymbol{w}||$ from the origin. The origin corresponds to the native protein. All points $\{\boldsymbol{c}_N - \boldsymbol{c}_D\}$ are located on one side of the hyperplane away from the origin, therefore satisfying the inequality requirement. A linear scoring function $\boldsymbol{w}$ such as the one represented by the straight line here can have perfect discrimination. (**c**) In the second toy problem, a set of inequalities are represented by a set of straight lines according to the first geometric view. A subset of the inequalities requires that the weight vector $\boldsymbol{w}$ be located in the shaded convex polygon on the left, but another subset of inequalities requires that $\boldsymbol{w}$ be located in the dashed convex polygon on the top. Since these two polygons do not intersect, there is no weight vector $\boldsymbol{w}$ that can satisfy all inequality requirements. That is, no linear scoring function can classify these decoys from native protein. (**d**) According to the second geometric view, no hyperplane can separate all points $\{\boldsymbol{c}_N - \boldsymbol{c}_D\}$ from the origin. But a nonlinear curve formed by a mixture of Gaussian kernels can have perfect separation of all vectors $\{\boldsymbol{c}_N - \boldsymbol{c}_D\}$ from the origin: It has perfect discrimination.

The second geometric view looks very different from the first view. However, the second view is dual and mathematically equivalent to the first geometric view. In the first view, a point $\boldsymbol{c}_N - \boldsymbol{c}_D$ determined by the structure–decoy pair $c_N = (\boldsymbol{s}_N, \boldsymbol{a}_N)$ and $c_D = (\boldsymbol{s}_N, \boldsymbol{a}_D)$ corresponds to a hyperplane representing an inequality, and a solution weight vector $\boldsymbol{w}$ corresponds to a point located in the final convex polyhedron. In the second view, each structure–decoy pair is represented as a point $\boldsymbol{c}_N - \boldsymbol{c}_D$ in $\mathbb{R}^d$, and the solution weight vector $\boldsymbol{w}$ is represented by a hyperplane separating all the points $\mathcal{C} = \{\boldsymbol{c}_N - \boldsymbol{c}_D\}$ from the origin.

2.4.2 Optimal Linear Potential Function

Several optimization methods have been applied to find the weight vector $\boldsymbol{w}$ of linear scoring function. The Rosenblatt perceptron method works by iteratively updating an initial weight vector $\boldsymbol{w}_0$ [84,124]. Starting with a random vector (e.g., $\boldsymbol{w}_0 = \boldsymbol{0}$), one tests each native protein and its decoy structure. Whenever the relationship $\boldsymbol{w} \cdot (\boldsymbol{c}_N - \boldsymbol{c}_D) + b < 0$ is violated, one updates $\boldsymbol{w}$ by adding to it a scaled violating vector $\eta \cdot (\boldsymbol{c}_N - \boldsymbol{c}_D)$. The final weight vector is therefore a linear combination of protein and decoy count vectors:

$$\boldsymbol{w} = \sum \eta(\boldsymbol{c}_N - \boldsymbol{c}_D) = \sum_{N \in \mathcal{N}} \alpha_N \boldsymbol{c}_N - \sum_{D \in \mathcal{D}} \alpha_D \boldsymbol{c}_D. \tag{2.37}$$

Here $\mathcal{N}$ is the set of native proteins, and $\mathcal{D}$ is the set of decoys. The set of coefficients $\{\alpha_N\} \cup \{\alpha_D\}$ gives a dual form representation of the weight vector $\boldsymbol{w}$, which is an expansion of the training examples including both native and decoy structures.

According to the first geometric view, if the final convex polyhedron $\mathcal{P}$ is nonempty, there can be an infinite number of choices of $\boldsymbol{w}$, all with perfect discrimination. But how do we find a weight vector $\boldsymbol{w}$ that is optimal? This depends on the criterion for optimality. For example, one can choose the weight vector $\boldsymbol{w}$ that minimizes the variance of score gaps between decoys and natives:

$$\arg_{\boldsymbol{w}} \min \frac{1}{|\mathcal{D}|} \sum \left(\boldsymbol{w} \cdot (\boldsymbol{c}_N - \boldsymbol{c}_D)\right)^2 - \left[\frac{1}{|\mathcal{D}|} \sum_D \left(\boldsymbol{w} \cdot (\boldsymbol{c}_N - \boldsymbol{c}_D)\right)\right]^2$$

as used in reference 118, or minimizing the Z-score of a large set of native proteins, or minimizing the Z-score of the native protein and an ensemble of decoys [21,85], or maximizing the ratio R between the width of the distribution of the score and the average score difference between the native state and the unfolded ones [42,45]. Effective linear sum scoring functions can be obtained by using perceptron learning and other optimization techniques [27,34,42,118,124]

There is another optimality criterion according to the second geometric view [48]. We can choose the hyperplane $(\boldsymbol{w}, b)$ that separates the set of points $\{\boldsymbol{c}_N - \boldsymbol{c}_D\}$ with the largest distance to the origin. Intuitively, we want to characterize proteins with a region defined by the training set points $\{\boldsymbol{c}_N - \boldsymbol{c}_D\}$. It is desirable to define this region such that a new unseen point drawn from the same protein distribution as $\{\boldsymbol{c}_N - \boldsymbol{c}_D\}$ will have a high probability to fall within the defined region. Nonprotein points following a different distribution, which is assumed to be centered around the origin when no *a priori* information is available, will have a high probability to fall outside the defined region. In this case, we are more interested in modeling the region or support of the distribution of protein data, rather than estimating its density distribution function. For linear scoring function, regions are half-spaces defined by hyperplanes, and the optimal hyperplane $(\boldsymbol{w}, b)$ is then the one with maximal distance to the origin. This is related to the novelty detection problem and single-class support vector machine studied in statistical learning theory [105,121,122]. In our case, any nonprotein points will need to be detected as outliers from the protein distribution characterized by $\{\boldsymbol{c}_N - \boldsymbol{c}_D\}$. Among all linear functions derived from the same set of native proteins and decoys, an optimal weight vector $\boldsymbol{w}$ is likely to have the least amount of mislabelings. The optimal weight vector $\boldsymbol{w}$ therefore can be found by solving the following quadratic programming problem:

$$\text{Minimize} \quad \frac{1}{2}||\boldsymbol{w}||^2 \tag{2.38}$$

$$\text{subject to} \quad \boldsymbol{w} \cdot (\boldsymbol{c}_N - \boldsymbol{c}_D) + b < 0 \text{ for all } N \in \mathcal{N} \text{ and } D \in \mathcal{D}. \tag{2.39}$$

The solution maximizes the distance $b/||\boldsymbol{w}||$ of the plane $(\boldsymbol{w}, b)$ to the origin. We obtained the solution by solving the following support vector machine problem:

$$\begin{aligned} &\text{Minimize} && \tfrac{1}{2}\|\boldsymbol{w}\|^2 \\ &\text{subject to} && \boldsymbol{w} \cdot \boldsymbol{c}_N + d \leq -1 \\ &&& \boldsymbol{w} \cdot \boldsymbol{c}_D + d \geq 1, \end{aligned} \tag{2.40}$$

where $d > 0$. Note that a solution of Problem (2.40) satisfies the constraints in Inequalities (2.39), since subtracting the second inequality here from the first inequality in the constraint conditions of (2.40) will give us $\boldsymbol{w} \cdot (\boldsymbol{c}_N - \boldsymbol{c}_D) + 2 \leq 0$.

2.4.3 Optimal Nonlinear Potential Function

It is possible that the linear weight vector $\boldsymbol{w}$ does not exist; that is, the final convex polyhedron $\mathcal{P} = \bigcap_{i=1}^{n} \mathcal{P}_i$ may be an empty set. This occurs if a large number of native protein structures are to be simultaneously stabilized against a large number of decoy conformations, and no such potential functions in the linear functional form can be found [118,125].

According to our geometric pictures, there are two possible scenarios. First, for a specific native protein i, there may be severe restriction from some inequality constraints, which makes $\mathcal{P}_i$ an empty set. Some decoys are very difficult to discriminate due to perhaps deficiency in protein representation. In these cases, it is impossible to adjust the weight vector so the native protein has a lower score than the sequence decoy. Figure 2.4c shows a set of inequalities represented by straight lines according to the first geometric view. In this case, there is no weight vector that can satisfy all these inequality requirements. That is, no linear scoring function can classify all decoys from native protein. According to the second geometric view (Fig. 2.4d), no hyperplane can separate all points $\{c_N - c_D\}$ from the origin, which corresponds to the native structures.

Second, even if a weight vector w can be found for each native protein — that is, w is contained in a nonempty polyhedron, it is still possible that the intersection of n polyhedra is an empty set; that is, no weight vector can be found that can discriminate all native proteins against the decoys simultaneously. Computationally, the question of whether a solution weight vector w exists can be answered unambiguously in polynomial time [54]. If a large number (e.g., hundreds) of native protein structures are to be simultaneously stabilized against a large number of decoy conformations (e.g., tens of millions), no such potential functions can be found computationally [118,125]. Similar conclusion is drawn in a study for protein design, where it was found that no linear potential function can simultaneously discriminate a large number of native proteins from sequence decoys [48].

A fundamental reason for such failure is that the functional form of linear sum is too simplistic. It has been suggested that additional descriptors of protein structures such as higher order interactions (e.g., three-body or four-body contacts) should be incorporated in protein description [12,92,136]. Functions with polynomial terms using up to 6 degrees of Chebyshev expansion have also been used to represent pairwise interactions in protein folding [32].

We now discuss an alternative approach. Let us still limit ourselves to pairwise contact interactions, although it can be naturally extended to include three or four body interactions [75]. We can introduce a nonlinear potential function or scoring function analogous to the dual form of the linear function in Eq. (2.37), which takes the following form:

$$H(f(s,a)) = H(c) = \sum_{D \in \mathcal{D}} \alpha_D K(c, c_D) - \sum_{N \in \mathcal{N}} \alpha_N K(c, c_N), \tag{2.41}$$

where $\alpha_D \geq 0$ and $\alpha_N \geq 0$ are parameters of the scoring function to be determined, and $c_D = f(s_N, a_D)$ from the set of decoys $\mathcal{D} = \{(s_N, a_D)\}$ is the contact vector of a sequence decoy D mounted on a native protein structure s_N, and $c_N = f(s_N, a_N)$ from the set of native training proteins $\mathcal{N} = \{(s_N, a_N)\}$ is the contact vector of a native sequence a_N mounted on its native structure s_N. In the study of Hu et al. [48], all decoy sequence $\{a_D\}$ were taken from real proteins possessing different fold structures. The difference of this functional form from linear function in Eq. (2.37) is that a kernel function $K(x, y)$ replaces the linear term. A convenient kernel function K is

$$K(x, y) = e^{-||x-y||^2/2\sigma^2} \qquad \text{for any vectors } x \text{ and } y \in \mathcal{N} \bigcup \mathcal{D},$$

where σ^2 is a constant. Intuitively, the surface of the scoring function has smooth Gaussian hills of height α_D centered on the location c_D of decoy protein D and has smooth Gaussian cones of depth α_N centered on the location c_N of native structures N. Ideally, the value of the scoring function will be -1 for contact vectors c_N of native proteins and will be $+1$ for contact vectors c_D of decoys.

2.4.4 Deriving Optimal Nonlinear Scoring Function

To obtain the nonlinear scoring function, our goal is to find a set of parameters $\{\alpha_D, \alpha_N\}$ such that $H(f(\boldsymbol{s}_N, \boldsymbol{a}_N))$ has value close to -1 for native proteins, and the decoys have values close to $+1$. There are many different choices of $\{\alpha_D, \alpha_N\}$. We use an optimality criterion originally developed in statistical learning theory [18,105,120]. First, we note that we have implicitly mapped each structure and decoy from $\mathbb{R}^{210}$ through the kernel function of $K(\boldsymbol{x}, \boldsymbol{y}) = e^{-||\boldsymbol{x}-\boldsymbol{y}||^2/2\sigma^2}$ to another space with dimension as high as tens of millions. Second, we then find the hyperplane of the largest margin distance separating proteins and decoys in the space transformed by the nonlinear kernel. That is, we search for a hyperplane with equal and maximal distance to the closest native proteins and the closest decoys in the transformed high-dimensional space. Such a hyperplane can be found by obtaining the parameters $\{\alpha_D\}$ and $\{\alpha_N\}$ from solving the following Lagrange dual form of quadratic programming problem:

$$\begin{array}{ll} \text{Maximize} & \sum_{i \in \mathcal{N} \cup \mathcal{D},} \alpha_i - \frac{1}{2} \sum_{i,j \in \mathcal{N} \cup \mathcal{D}} y_i y_j \alpha_i \alpha_j e^{-||\boldsymbol{c}_i - \boldsymbol{c}_j||^2/2\sigma^2} \\ \text{subject to} & 0 \leq \alpha_i \leq C, \end{array}$$

where C is a regularizing constant that limits the influence of each misclassified protein or decoy [18,105,120–122], $y_i = -1$ if i is a native protein, and $y_i = +1$ if i is a decoy. These parameters lead to optimal discrimination of an unseen test set [18,105,120–122]. When projected back to the space of $\mathbb{R}^{210}$, this hyperplane becomes a nonlinear surface. For the toy problem of Fig. 2.4, Fig. 2.4d shows that such a hyperplane becomes a nonlinear curve in $\mathbb{R}^2$ formed by a mixture of Gaussian kernels. It separates perfectly all vectors $\{\boldsymbol{c}_N - \boldsymbol{c}_D\}$ (black and green) from the origin. That is, a nonlinear scoring function can have perfect discrimination.

2.4.5 Optimization Techniques

The techniques that have been used for optimizing potential function include perceptron learning, linear programming, gradient descent, statistical analysis, and support vector machine [8,9,48,118,125,128]. These are standard techniques that can be found in optimization and machine learning literature. For example, there are excellent linear programming solvers based on the simplex method, as implemented in CLP, GLPK, and LP_SOLVE [11], and based on the interior point method as implemented in the BPMD [83], the HOPDM and the PCx packages [23]. We neglect the details of these techniques and point readers to the excellent treatises of [96,119].

2.5 APPLICATIONS

Knowledge-based potential function has been widely used in the study of protein structure prediction, protein folding, and protein–protein interaction. In this section we discuss briefly some of these applications.

2.5.1 Protein Structure Prediction

Protein structure prediction is a complex task that involves two major components: sampling the conformational space and recognizing the near-native structures from the ensemble of sampled conformations.

In protein structure prediction, methods for conformational sampling generates a large number of candidate protein structures. These are often called *decoys*. Among these decoys, only a few are near-native structures that are very similar to the native structure. Many decoy sets have been developed which are used as objective benchmarks to test if an knowledge-based potential function can successfully identify the native and near native structures. For example, Park and Levitt (1996) constructed a 4-state-reduced decoy set. This decoy test set contains native and near-native conformations of seven sequences, along with about 650 misfolded structures for each sequence. In a successful study of structure predictions, an knowledge-based potential function then can be used to discriminate the near-native structures from all other decoys [7]. Furthermore, a knowledge-based potential function can not only be applied at the end of the conformation sampling to recognize near-native structures, but can also be used during conformation generation to guide the efficient sampling of protein structures [45,52].

2.5.2 Protein–Protein Docking Prediction

Knowledge-based potential function can also be used to study protein–protein interactions. Here we give an example of predicting the binding surface of antibody or antibody related proteins (e.g., Fab fragment, T-cell receptor) [74]. When docking two proteins together, we say a *cargo* protein is docked to a fixed *seat* protein. To determine the binding surfaces on the cargo protein, we can examine all possible surface patches on the unbound structure of cargo protein as candidate binding interfaces. The alpha knowledge-based potential function is then used to identify native or near-native binding surfaces. To evaluate the performance of the potential function, we assume the knowledge of the binding interface on the seat protein. We further assume that the degree of near neighbors for interface residues is known.

We first partition the surface of the unbound cargo protein into candidate surface patches; each has the same size as the native binding surface of m residues. A candidate surface patch is generated by starting from a surface residue on the cargo protein, and following alpha edges on the boundary of the alpha shape by breadth-first search, until m residues are found (Fig. 2.5). We construct n candidate surface patches by starting in turn from each of the n surface residue on the cargo protein. Because each surface residue is the center of one of the n candidate surface patch, the set of candidate surface patches cover exhaustively the whole protein binding interface.

Second, we assume that a candidate surface patch on the cargo protein has the same set of contacts as that of the native binding surface. The degree of near neighbors for each hypothetical contacting residue pair is also assumed to be the same. We replace the m residues of the native surface with the m residues from the candidate surface patch. There are $\frac{m!}{\prod_{i=1}^{20} m_i!}$ different ways to permute the m residues of the candidate

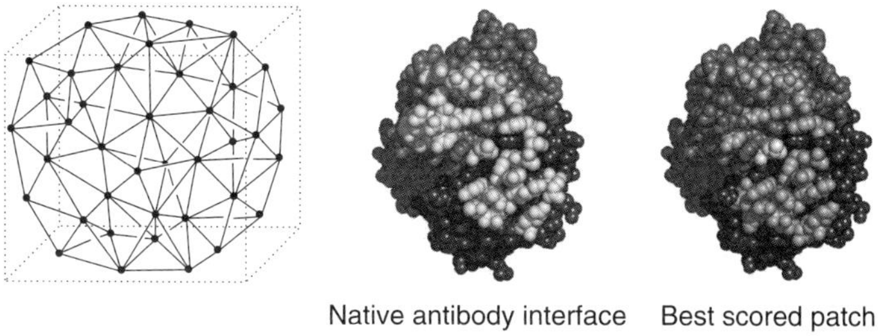

(a) (b)

FIGURE 2.5 Recognition of binding surface patch of protein targets using the geometric potential function. **(a)** Boundary of alpha shape for a *cargo* protein. Each node represents a surface residue, and each edge represents the alpha edge between two surface residues. A candidate surface patch is generated by starting from a surface residue on the cargo protein, and following alpha edges on the boundary of the alpha shape by breadth-first search, until m residues are included. **(b)** Native interface and the surface patch with the best score on the antibody of the protein complex. Only heavy chain (in red) and light chain (in blue) of the antibody are drawn. The antigen is omitted from this illustration for clarity. The best scored surface patch (in green) resembles the native interface (in yellow): 71% residues from this surface patch are indeed on the native binding interface. The residue in white is the starting residue used to generate this surface patch with the best score.

surface patch, where m_i is the number of residue type i on the candidate surface patch. A typical candidate surface patch has about 20 residues; therefore the number of possible permutation is very large. For each candidate surface patch, we take a sample of 5000 random permutations. For a candidate surface patch SP_i, we assume that the residues can be organized so that they can interact with the binding partner at the lowest energy. Therefore, the binding energy $E(SP_i)$ is estimated as

$$E(SP_i) = \min_k E(SP_i)_k, \qquad k = 1, \ldots, 5000.$$

Here $E(SP_i)_k$ is calculated based on the residue-level packing and distance-dependent potential for the kth permutation. The value of $E(SP_i)$ is used to rank the candidate surface patches.

We can assess the statistical potential by taking antibody/antigen protein in turn as the seat protein and taking the antigen/antibody as cargo protein. The native interface on the seat protein is fixed. We then test if our statistical potential can discriminate native surface patch on the cargo protein from the set of candidate surface patches. We can also test if the best scored patch resembles the native patch. An example of the predicted antigen–binding interface of T02 is shown in Fig. 2.5b. For five out of the seven protein complexes, the native patches on both the antibody and the antigen are successfully predicted (Table 2.2). Over 50% of the residues from the best scored patch overlaps with corresponding native patch. The statistical potential does not work as well for T04 and T05, because the antibodies of these two complexes do not use their CDR domains to recognize the antigens as an antibody usually does, and

TABLE 2.2 Recognition of Native Binding Surface of CAPRI Targets by Alpha Potential Function

		Antibody[a]		Antigen	
Target	Complex	Rank^{b}_{native}	Overlap[c]	Rank_{native}	Overlap
T02	Rotavirus VP6-Fab	1/283[d]	0.71	1/639	0.68
T03	Flu hemagglutinin-Fab	1/297	0.56	1/834	0.71
T04	α-amylase-camelid Ab VH 1	56/89	0.60	102/261	0.03
T05	α-amylase-camelid Ab VH 2	23/90	0.57	57/263	0.25
T06	α-amylase-camelid Ab VH 3	1/88	0.70	1/263	0.62
T07	SpeA superantigen TCRβ	1/172	0.57	1/143	0.61
T13	SAG1–antibody complex	1/286	0.64	1/249	0.69

[a] "Antibody": Different surface patches on the antibody molecule are evaluated by the scoring function, while the native binding surface on the antigen remains unchanged. "Antigen": similarly defined as "Antibody".

[b] Ranking of the native binding surface among all candidate surface patches.

[c] Fraction of residues from the best candidate surface patch that overlap with residues from the native binding surface patch.

[d] The first number is the rank of native binding surface and the second number is the number of total candidate surface patches.

such examples are not present in the dataset of the 34 antibody–antigen complexes, based on which the alpha potential function was obtained.

2.5.3 Protein Design

Protein design aims to identify sequences compatible with a given protein fold but incompatible to any alternative folds [58,59]. The goal is to design a novel protein that may not exist in nature but has enhanced or novel biological function. Several novel proteins have been successfully designed [24,46,66,79]. The problem of protein design is complex, because even a small protein of just 50 residues can have an astronomical number of sequences (10^{65}). This clearly precludes exhaustive search of the sequence space with any computational or experimental method. Instead, protein design methods rely on potential functions for biasing the search towards the feasible regions that encode protein sequences. To select the correct sequences and to guide the search process, a design potential function is critically important. Such a scoring function should be able to characterize the global fitness landscape of many proteins simultaneously.

Here we briefly describe the application of the optimal nonlinear design potential function discussed in Section 2.4.3 [48] in protein design. The aim is to solve a simplified protein sequence design problem, namely, to distinguish each native sequence for a major portion of representative protein structures from a large number of alternative decoy sequences, each a fragment from proteins of different fold.

To train the nonlinear potential function, a list of 440 proteins was compiled from the WHATIF98 database [125]. Using gapless threading [81], a set of 14,080,766 sequence decoys was obtained. The entries in the WHATIF99 database that are not present in WHATIF98 are used as a test set. After cleaningup, the test set consists of 194 proteins and 3,096,019 sequence decoys.

To test the design scoring functions for discriminating native proteins from sequence decoys, we take the sequence $\boldsymbol{a}$ from the conformation–sequence pair $(\boldsymbol{s}_N, \boldsymbol{a})$ for a protein with the lowest score as the predicted sequence. If it is not the native sequence $\boldsymbol{a}_N$, the discrimination failed and the design scoring function does not work for this protein.

The nonlinear design scoring function is capable of discriminating all of the 440 native sequences. In contrasts, no linear scoring function can succeed in this task. The nonlinear potential function also works well for the test set, where it succeeded in correctly identifying 93.3% (181 out of 194) of native sequences in the independent test set of 194 proteins. This compares favorably with results obtained using optimal linear folding scoring function taken as reported in reference 118, which succeeded in identifying 80.9% (157 out of 194) of this test set. It also has a better performance than optimal linear scoring function based on calculations using parameters reported in reference 8, which succeeded in identifying 73.7% (143 out of 194) of proteins in the test set. The Miyazawa–Jernigan statistical potential succeeded in identifying 113 native proteins out of 194) (success rate 58.2%).

2.5.4 Protein Stability and Binding Affinity

Because the stability of protein in the native conformation is determined by the distribution of the full ensemble of conformations, namely, the partition function $Z(\boldsymbol{a})$ of the protein sequence $\boldsymbol{a}$, care must be taken when using statistical potentials to compare the stabilities of different protein sequences adopting the same given conformation as in protein design [88,112]. This issue is discussed in some detail in section 2.6.1.

Nevertheless, it is expected that statistical potential should work well in estimating protein stability changes upon mutations, as the change in partition functions of the protein sequence is small. In most such studies and those using physics-based empirical potential [14]), good correlation coefficient (0.6–0.8) between predicted and measured stability change can be achieved [14,38,39,43,47,137].

Several studies have shown that statistical potentials can also be used to predict quantitative binding free energy of protein–protein or protein–ligand interactions [26,78,86,91,134]. In fact, Xu et al. showed that a simple number count of hydrophilic bridges across the binding interface is strongly correlated with binding free energies of protein–protein interaction [129]. This study suggests that binding free energy may be predicted successfully by number counts of different types of interfacial contacts defined using some distance threshold. Such number count studies provide a useful benchmark to quantify the improvement in predicting binding free energy when using statistical potentials for different protein–protein and protein–ligand complexes. Similar to prediction of protein stability change upon mutation, knowledge based potential function also played an important role in a successful study of predicting binding free energy changes upon mutation [62,63].

2.6 DISCUSSION AND SUMMARY

2.6.1 Knowledge-Based Statistical Potential Functions

The statistical potential functions are often derived based on several assumptions: (a) Protein energetics can be decomposed into pairwise interactions; (b) interactions are independent from each other; (c) the partition function in native proteins Z and in reference states Z' are approximately equal; (d) the probability of occupancy of a state follows the Boltzmann distribution. These assumptions may be unrealistic, which raises questions about the validity of the statistical potential functions: Can statistical potential functions provide energy-like quantities such as the folding free energy of a protein or the binding free energy of a protein–protein complex [117]? Can statistical potential functions correctly recognize the native structures from alternative conformations?

The Assumptions of Statistical Knowledge-Based Potential Functions. From Eq. (2.5), we can obtain the potential function $H(\boldsymbol{c})$ by estimating the probability $\pi(\boldsymbol{c})$. However, we need a number of assumptions for this approach to work. We need the independency assumption to have

$$\pi(\boldsymbol{c}) = \prod_i \pi(c_i) = \prod_i \prod_{c_i} \pi_i,$$

where c_i is the number of occurrences of ith structural feature (e.g., the number of a specific residue pair contact) and π_i is the probability of the ith structural feature in the database. That is, we have to assume that the distribution of a specific structural feature is independent and not influenced by any other features and is of no consequence for the distribution of other features as well. We also need to assume that $\boldsymbol{c}$ provides an adequate characterization of protein interactions, and the functional form of $\boldsymbol{w} \cdot \boldsymbol{c}$ provides the correct measurement of the energy of the interactions. We further need to assume that the energy for a protein–solvent system is decomposable; that is, the overall energy can be partitioned into many basic energy terms, such as pairwise interactions and desolvation energies. Moreover, the partition functions Z' in a chosen reference state are approximately equal to the partition functions Z in native proteins. These assumptions go along with the assumption that their structural features contained in the protein database are correctly sampled under the Boltzmann distribution. For any protein descriptor, we have

$$\pi_i \propto \exp(-w_i).$$

To calculate π_i in practice, we have to rely on another assumption that all protein structures are crystallized at the same temperature. Therefore, the distribution π_i is reasonably similar for all proteins in the database, and hence the frequency counts of protein descriptors in different protein structures can be combined by simple summation with equal weight.

Clearly, none of these assumptions are strictly true. However, the success of many applications of using the statistical knowledge-based potentials indicate that they do capture many important properties of proteins. The questions for improving the statistical potential function are, How seriously is each of these assumptions violated, and to what extent does it affect the validity of the potential function? A few assumptions specific to a particular potential function (such as the coordination and solvation assumptions for the Miyazawa–Jernigan's reaction model) have been described earlier. Here we discuss several assumptions in details below.

Interactions Are Not Independent. Using a HP (hydrophobic–polar) model on two-dimensional lattice, Thomas and Dill (1996) tested the accuracy of Miyazawa–Jernigan contact potentials and Sippl's distance-dependent potentials. In the HP model, a peptide chain contains only two types of monomer: H and P. The true energies are set as $H_{(H,H)} = -1$, $H_{(H,P)} = 0$ and $H_{(P,P)} = 0$. Monomers are in contact if they are nonbonded nearest neighbors on the lattice. The conformational space was exhaustively searched for all sequences with the chain length from 11 to 18. A sequence is considered to have a native structure if it has a unique ground energy state. All native structures were collected to build a structure database, from which the statistical potentials are extracted by following the Miyazawa–Jernigan or the Sippl method. The extracted energies are denoted as $e_{(H,H)}$, $e_{(H,P)}$, and $e_{(P,P)}$.

It was found that neither of these two methods can extract the correct energies. All extracted energies by these two methods depend on chain length, while the true energies do not. Using the Miyazawa–Jernigan's method, the (H, H) contact is correctly determined as dominant and attractive. However, the estimated values for $e_{(H,P)}$ and $e_{(P,P)}$ are not equal to zero, whereas the true energies $H_{(H,P)}$ and $H_{(P,P)}$ are equal to zero. Using Sippl's method, the extracted potentials erroneously show a distance dependence; that is, (H, H) interactions are favorable in short distance but unfavorable in long distance, and conversely for (P, P) interactions, whereas the true energies in the HP model only exist between a first-neighbor (H, H) contact and become zero for all the interactions separated by two or more lattice units.

These systematic errors result from the assumption that the pairwise interactions are independent, and thus the volume exclusion in proteins can be neglected [117]. However, (H, H) interactions indirectly affect the observed frequencies of (H, P) and (P, P) interactions. First, in both contact and distance-dependent potentials, because only a limited number of inter-residue contacts can be made within the restricted volume at a given distance, the high density of (H, H) pairs at short distances is necessarily coupled with the low density (relative to reference state) of (H, P) and (P, P) pairs at the same distances, especially at the distance of one lattice unit. As a result, the extracted (H, P) and (P, P) energies are erroneously unfavorable at short distance. Second, for distance-dependent potentials, the energy of a specific type of pair interaction at a given distance is influenced by the same type of pair at different distances. For example, the high density of (H, H) pairs at short distances causes a compensating depletion (relative to the uniform density reference state) at certain longer distances, and conversely for (H, P) and (P, P) interactions. Admittedly this study was carried out using models of short chain lengths and a simple alphabet of residues where the foldable sequences may be very homologous, hence the observed artifacts are profound; the deficiencies of the statistical potentials revealed in this study, such as the excluded volume effect, are likely to be significant in potential functions derived from real proteins.

Pairwise Interactions Are Not Additive. Interactions stabilizing proteins are often modeled by pairwise contacts at atom or residue level. An assumption associated with this approach is the additivity of pairwise interactions, namely, the total energy or fitness score of a protein is the linear sum of all of its pairwise interactions.

However, the nonadditivity effects have been clearly demonstrated in cluster formation of hydrophobic methane molecules both in experiment [10] and in simulation [22,98,108,109]. Protein structure refinement will likely require higher order interactions [12]. Some three-body contacts have been introduced in several studies [30,40,41,101], where physical models explicitly incorporating three-body interactions are developed. In addition, several studies of Delaunay four-body interactions clearly showed the importance of including higher-order interactions in explaining the observed frequency distribution of residue contacts [19,35,65,92,111,136].

Li and Liang (2005) introduced a geometric model based on the Delaunay triangulation and alpha shape to collect three-body interactions in native proteins. A nonadditivity coefficient $\nu_{(i,j,k)}$ is introduced to compare the three-body potential energy $e_{(i,j,k)}$ with the summation of three pairwise interactions $e_{i,j}$, $e_{(i,k)}$, and $e_{(j,k)}$:

$$\nu_{(i,j,k)} = \exp[-e_{(i,j,k)}] / \exp[-(e_{(i,j)} + e_{(i,k)} + e_{(j,k)})].$$

There are three possibilities: (1) $\nu = 1$: interaction of a triplet type is additive in nature and can be well approximated by the sum of three pairwise interactions; (2) $\nu > 1$: Three-body interactions are cooperative and their association is more favorable than three independent pairwise interactions; (3) $\nu < 1$: three-body interactions are anti-cooperative.

After systematically quantifying the nonadditive effects of all 1540 three-body contacts, it was found that hydrophobic interactions and hydrogen-bonding interactions make nonadditive contributions to protein stability, but the nonadditive nature depends on whether such interactions are located in protein interior or on protein surface. When located in interior, many hydrophobic interactions such as those involving alkyl residues are anti-cooperative, namely $\nu < 1$. Salt-bridge and regular hydrogen-bonding interactions such as those involving ionizable residues and polar residues are cooperative in interior. When located on protein surface, these salt-bridge and regular hydrogen-bonding interactions are anti-cooperative with $\nu < 1$, and hydrophobic interactions involving alkyl residues become cooperative [75].

Sequence Dependency of the Partition Function $Z(a)$. We can obtain the total effective energy $\Delta E(s, a)$ given a structure conformation s and its amino acid sequence

$\boldsymbol{a}$ from Eq. (2.6):

$$\begin{aligned}\Delta H(f(s, \boldsymbol{a})) &= \Delta H(\boldsymbol{c}) = \sum_i \Delta H(c_i) \\ &= -kT \sum_{c_i} \ln\left(\frac{\pi(c_i)}{\pi'(c_i)}\right) - kT \ln\left(\frac{Z(\boldsymbol{a})}{Z'(\boldsymbol{a})}\right),\end{aligned} \tag{2.42}$$

where c_i is the total number count of the occurrence of the ith descriptor — for example, the total number of ith type of pairwise contact. The summation involving $Z(\boldsymbol{a})$ and $Z'(\boldsymbol{a})$ is ignored during the evaluation of $\Delta H(c_i)$ by assuming $Z(\boldsymbol{a}) \approx Z'(\boldsymbol{a})$.

It is clear that both $Z(\boldsymbol{a})$ and $Z'(\boldsymbol{a})$ do not depend on the particular structural conformation $\boldsymbol{s}$. Therefore, the omission of the term of the partition functions $-kT \ln\left(\frac{Z(\boldsymbol{a})}{Z'(\boldsymbol{a})}\right)$ will not affect the rank ordering of energy values of different conformations (i.e. , decoys) for the same protein sequence. On the other hand, it is also clear that both $Z(\boldsymbol{a})$ and $Z'(\boldsymbol{a})$ depend on the specific sequence $\boldsymbol{a}$ of a protein. Therefore, there is no sound theoretical basis to compare the stabilities between different proteins using the same knowledge-based potential function, unless the ratio of $Z(\boldsymbol{a})/Z'(\boldsymbol{a})$ for each individual sequence is known and is included during the evaluation [87,104,112]. Notably, DFIRE and other statistical energy functions have been successfully used to predict binding affinities across different protein–protein/peptide complexes. Nevertheless, the theoretical basis is not certain either, because the values of partition function Z(a)s for different protein complexes can be very different. It remains to be seen whether a similarly successful prediction of binding affinities can be achieved just by using the number of native interface contacts at some specific distance — interval, that is, the packing density along the native interface. This omission is probably not seriously detrimental for the problem of predicting free energy change of a protein monomer or binding free energy change of a protein–protein complex upon point mutations, because the distribution of the ensemble of protein conformations may not change significantly after one or several point mutations.

Evaluating Potential Function. The measure used for performance evaluation of potential functions is important. For example, the z-score of native protein among decoys is widely used as an important performance statistic. However, the z-score strongly depends on the properties of the decoy set. Imagine we have access to the true energy function. If a decoy set has a diverse distribution in true energy values, the z-score of the native structure will not be very large. However, this should not suggest that a knowledge-based energy function that gives a larger z-score for native protein is better than the true energy function. Alternative measures may provide more accurate or useful performance evaluation. For example, the correlation r of energy value and CRMSD may be helpful in protein structure prediction. Since a researcher has no access to the native structure, (s)he has to rely on the guidance of an energy function to search for better structures with lower CRMSD to the unknown native structure. For this purpose, a potential function with a large r will be very useful.

Perhaps the performance of a potential function should be judged not by a single statistic but comprehensively by a number of measures.

2.6.2 Relationship of Knowledge-Based Energy Functions and Further Development

The Miyazawa–Jernigan contact potential is the first widely used knowledge-based potential function. Because it is limited by the simple spatial description of a cutoff distance, it cannot capture the finer spatial details. Several distance-dependent potentials have been developed to overcome this limitation and in general have better performance [80,104,137]. A major focus of works in this area is the development of models for the reference state. For example, the use of the ideal gas as reference state in the potential function DFIRE significantly improves the performance in folding and docking decoy discrimination [132].

Because protein surface, interior, and protein–protein interface are packed differently, the propensity of the same pairwise interaction can be different depending on whether the residues are solvent-exposed or are buried. The contact potential of Simons et al. considers two types of environment, namely, buried and nonburied environments separately [110]. The geometric potential function [74] described in Section 2.3.5 incorporates both dependencies on distance and fine-graded local packing, resulting in significant improvement in performance. Knowledge-based potential has also been developed to account for the loss of backbone, side-chain, and translational entropies in folding and binding [4,68].

Another emphasis of recent development of potential function is the orientational dependency of pairwise interaction [16,17,64,89]. Kortemme et al. developed an orientation-dependent hydrogen-bonding potential, which improved prediction of protein structure and specific protein–protein interactions [64]. Miyazawa and Jernigan developed a fully anisotropic distance-dependent potential, with drastic improvements in decoy discrimination over the original Miyazawa–Jernigan contact potential [89].

Computational Efficiency. Given current computing power, all potential functions discussed above can be applied to large-scale discrimination of native or near-native structures from decoys. For example, the geometric potential requires complex computation of the Delaunay tetrahedrization and alpha shape of the molecule. Nevertheless, the time complexity is only $\mathcal{O}(N \log N)$, where N is the number of residues for residual-level potentials or atoms for atom-level potentials. For comparison, a naive implementation of contact computing without the use of proper data structure such as a quad-tree or k–d tree is $\mathcal{O}(N^2)$.

In general, atom-level potentials have better accuracy in recognizing native structures than residue-level potentials, and is often preferred for the final refinement of predicted structures, but it is computationally too expensive to be applicable in every step of a folding or sampling computation.

Potential Function for Membrane Protein. The potential functions we have discussed so far are based on the structures of soluble proteins. Membrane proteins are located in a very different physicochemical environment. They also have different amino acid composition, and they fold differently. Potential functions developed for soluble proteins are therefore not applicable to membrane proteins. For example, Cys–Cys has the strongest pairing propensity because of the formation of disulfide bond. However, Cys–Cys pairs rarely occur in membrane proteins. This and other differences in pairwise contact propensity between membrane and soluble proteins are discussed in reference 2.

Nevertheless, the physical models underlying most potential functions developed for soluble proteins can be modified for membrane proteins [1–3,49,97]. For example, Sale et al. used the MHIP potential developed in reference 2 to predict optimal bundling of TM helices. With the help of 27 additional sparse distance constraints from experiments reported in the literature, these authors succeeded in predicting the structure of dark-adapted rhodopsin to within 3.2 Å of the crystal structure [103].

The development of empirical potential function for β-barrel membrane proteins based on the reference state using the internal random model or the permutation model enabled successes in high-resolution structure predictions of the transmembrane regions of β-barrel membrane proteins, including those of novel architecture and those from eukaryotic cells [49,53,77,94]. The empirical potential function has also been successfully applied to predict the oligomerization states [93], in identification of the protein–protein interaction interfaces [93], and in discovery of mechanisms of stabilization of β-barrel membrane proteins [93]. The stability calculation based on such empirical potential function has also been successfully applied to design the oligomerization state of the OmpF protein [36], as well as in identifying the biochemical mechanism of the VDAC proteins [37].

2.6.3 Optimized Potential Function

The knowledge-based potential function derived by optimization has a number of characteristics that are distinct from statistical potential. We discuss in detail below.

Training Set for Optimized Potential Function. Unlike statistical potential functions where each native protein in the database contribute to the knowledge-based scoring function, only a subset of native proteins contribute. In an optimized potential function, in addition, a small fraction of decoys also contribute to the scoring function. In the study of Hu et al. [48], about 50% of native proteins and $< 0.1\%$ of decoys from the original training data of 440 native proteins and 14 million sequence decoys contribute to the potential function.

As illustrated in the second geometric views, the discrimination of native proteins occurs at the boundary surface between the vector points and the origin. It does not help if the majority of the training data are vector points away from the boundary surface. This implies the need for optimized potential to have appropriate training data. If no *a priori* information is known, it is likely that many decoys (>millions) will be needed to accurately define the discrimination boundary surface, because of the usually large dimension of the descriptors for proteins. However, this imposes a significant computational burden.

Various strategies have been developed to select only the most relevant vector points. One may only need to include the most difficult decoys during training, such as decoys with lower energy than native structures, decoys with lowest absolute energies, and decoys already contributing to the potential function in previous iteration [48,84,118]. In addition, an iterative training process is often necessary [48,84,118].

Reduced Nonlinear Potential Function. The use of nonlinear terms for potential function involves large datasets, because they are necessary *a priori* to define accurately the discrimination surface. This demands the solution of a huge optimization problem. Moreover, the representation of the boundary surface using a large basis set requires expensive computing time for the evaluation of a new unseen contact vector $\boldsymbol{c}$. To overcome these difficulties, nonlinear potential function needs to be further simplified.

One simple approach is to use alternative optimal criterion, for example, by minimizing the distance expressed in 1-norm instead of the standard 2-norm Euclidean distance. The resulting potential function will automatically have reduced terms. Another promising approach is to use rectangle kernels [130].

Potential Function by Optimal Regression. Optimized potential functions are often derived based on decoy discrimination, which is a form of binary classification. Here we suggest a conceptual improvement that can significantly improve the development of optimized potential functions. If we can measure the thermodynamic stabilities of all major representative proteins under identical experimental conditions (e.g., temperature, pH, salt concentration, and osmolarity), we can attempt to develop potential functions with the objective of minimizing the regression errors of fitted energy values and measured energy values. The resulting energy surface will then provide quantitative information about protein stabilities. However, the success of this strategy will depend on coordinated experimental efforts in protein thermodynamic measurements. The scale of such efforts may need to be similar to that of genome sequencing projects and structural genomics projects.

2.6.4 Data Dependency of Knowledge-Based Potentials

There are many directions to improve knowledge-based potential functions. Often it is desirable to include additional descriptors in the energy functions to more accurately account for solvation, hydrogen bonding, backbone conformation (e.g., ϕ and ψ angles), and side-chain entropies. Furthermore, potential functions with different descriptors and details may be needed for different tasks (e.g., backbone prediction versus structure refinement [99]).

An important issue in both statistical potential and optimized potential is their dependency on the amount of available training data and possible bias in such data.

For example, whether a knowledge-based potential derived from a bias dataset is applicable to a different class of proteins is the topic of several studies [56,133]. Zhang et al. further study the effect of database choice on statistical potential [133]. In addition, when the amount of data is limited, overfitting is a serious problem if too many descriptors are introduced in either of the two types of potential functions. For statistical potential, hierarchical hypothesis testing should help to decide whether additional terms are warranted. For optimized potential, cross-validation will help to uncover possible overfitting [48].

Summary. In this chapter, we discussed the general framework of developing knowledge-based potential functions in terms of molecular descriptors, functional form, and parameter calculations. We also discussed the underlying thermodynamic hypothesis of protein folding. With the assumption that frequently observed protein features in a database of structures correspond to a low energy state, frequency of observed interactions can be converted to energy terms. We then described in detail the models behind the Miyazawa–Jernigan contact potential, distance-dependent potentials, and geometric potentials. We also discussed how to weight sample structures of varying degree of sequence similarity in the structural database. In the section on the optimization method, we describe general geometric models for the problem of obtaining optimized knowledge-based potential functions, as well as methods for developing optimal linear and nonlinear potential functions. This is followed by a brief discussion of several applications of the knowledge-based potential functions. Finally, we point out general limitations and possible improvements for the statistical and optimized potential functions.

Remark. Anfinsen's thermodynamic hypothesis can be found in references 5 and 6. More technical details of the Miyazawa–Jernigan contact potential are described in references 87 and 88. The distance dependent potential function was first proposed by Sippl [112], with further development described in references 80 and 104. The development of geometric potentials can be found in [19,65,73,82,136]. The gas-phase approximation of the reference state is discussed in reference 137. Thomas and Dill offered insightful comments about the deficiency of knowledge-based statistical potential functions [117]. The development of optimized linear potential functions can be found in references 84, 118 and 125. The geometric view for designing the optimized potential function and the nonlinear potential function are based on the results in reference 48.

REFERENCES

1. L. Adamian, R. Jackups, T. A. Binkowski, and J. Liang. Higher-order interhelical spatial interactions in membrane proteins. *J. Mol. Biol.*, **327**:251–272, 2003.
2. L. Adamian and J. Liang. Helix–helix packing and interfacial pairwise interactions of r esidues in membrane proteins. *J. Mol. Biol.*, **311**:891–907, 2001.

3. L. Adamian and J. Liang. Interhelical hydrogen bonds and spatial motifs in membrane proteins: Polar clamps and serine zippers. *Proteins*, **47**:209–218, 2002.
4. L. M. Amzel. Calculation of entropy changes in biological processes: Folding, binding, and oligomerization. *Methods Enzymol*, **323**:167–77, 2000.
5. C. Anfinsen, E. Haber, M. Sela, and F. White. The kinetics of formation of native ribonuclease during oxidation of the reduced polypeptide chain. *Proc. Natl. Acad. Sci., USA*, **47**:1309–1314, 1961.
6. C. B. Anfinsen. Principles that govern the folding of protein chains. *Science*, **181**:223–230, 1973.
7. Park B. and Levitt M. Energy functions that discriminate x-ray and near-native folds from well-constructed decoys. *J. Mol. Biol.*, **258**:367–392, 1996.
8. U. Bastolla, J. Farwer, E. W. Knapp, and M. Vendruscolo. How to guarantee optimal stability for most representative structurs in the protein data bank. *Proteins*, **44**:79–96, 2001.
9. U. Bastolla, M. Vendruscolo, and E. W. Knapp. A statistical mechanical method to optimize energy functions for protein folding. *Proc. Natl. Acad. Sci. USA*, **97**:3977–3981, 2000.
10. A. Ben-Naim. Statistical potentials extracted from protein structures: Are these meaningful potentials? *J. Chem. Phys.*, **107**:3698–3706, 1997.
11. M Berkelaar. LP_Solve package. 2004.
12. M. R. Betancourt and D. Thirumalai. Pair potentials for protein folding: Choice of reference states and sensitivity of predicted native states to variations in the interaction schemes. *Protein Sci.*, **8**:361–369, 1999.
13. J. R. Bienkowska, R. G. Rogers, and T. F. Smith. Filtered neighbors threading. *Proteins*, **37**:346–359, 1999.
14. A. J. Bordner and R. A. Abagyan. Large-scale prediction of protein geometry and stability changes for arbitrary single point mutations. *Proteins*, **57**(2):400–413, 2004.
15. B. R. Brooks, C.L. Brooks, 3rd, A. D. Mackerell, Jr. L. Nilsson, R. J. Petrella, B. Roux, Y. Won, G. Archontis, C. Bartels, S. Boresch, A. Caflisch, L. Caves, Q. Cui, A.R. Dinner, M. Feig, S. Fischer, J. Gao, M. Hodoscek, W. Im, K. Kuczera, T. Lazaridis, J. Ma, V. Ovchinnikov, E. Paci, R. W. Pastor, C. B. Post, J. Z. Pu, M. Schaefer, B. Tidor, R. M. Venable, H. L. Woodcock, X. Wu, W. Yang, D. M. York, and M. Karplus. CHARMM: The biomolecular simulation program. *J. Comput. Chem.*, **30**:1545–1614, 2009.
16. N. V. Buchete, J. E. Straub, and D. Thirumalai. Anisotropic coarse-grained statistical potentials improve the ability to identify nativelike protein structures. *J. Chem. Phys.*, **118**:7658–7671, 2003.
17. N. V. Buchete, J. E. Straub, and D. Thirumalai. Orientational potentials extracted from protein structures improve native fold recognition. *Protein Sci.*, **13**:862–874, 2004.
18. C. J. C. Burges. A tutorial on support vector machines for pattern recognition. *Knowledge Discovery and Data Mining*, **2**(2):121–167, 1998.
19. C. W. Carter, Jr., B. C. LeFebvre, S. A. Cammer, A. Tropsha, and M. H. Edgell. Four-body potentials reveal protein-specific correlations to stability changes caused by hydrophobic core mutations. *J. Mol. Biol.*, **311**(4):625–638, 2001.

20. H. S. Chan and K. A. Dill. Origins of structure in globular proteins. *Proc. Natl. Acad. Sci. USA*, **87**(16):6388–6392, 1990.
21. T. L. Chiu and R. A. Goldstein. Optimizing energy potentials for success in protein tertiary structure prediction. *Folding Des.*, **3**:223–228, 1998.
22. C. Czaplewski, S. Rodziewicz-Motowidlo, A. Liwo, D. R. Ripoll, R. J. Wawak, and H. A. Scheraga. Molecular simulation study of cooperativity in hydrophobic association. *Protein Sci.*, **9**:1235–1245, 2000.
23. J. Czyzyk, S. Mehrotra, M. Wagner, and S. Wright. PCx package. 2004.
24. B. I. Dahiyat and S. L. Mayo. *De Novo* protein design: Fully automated sequence selection. *Science*, **278**:82–87, 1997.
25. J. M. Deutsch and T. Kurosky. New algorithm for protein design. *Phys. Rev. Lett.*, **76**(2):323–326, 1996.
26. R. S. DeWitte and E. I Shakhnovich. SMoG: de novo design method based on simple, fast and accurate free energy estimates. 1. Methodology and supporting evidence. *J. Am. Chem. Soc.*, **118**:11733–11744, 1996.
27. J. R. Dima, R. I. Banavar, and A. Maritan. Scoring functions in protein folding and design. *Protein Sci.*, **9**:812–819, 2000.
28. H. Dobbs, E. Orlandini, R. Bonaccini, and F. Seno. Optimal potentials for predicting inter-helical packing in transmembrane proteins. *Proteins*, **49**(3):342–349, 2002.
29. Y. Duan and P. A. Kollman. Pathways to a protein folding intermediate observed in a 1-microsecond simulation in aqueous solution. *Science*, **282**(5389):740–744, 1998.
30. M. P. Eastwood and P. G. Wolynes. Role of explicitly cooperative interactions in protein folding funnels: A simulation study. *J Chem Phys*, **114**(10):4702–4716, 2001.
31. H. Edelsbrunner. *Algorithms in Combinatorial Geometry*. Springer-Verlag, Berlin, 1987.
32. B. Fain, Y. Xia, and M. Levitt. Design of an optimal Chebyshev-expanded discrimination function for globular proteins. *Protein Sci.*, **11**:2010–2021, 2002.
33. A. V. Finkelstein, A. Ya. Badretdinov, and A. M. Gutin. Why do protein architectures have boltzmann-like statistics? *Proteins*, **23**(2):142–50, 1995.
34. M. S. Friedrichs and P. G. Wolynes. Toward protein tertiary structure recognition by means of associative memory hamiltonians. *Science*, **246**:371–373, 1989.
35. H. H. Gan, A. Tropsha, and T. Schlick. Lattice protein folding with two and four body statistical potentials. *Proteins*, **43**(2):161–174, 2001.
36. D. Gessmann, F. Mager, H. Naveed, T. Arnold, S. Weirich, D. Linke, J. Liang, and S. Nussberger. Improving the resistance of a eukaryotic beta-barrel protein to thermal and chemical perturbations. *J. Mol. Biol.*, **413**(1):150–161, 2011.
37. S. Geula, H. Naveeds, J. Liangs, and V. Shoshan-Barmatz. Structure-based analysis of VDAC1 protein: Defining oligomer contact sites. *J. Biol. Chem.*, **287**(3):2179–2190, 2012.
38. D. Gilis and M. Rooman. Stability changes upon mutation of solvent-accessible residues in proteins evaluated by database-derived potentials. *J. Mol. Biol.*, **257**(5):1112–26, 1996.
39. D. Gilis and M. Rooman. Predicting protein stability changes upon mutation using database-derived potentials: solvent accessibility determines the importance of local versus non-local interactions along the sequence. *J. Mol. Biol.*, **272**(2):276–290, 1997.

40. A. Godzik, A. Kolinski, and J. Skolnick. Topology fingerprint approach to the inverse protein folding problem. *J. Mol. Biol.*, **227**(1):227–238, 1992.

41. A. Godzik and J. Skolnick. Sequence-structure matching in globular proteins: Application to supersecondary and tertiary structure determination. *Proc. Natl. Acad. Sci., USA*, **89**(24):12098–12102, 1992.

42. R. Goldstein, Z. A. Luthey-Schulten, and P. G. Wolynes. Protein tertiary structure recognition using optimized hamiltonians with local interactions. *Proc. Natl. Acad. Sci. USA*, **89**:9029–9033, 1992.

43. R. Guerois, J. E. Nielsen, and L. Serrano. Predicting changes in the stability of proteins and protein complexes: a study of more than 1000 mutations. *J. Mol. Biol.*, **320**(2):369–387, 2002.

44. M. H. Hao and H. A. Scheraga. How optimization of potential functions affects protein folding. *Proc. Natl. Acad. Sci., USA*, **93**(10):4984–4989, 1996.

45. M.-H. Hao and H. A. Scheraga. Designing potential energy functions for protein folding. *Curr. Opin. Struct. Biol.*, **9**:184–188, 1999.

46. R. B. Hill, D. P. Raleigh, A. Lombardi, and W. F. DeGrado. *De novo* design of helical bundles as models for understanding protein folding and function. *Acc. Chem. Res.*, **33**:745–754, 2000.

47. C. Hoppe and D. Schomburg. Prediction of protein thermostability with a direction- and distance-dependent knowledge-based potential. *Protein Sci.*, **14**:2682–2692, 2005.

48. C. Hu, X. Li, and J. Liang. Developing optimal non-linear scoring function for protein design. *Bioinformatics*, **20**(17):3080–3098, 2004.

49. R. Jackups, Jr. and J. Liang. Interstrand pairing patterns in beta-barrel membrane proteins: The positive-outside rule, aromatic rescue, and strand registration prediction. *J. Mol. Biol.*, **354**(4):979–993, 2005.

50. R. Janicke. Folding and association of proteins. *Prog. Biophys. Mol. Biol.*, **49**:117–237, 1987.

51. J. Janin, K. Henrick, J. Moult, L. T. Eyck, M. J. Sternberg, S. Vajda, I. Vakser, and S. J. Wodak. CAPRI: A Critical Assessment of PRedicted Interactions. *Proteins*, **52**(1):2–9, 2003.

52. R. L. Jernigan and I. Bahar. Structure-derived potentials and protein simulations. *Curr. Opin. Struct. Biol.*, **6**:195–209, 1996.

53. Ronald Jackups, Jr., and Jie Liang. Combinatorial analysis for sequence and spatial motif discovery in short sequence fragments. *IEEE/ACM Trans. Comput. Biol. Bioinform.*, **7**(3):524–536, 2010.

54. N. Karmarkar. A new polynomial-time algorithm for linear programming. *Combinatorica*, **4**:373–395, 1984.

55. M. Karplus and G. A. Petsko. Molecular dynamics simulations in biology. *Nature*, **347**:631–639, 1990.

56. J. Khatun, S. D. Khare, and N. V. Dokholyan. Can contact potentials reliably predict stability of proteins? *J. Mol. Biol.*, **336**:1223–1238, 2004.

57. J. A. Kocher, M. J. Rooman, and S. J. Wodak. Factors influencing the ability of knowledge-based potentials to identify native sequence–structure matches. *J. Mol. Biol.*, **235**:1598–1613, 1994.

58. P. Koehl and M. Levitt. *De novo* protein design. I. In search of stability and specificity. *J. Mol. Biol.*, **293**:1161–1181, 1999.

59. P. Koehl and M. Levitt. *De novo* protein design. II. Plasticity of protein sequence. *J. Mol. Biol.*, **293**:1183–1193, 1999.

60. K. K. Koretke, Z. Luthey-Schulten, and P. G. Wolynes. Self-consistently optimized statistical mechanical energy functions for sequence structure alignment. *Protein Sci.*, **5**:1043–1059, 1996.

61. K. K. Koretke, Z. Luthey-Schulten, and P. G. Wolynes. Self-consistently optimized energy functions for protein structure prediction by molecular dynamics. *Proc. Natl. Acad. Sci., USA*, **95**(6):2932–2937, 1998.

62. T. Kortemme and D. Baker. A simple physical model for binding energy hot spots in protein–protein complexes. *Proc. Natl. Acad. Sci., USA*, **99**:14116–14121, 2002.

63. T. Kortemme, D. E. Kim, and D. Baker. Computational alanine scanning of protein–protein interfaces. *Sci. STKE*, **2004**:pl2, 2004.

64. T. Kortemme, A. V. Morozov, and D. Baker. An orientation-dependent hydrogen bonding potential improves prediction of specificity and structure for proteins and protein–protein complexes. *J. Mol. Biol.*, **326**:1239–1259, 2003.

65. B. Krishnamoorthy and A. Tropsha. Development of a four-body statistical pseudo-potential to discriminate native from non-native protein conformations. *Bioinformatics*, **19**(12):1540–1548, 2003.

66. B. Kuhlman, G. Dantas, G. C. Ireton, G. Varani, B. L. Stoddard, and D. Baker. Design of a novel globular protein fold with atomic-level accuracy. *Science*, **302**:1364–1368, 2003.

67. T. Lazaridis and M. Karplus. Effective energy functions for protein structure prediction. *Curr. Opin. Struct. Biol.*, **10**:139–145, 2000.

68. K. H. Lee, D. Xie, E. Freire, and L. M. Amzel. Estimation of changes in side chain configurational entropy in binding and folding: General methods and application to helix formation. *Proteins*, **20**:68–84, 1994.

69. C. M. R. Lemer, M. J. Rooman, and S. J. Wodak. Protein-structure prediction by threading methods—evaluation of current techniques. *Proteins*, **23**:337–355, 1995.

70. M. Levitt and A. Warshel. Computer simulation of protein folding. *Nature*, **253**:694–698, 1975.

71. H. Li, R. Helling, C. Tang, and N. Wingreen. Emergence of preferred structures in a simple model of protein folding. *Science*, **273**:666–669, 1996.

72. H. Li, C. Tang, and N. S. Wingreen. Nature of driving force for protein folding: A result from analyzing the statistical potential. *Phys. Rev. Lett.*, **79**:765–768, 1997.

73. X. Li, C. Hu, and J. Liang. Simplicial edge representation of protein structures and alpha contact potential with confidence measure. *Proteins*, **53**:792–805, 2003.

74. X. Li and J. Liang. Computational design of combinatorial peptide library for modulating protein–protein interactions. *Pacific Symp. Biocomput.*, 28–39, 2005.

75. X. Li and J. Liang. Geometric cooperativity and anti-cooperativity of three-body interactions in native proteins. *Proteins*, **60**:46–65, 2005.

76. J. Liang and K. A. Dill. Are proteins well-packed? *Biophys. J.*, **81**:751–766, 2001.

77. J. Liang, H. Naveed, D. Jimenez-Morales, L. Adamian, and M. Lin. Computational studies of membrane proteins: Models and predictions for biological understanding. *Biochim. Biophys. Acta—Biomembranes*, **1818**(4):927–941, 2012.
78. S. Liu, C. Zhang, H. Zhou, and Y. Zhou. A physical reference state unifies the structure-derived potential of mean force for protein folding and binding. *Proteins*, **56**:93–101, 2004.
79. L. L. Looger, M. A. Dwyer, J. J. Smith, and H. W. Hellinga. Computational design of receptor and sensor proteins with novel functions. *Nature*, **423**:185–190, 2003.
80. H. Lu and J. Skolnick. A distance-dependent atomic knowledge-based potential for improved protein structure selection. *Proteins*, **44**:223–232, 2001.
81. V. N. Maiorov and G. M. Crippen. Contact potential that pecognizes the correct folding of globular proteins. *J. Mol. Biol.*, **227**:876–888, 1992.
82. B. J. McConkey, V. Sobolev, and M. Edelman. Discrimination of native protein structures using atom-atom contact scoring. *Proc. Natl. Acad. Sci. USA*, **100**(6):3215–3220, 2003.
83. C. S. Mészáros. Fast Cholesky factorization for interior point methods of linear programming. *Comp. Math. Appl.*, **31**:49 – 51, 1996.
84. C. Micheletti, F. Seno, J. R. Banavar, and A. Maritan. Learning effective amino acid interactions through iterative stochastic techniques. *Proteins*, **42**(3):422–431, 2001.
85. L. A. Mirny and E. I. Shakhnovich. How to derive a protein folding potential? A new approach to an old problem. *J. Mol. Biol.*, **264**:1164–1179, 1996.
86. B. O. Mitchell, R. A. Laskowski, A. Alex, and J. M. Thornton. BLEEP: potential of mean force describing protein–ligand interactions: II. Calculation of binding energies and comparison with experimental data. *J. Comp. Chem.*, **20**:1177–1185, 1999.
87. S. Miyazawa and R. L. Jernigan. Estimation of effective interresidue contact energies from protein crystal structures: Quasi-chemical approximation. *Macromolecules*, **18**:534–552, 1985.
88. S. Miyazawa and R. L. Jernigan. Residue-residue potentials with a favorable contact pair term and an unfavorable high packing density term. *J. Mol. Biol.*, **256**:623–644, 1996.
89. S. Miyazawa and R. L. Jernigan. How effective for fold recognition is a potential of mean force that includes relative orientations between contacting residues in proteins? *J. Chem. Phys.*, **122**:024901, 2005.
90. F. A. Momany, R. F. McGuire, A. W. Burgess, and H. A. Scheraga. Energy parameters in polypeptides. VII. Geometric parameters, partial atomic charges, nonbonded interactions, hydrogen bond interactions, and intrinsic torsional potentials for the naturally occurring amino acids. *J. Phys. Chem.*, **79**(22):2361–2381, 1975.
91. I. Muegge and Y. C. Martin. A general and fast scoring function for protein-ligand interactions: A simplified potential approach. *J. Med. Chem.*, **42**:791–804, 1999.
92. P. J. Munson and R. K. Singh. Statistical significane of hierarchical multi-body potential based on delaunay tessellation and their application in sequence–structure alignment. *Protein Sci.*, **6**:1467–1481, 1997.
93. H. Naveed, R. Jackups, Jr., and J. Liang. Predicting weakly stable regions, oligomerization state, and protein–protein interfaces in transmembrane domains of outer membrane proteins. *Proc. Nat. Acad. Sci. USA*, **106**(31):12735–12740, 2009.

94. H. Naveed, Y. Xu, R. Jackups, and J. Liang. Predicting three-dimensional structures of transmembrane domains of IŠ-barrel membrane proteins. *J. Am. Chem. Soc.*, **134**(3):1775–1781, 2012.

95. K. Nishikawa and Y. Matsuo. Development of pseudoenergy potentials for assessing protein 3-D-1-D compatibility and detecting weak homologies. *Protein Eng.*, **6**:811–820, 1993.

96. C. H. Papadimitriou and K. Steiglitz. *Combinatorial optimization: Algorithms and complexity*. Dover, 1998.

97. Y. Park, M. Elsner, R. Staritzbichler, and V. Helms. Novel scoring function for modeling structures of oligomers of transmembrane alpha-helices. *Proteins*, **57**(3):577–585, 2004.

98. J. A. Rank and D. Baker. A desolvation barrier to hydrophobic cluster formation may contribute to the rate-limiting step in protein folding. *Protein Sci.*, **6**(2):347–354, 1997.

99. C. A. Rohl, C. E. Strauss, K. M. Misura, and D. Baker. Protein structure prediction using rosetta. *Methods Enzymol.*, **383**:66–93, 2004.

100. C. A. Rohl, C. E. M. Strauss, K. M. S. Misura, and D. Baker. *Protein Structure Prediction Using Rosetta*, Vol. 383 of *Methods in Enzymology*, pp. 66–93. Elsevier, Department of Biochemistry and Howard Hughes Medical Institute, University of Washington, Seattle, Washington, 2004.

101. A. Rossi, C. Micheletti, F. Seno, and A. Maritan. A self-consistent knowledge-based approach to protein design. *Biophys. J.*, **80**(1):480–490, 2001.

102. W. P. Russ and R. Ranganathan. Knowledge-based potential functions in protein design. *Curr. Opin. Struct. Biol.*, **12**:447–452, 2002.

103. K. Sale, J. L. Faulon, G. A. Gray, J. S. Schoeniger, and M. M. Young. Optimal bundling of transmembrane helices using sparse distance constraints. *Protein Sci.*, **13**(10):2613–2627, 2004.

104. R. Samudrala and J. Moult. An all-atom distance-dependent conditional probability discriminatory function for protein structure prediction. *J. Mol. Biol.*, **275**:895–916, 1998.

105. B. Schölkopf and A. J. Smola. *Learning with kernels: Support Vector Machines, Regularization, Optimization, and Beyond.* MIT Press, Cambridge, MA, 2002.

106. E. I. Shakhnovich. Proteins with selected sequences fold into unique native conformation. *Phys. Rev. Lett.*, **72**:3907–3910, 1994.

107. E. I. Shakhnovich and A. M. Gutin. Engineering of stable and fast-folding sequences of model proteins. *Proc. Natl. Acad. Sci. USA*, **90**:7195–7199, 1993.

108. S. Shimizu and H. S. Chan. Anti-cooperativity in hydrophobic interactions: A simulation study of spatial dependence of three-body effects and beyond. *J. Chem. Phys.*, **115**(3):1414–1421, 2001.

109. S. Shimizu and H. S. Chan. Anti-cooperativity and cooperativity in hydrophobic interactions: Three-body free energy landscapes and comparison with implicit-solvent potential functions for proteins. *Proteins*, **48**:15–30, 2002.

110. K. T. Simons, I. Ruczinski, C. Kooperberg, B. Fox, C. Bystroff, and D. Baker. Improved recognition of native-like protein structures using a combination of sequence-dependent and sequence-independent features of proteins. *Proteins*, **34**:82–95, 1999.

111. R. K. Singh, A. Tropsha, and I. I. Vaisman. Delaunay tessellation of proteins: Four body nearest-neighbor propensities of amino acid residues. *J. Comput. Biol.*, **3**(2):213–221, 1996.

112. M. J. Sippl. Calculation of conformational ensembles from potentials of the main force. *J. Mol. Biol.*, **213**:167–180, 1990.

113. M. J. Sippl. Boltzmann's principle, knowledge-based mean fields and protein folding. an approach to the computational determination of protein structures. *J. Comput. Aided Mol. Des.*, **7**(4):473–501, 1993.

114. M. J. Sippl. Knowledge-based potentials for proteins. *Curr. Opin. Struct. Biol.*, **5**(2):229–235, 1995.

115. S. Tanaka and H. A. Scheraga. Medium- and long-range interaction parameters between amino acids for predicting three-dimensional structures of proteins. *Macromolecules*, **9**:945–950, 1976.

116. P. D. Thomas and K. A. Dill. An iterative method for extracting energy-like quantities from protein structures. *PNAS*, **93**(21):11628–11633, 1996.

117. P. D. Thomas and K. A. Dill. Statistical potentials extracted from protein structures: How accurate are they? *J. Mol. Biol.*, **257**:457–469, 1996.

118. D. Tobi, G. Shafran, N. Linial, and R. Elber. On the design and analysis of protein folding potentials. *Proteins*, **40**:71–85, 2000.

119. R. J. Vanderbei. *Linear Programming: Foundations and Extensions*. Kluwer Academic Publishers, Dordrecht, 1996.

120. V. Vapnik. *The Nature of Statistical Learning Theory*. Springer, New York, 1995.

121. V. Vapnik and A. Chervonenkis. A note on one class of perceptrons. *Automation and Remote Control*, **25**, 1964.

122. V. Vapnik and A. Chervonenkis. *Theory of Pattern Recognition [in Russian]*. Nauka, Moscow, 1974. (German Translation: W. Wapnik & A. Tscherwonenkis, *Theorie der Zeichenerkennung*, Akademie–Verlag, Berlin, 1979.)

123. E. Venclovas, A. Zemla, K. Fidelis, and J. Moult. Comparison of performance in successive CASP experiments. *Proteins*, **45**:163–170, 2003.

124. M. Vendruscolo and E. Domanyi. Pairwise contact potentials are unsuitable for protein folding. *J. Chem. Phys.*, **109**:11101–11108, 1998.

125. M. Vendruscolo, R. Najmanovich, and E. Domany. Can a pairwise contact potential stabilize native protein folds against decoys obtained by threading? *Proteins*, **38**:134–148, 2000.

126. S. J. Wodak and M. J. Rooman. Generating and testing protein folds. *Curr. Opin. Struct. Biol.*, **3**:247–259, 1993.

127. P. G. Wolynes, J. N. Onuchic, and D. Thirumalai. Navigating the folding routes. *Science*, **267**:1619–1620, 1995.

128. Y. Xia and M. Levitt. Extracting knowledge-based energy functions from protein structures by error rate minimization: Comparison of methods using lattice model. *J Chem. Phys.*, **113**:9318–9330, 2000.

129. D. Xu, S. L. Lin, and R. Nussinov. Protein binding versus protein folding: The role of hydrophilic bridges in protein associations. *J. Mol. Biol.*, **2651**:68–84, 1997.

130. Y. Xu, C. Hu, Y. Dai, and J. Liang. On simplified global nonlinear function for fitness landscape: A case study of inverse protein folding. *PLOS ONE*, **9**(8):e104403, 2014.

131. C. Zhang and S. H. Kim. Environment-dependent residue contact energies for proteins. *PNAS*, **97**(6):2550–2555, 2000.

132. C. Zhang, S. Liu, H. Zhou, and Y. Zhou. An accurate, residue-level, pair potential of mean force for folding and binding based on the distance-scaled, ideal-gas reference state. *Protein Sci.*, **13**:400–411, 2004.

133. C. Zhang, S. Liu, H. Zhou, and Y. Zhou. The dependence of all-atom statistical potentials on structural training database. *Biophys. J.*, **86**(6):3349–3358, 2004.

134. C. Zhang, S. Liu, Q. Zhu, and Y. Zhou. A knowledge-based energy function for protein–ligand, protein–protein, and protein–DNA complexes. *J. Med. Chem.*, **48**:2325–2335, 2005.

135. C. Zhang, G. Vasmatzis1, J. L. Cornette, and C. DeLisi. Determination of atomic desolvation energies from the structures of crystallized proteins. *J. Mol. Biol.*, **267**:707–726, 1997.

136. W. Zheng, S. J. Cho, I. I. Vaisman, and A. Tropsha. A new approach to protein fold recognition based on Delaunay tessellation of protein structure. In R. B. Altman, A. K. Dunker, L. Hunter, and T. E. Klein, editors, *Pacific Symposium on Biocomputing'97*, p. 486–497. World Scientific, Singapore, 1997.

137. H. Y. Zhou and Y. Q. Zhou. Distance-scaled, finite ideal-gas reference state improves structure-derived potentials of mean force for structure selection and stability prediction. *Protein Sci.*, **11**:2714–2726, 2002.

EXERCISES

2.1 To capture higher-order interactions in proteins, one can construct the three-body propensity function. The propensity $P(i, j, k)$ for residues of type i, j, k to interact can be modeled as the odds ratio of the observed probability $q(i, j, k)$ of a three-body (triple) atomic contacts involving residue i, j, and k and the expected probability $p(i, j, k)$ $P(i, j, k) \equiv \frac{q(i,j,k)}{p(i,j,k)}$. To compute the observed probability $q(i, j, k)$, we can use $q(i, j, k) = a(i, j, k)/\sum_{i',j',k'} a(i', j', k')$, where $a(i, j, k)$ is the number count of atomic contacts among residue types i, j, and k, and $\sum_{i',j',k'} a(i', j', k')$ is the total number of all atomic three-body contacts. For the random probability $p(i, j, k)$, let us assume it is the probability that three atoms are picked from a residue of type i, a residue of type j, and a residue of type k, when chosen randomly and independently from the pooled database of protein structures. Denote the number of interacting residues of type i as N_i, the number of atoms residue of type i has as n_i, and the total number of interacting atoms as n.

(a) Assume that all three interacting residues are of different types—for example, $i \neq j \neq k$. What is the probability that we first pick up an atom from a residue of type i, then an atom from a residue of type j, and with the third atom picked up to be from a residue of type k?

(**b**) Now consider all other possible sequences of picking up an atom each from an i, j, and k residue type. Write down the formula for $p(i, j, k)$.

(**c**) When two of the three interacting residues are of the same type, that is, $i = j \neq k$, what is the formula for $p(i, j, k)$?

(**d**) When all three residues are of the same type, that is, $i = j = k$, what is the formula for $p(i, j, k)$?

2.2 β-barrrel membrane proteins are found in a large number of pathogeneic gram-negative bacteria. Their transmembrane (TM) segments are β-strands. We can obtain the empirical propensity $P(X, Y)$ for interacting pairs of residue types X and Y on neighboring β-strands as $P(X, Y) = f_{\text{obs}}(X, Y)/\mathbb{E}[f(X, Y)]$, where $f_{\text{obs}}(X, Y)$ is the observed count of X–Y contacts in the strand pair, and $\mathbb{E}[f(X, Y)]$ is the expected count of X–Y contacts in a null model.

As the TM strands are short, there is strong coupling between presence and absence of residues residing on the same strand. Commonly used techniques such as the χ^2-distribution, in which normality is assumed, or the Bernoulli model, in which residues are drawn with replacement, are not valid. One can use the *permutation model* or the *internally random model*, in which residues within each of the two interacting strands are permuted exhaustively and independently, and hence are drawn without replacement. Each permutation is assumed to occur with equal probability. In this model, an X–Y contact forms if in a permuted strand pair two interacting residues happen to be of type X and type Y. $\mathbb{E}[f(X, Y)]$ is then the expected number of X–Y contacts in the strand pairs.

(**a**) We first examine the simpler cases when X is the same as Y, that is, X–X pairs. Let x_1 be the number of residues of type X in the first strand, x_2 the number of residues of type X in the second strand, and l the common length of the strand pair. We randomly select residues from one strand to pair up with residues from the other strand. We wish to know the probability of exactly $i = f(X,\ X)$ number of $X - X$ contacts. How many ways are there to place the x_2 residues of type X in the second strand?

(**b**) How many ways are there to have each of the i residues to be paired with one of the x_1 residues of type X on the first strand?

(**c**) How many ways are there to have each of the $x_2 - i$ residues be paired with one of the $l - x_1$ non-X residues?

(**d**) What is the probability $\mathbb{P}_{XX}(i)$ of $i = f(X, X)$ number of X–X contacts in a strand pair? What type of distribution is this?

(**e**) What is the expected number of $(X,\ X)$ contacts?

(**f**) If the two contacting residues are not of the same type, that is, $X \neq Y$, what is the probability $\mathbb{P}_{XY}(i)$ of $i = f(X, Y)$ number of X–Y contacts in a strand pair? (*Hint*: Consider $f(X, Y | X \in s_1, Y \in s_2)$ for the case where the type X residues are in the first strand s_1 and type Y in the second strand s_2, and $f(X, Y | X \in s_2, Y \in s_1)$ for the other case where the type Y residues are in s_1 and type X in s_2.)

(g) What is the probability $\mathbb{P}_{XY}(m)$ that there are a total of $i + j = m$ X–Y contacts? Note the complication that the variables $f(X, Y|X \in s_1, Y \in s_2)$ and $f(X, Y|X \in s_2, Y \in s_1)$ are dependent; that is, the placement of an X–Y pair may affect the probability of a Y–X pair in the same strand pair.

2.3 For a potential function in the form of weighted linear sum of interactions, show proof that a decoy always has energy values higher than the native structure by at least an amount of $b > 0$, that is,

$$\boldsymbol{w} \cdot (\boldsymbol{c}_D - \boldsymbol{c}_N) > b \qquad \text{for all } \{(\boldsymbol{c}_D - \boldsymbol{c}_N) | D \in \mathcal{D} \text{ and } N \in \mathcal{N}\} \tag{2.43}$$

if and only if the origin $\mathbf{0}$ is not contained within the convex hull of the set of points $\{(\boldsymbol{c}_D - \boldsymbol{c}_N) | D \in \mathcal{D} \text{ and } N \in \mathcal{N}\}$, namely, the smallest convex body that contain all $\{(\boldsymbol{c}_D - \boldsymbol{c}_N)\}$. Note that by the definition of convexity, any point $\boldsymbol{x}$ inside or on the convex hull $\mathcal{A}$ can be expressed as a convex combination of points on the convex hull, namely

$$\boldsymbol{x} = \sum_{(\boldsymbol{c}_D - \boldsymbol{c}_N) \in \mathcal{A}} \lambda_{\boldsymbol{c}_D - \boldsymbol{c}_N} \cdot (\boldsymbol{c}_D - \boldsymbol{c}_N) \quad \text{and} \quad \sum \lambda_{\boldsymbol{c}_D - \boldsymbol{c}_N} = 1, \qquad \lambda_{\boldsymbol{c}_D - \boldsymbol{c}_N} > 0.$$

3

SAMPLING TECHNIQUES: ESTIMATING EVOLUTIONARY RATES AND GENERATING MOLECULAR STRUCTURES

3.1 INTRODUCTION

Many problem encountered in computational studies of biological system can be formulated as characterization of its ensemble properties, such as those of a population of molecules or cells. For example, we may need to evaluate the expected energy value of protein molecules in a population of conformations; or we may be interested in estimating the concentration of a key regulator that controls whether a cell switches to a different lifestyle. What lies at the heart of solving these problems is to arrive at an accurate estimation of the ensemble properties of the system. This often requires the application of Monte Carlo sampling techniques.

Denote the state of the biological system as $\boldsymbol{x}$, and assume that it follows some distribution $\pi(\boldsymbol{x})$. If the property associated with the system at state $\boldsymbol{x}$ can be expressed using a scalar function $f(\boldsymbol{x})$, our task is then to estimate

$$\mathbb{E}[f(\boldsymbol{x})] = \int_{\boldsymbol{x} \in \mathcal{R}} f(\boldsymbol{x})\pi(\boldsymbol{x})\, d\boldsymbol{x}, \tag{3.1}$$

where $\mathcal{R}$ is the region of the state space of interests. A key problem is to properly generate samples that follow the probability distribution $\pi(\boldsymbol{x})$. Below we discuss a few specific examples arising from studies in bioinformatics and systems biology.

Example 1. Evolution of Biomolecules. Sequence alignment is a widely used method to infer biological information about a biomolecule. A key ingredient for

Models and Algorithms for Biomolecules and Molecular Networks, First Edition. Bhaskar DasGupta and Jie Liang.

alignment is the scoring matrices used to evaluate similarity between sequences, which are derived based on an evolutionary model of the biomolecules [2,11,28].

Assuming that the evolutionary process of a protein molecule can be described by a continuous-time Markov process, we can derive scoring matrices from the underlying amino acid substitution rates of the evolutionary process [12,27,29,30]. Denote the 20×20 substitution rates in the matrix form as $\boldsymbol{Q} \in \mathbb{R}^{20} \times \mathbb{R}^{20}$, and assume that we have a set of multiple-aligned sequences S and a phylogenetic tree $\boldsymbol{T}$. Our task is then to estimate the distribution $\pi(\boldsymbol{Q}|S, \boldsymbol{T})$ that $\boldsymbol{Q}$ follows.

Following the Bayesian framework, the distribution $\pi(\boldsymbol{Q}|S, \boldsymbol{T})$ can be calculated as

$$\pi(\boldsymbol{Q}|S, \boldsymbol{T}) \propto \int P(S|\boldsymbol{T}, \boldsymbol{Q}) \cdot \pi(\boldsymbol{Q})\, d\boldsymbol{Q},$$

where the probability $P(S|\boldsymbol{TQ})$ of obtaining a set of aligned sequences for a given set of substitution rates $\boldsymbol{Q}$, and a phylogenetic tree $\boldsymbol{T}$ can be evaluated through the underlying Markovian model. Additional information of prior knowledge of the likely values of $\boldsymbol{Q}$ can be incorporated in the prior distribution $\pi(\boldsymbol{Q})$.

Example 2. Loop Entropy. Conformational entropy is an important factor that contributes to the stability of biomolecules. For example, loops located on the surface of a protein often are flexible, and their conformational entropy may influence the stability of the protein. Similarly, loops in different types of RNA secondary structures may influence the final structure and stability of RNA molecules [33].

The problem of estimating loop entropy can be formulated as follows. Let the conformation of a loop of length l be $\boldsymbol{x} = (x_0, \ldots, x_l)$, where $x_i \in \mathbb{R}^3$ is the coordinates of the ith residue or nucleotide. The positions of the first residue $\boldsymbol{x}_0$ and the last residue $\boldsymbol{x}_l$ are fixed. The probability $\pi(\boldsymbol{x})$ that a chain of length l with only one end fixed closes up so its lth residue is at the correct position of $\boldsymbol{x}_l$ gives the conformational entropy s of the loop:

$$s = \int \pi(\boldsymbol{x}) \ln \pi(\boldsymbol{x})\, d\boldsymbol{x},$$

The task is then to accurately estimate $\pi(\boldsymbol{x})$.

Example 3. Biological systems such as gene circuits and protein–protein interaction networks have often been fine-tuned through evolution to be robust against fluctuations in environmental conditions, such as nutrient concentration and damaging factors such as UV irradiation. Despite the intrinsic stochastic nature of many biological processes, they often behave consistently and reliably. However, when the system behavior deviates significantly from the norm, a cell often enters into an abnormal or emergent state, which may manifest as a disease state at organismic level.

We can formulate a simplified model of this problem. The state of a biological network system can be represented by the amount of relevant molecular species. Assume that there are m molecular species, and the vector $\boldsymbol{x}$ of their concentrations or copy numbers is $\boldsymbol{x} = (x_1, \ldots, x_m)$. Let the region $\mathcal{R}_n$ for $\boldsymbol{x}$ represent the region of normal states, and let $\mathcal{R}_d$ represent the region of a disease state. The probability that the system will be in the normal state at time t can be calculated as

$$\pi[\boldsymbol{x}(t) \in \mathcal{R}_n | \boldsymbol{x}(0)] = \int_{\boldsymbol{x}(t) \in \mathcal{R}_n} \pi[\boldsymbol{x}(t) | \boldsymbol{x}(0)]\, d\boldsymbol{x}(t).$$

Similarly, the probability that the system will be in the disease state is

$$\pi[\boldsymbol{x}(t) \in \mathcal{R}_d | \boldsymbol{x}(0)] = \int_{\boldsymbol{x}(t) \in \mathcal{R}_d} \pi[\boldsymbol{x}(t) | \boldsymbol{x}(0)]\, d\boldsymbol{x}(t).$$

We may also be interested in assessing the probability of rare event that the system moves from the normal state to a disease state. This can be calculated as

$$\pi[\boldsymbol{x}(t) \in \mathcal{R}_d | \boldsymbol{x}(0) \in \mathcal{R}_n] = \int_{\boldsymbol{x}(t) \in \mathcal{R}_d; \boldsymbol{x}(0) \in \mathcal{R}_n;} \pi[\boldsymbol{x}(t) | \boldsymbol{x}(0)]\, d\boldsymbol{x}(t).$$

The primitive for studying these problems is to assess the probability $\pi[\boldsymbol{x}(t)|\boldsymbol{x}(0)]$ that the system will be in state $\boldsymbol{x}(t)$ at time t, given an initial condition $\boldsymbol{x}(0)$.

3.2 PRINCIPLES OF MONTE CARLO SAMPLING

3.2.1 Estimation Through Sampling from Target Distribution

To estimate properties of a biological system, our task is to generate samples following the distribution $\pi(\boldsymbol{x})$ we are interested in, which is called the *target distribution*. For example, calculating

$$\mathbb{E}[f(\boldsymbol{x})] = \int_{\boldsymbol{x} \in \mathcal{R}} f(\boldsymbol{x}) \pi(\boldsymbol{x})\, d\boldsymbol{x}$$

depends on the ability to generate proper samples from the distribution $\pi(\boldsymbol{x})$. If we are able to generate m such samples, we can have our estimation as

$$\mathbb{E}[f(\boldsymbol{x})] \approx \frac{1}{m}[f(\boldsymbol{x}^{(1)}) + \cdots + f(\boldsymbol{x}^{(m)})]. \qquad (3.2)$$

This can be a very challenging task, as we may have difficulty in sampling sufficiently in high-dimensional space, a frequently encountered problem. As an example, a two-dimensional Ising model with 30×30 lattice sites has a magnetic spin in each site (Fig. 3.1). A spin σ can take either an *up* $(+1)$ or *down*(-1) position, which is denoted as $x_\sigma \in \{+1, -1\}$. The overall energy $H(\boldsymbol{x})$of the system is determined by the magnetic field h_σ that each spin σ experiences and the interactions that each spin has with its four neighbors:

$$H(\boldsymbol{x}) = -J \sum_{\sigma \sim \sigma'} x_\sigma x_{\sigma'} + \sum_{\sigma} h_\sigma x_\sigma.$$

Here J is the strength of interactions between neighboring spins, h_σ is the magnitude of the external magnetic field, and the state $\boldsymbol{x} = (x_1, \ldots, x_{900})$ can take any of 2^{900}

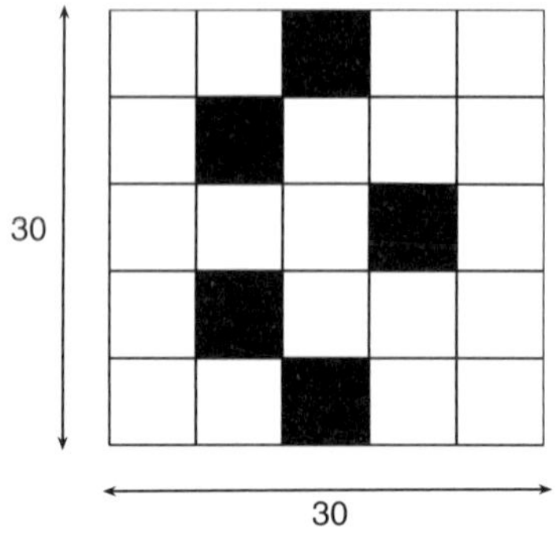

FIGURE 3.1 The Ising model of 30×30 size, with a total of $30 \times 30 = 900$ sites. The spin at each site can adopt either an ''up'' or a ''down'' state, with a total of 2^{900} possible configurations. Integration in this high-dimensional space of this simple example is challenging.

configurations. The task of estimating physical properties is to compute internal energy $< u > = \mathbb{E}[H(\boldsymbol{x})]$, the free energy $F = -kT \log[\sum_{\boldsymbol{x}} \exp(H(\boldsymbol{x})/kT)]$, and the specific heat $C = \frac{\partial <u>}{\partial T}$; all require summation in the 900-dimensional space, which is a challenging task.

There may be other difficulties in integrating Eq. (3.1). We may have good models of the underlying physical processes, but we may have no knowledge of the explicit form of the distribution function $\pi(\boldsymbol{x})$. Even if we know the functional form of $\pi(\boldsymbol{x})$, sampling from this distribution can be quite difficult. For example, the region $\mathcal{R}$ of interests where we need to sample from may be of very constrained in high-dimensional space, and it would be difficult to generate samples from it.

3.2.2 Rejection Sampling

Computing the probability $\pi(\boldsymbol{x})$ of a state $\boldsymbol{x}$ is challenging, but often we can evaluate the relative probability of a state $\boldsymbol{x}$, namely, $\pi(\boldsymbol{x})$ up to a constant, $c\pi(\boldsymbol{x})$, where the value of the constant c is unknown. We can generate samples from a trial distribution $g(\boldsymbol{x})$ (also called a *sampling* or a *proposal distribution*), such that it covers $c\pi(\boldsymbol{x})$ after multiplication by a constant M. That is, we have $Mg(\boldsymbol{x}) \geq c\pi(\boldsymbol{x})$ for all $\boldsymbol{x}$. We can use the *rejection sampling* method to obtain samples from the target distribution $\pi(\boldsymbol{x})$. This is summarized in Algorithm 3.2.2. The accepted samples following this procedure will correctly follow $\pi(\boldsymbol{x})$.

The principle behind rejection sampling is illustrated in Fig. 3.2. Note that $Mg(\boldsymbol{x})$ covers $c\pi(\boldsymbol{x})$ everywhere. After we draw a sample $\boldsymbol{x}$ from $g(\boldsymbol{x})$, we will only accept it with a probability. The acceptance ratio r, which is $\frac{c\pi(\boldsymbol{x})}{Mg(\boldsymbol{x})}$, will remove excessive samples beyond the target distribution $c\pi(\boldsymbol{x})$. The final accepted samples $\boldsymbol{x}$ will therefore follow the correct target distribution $\pi(\boldsymbol{x})$, up to a constant c.

Algorithm 3.2.2 Rejection Sampling

repeat
  Sample a candidate new state $\boldsymbol{x}$ from a proposal distribution function $g(\boldsymbol{x})$
  Compute the ratio $r = \frac{c\pi(\boldsymbol{x})}{Mg(\boldsymbol{x})}$
  Draw u from $\mathcal{U}[0, 1]$
  if $(u \leq r)$ **then**
    Accept sample $\boldsymbol{x}$
  else
    Reject $\boldsymbol{x}$
  end if
until convergency

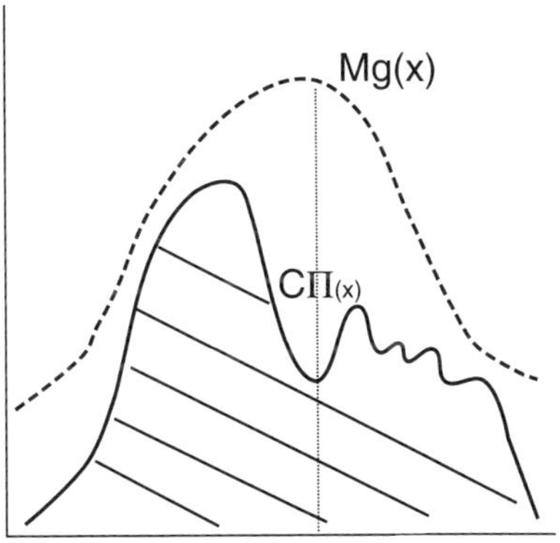

FIGURE 3.2 Illustration of rejection sampling. The target distribution up to a constant $c\pi(\boldsymbol{x})$ is covered everywhere by the sampling distribution $Mg(\boldsymbol{x})$. As samples are generated at $\boldsymbol{x}$ from $Mg(\boldsymbol{x})$, it is different from $c\pi(\boldsymbol{x})$. By accepting only a fraction of $\frac{c\pi(\boldsymbol{x})}{Mg(\boldsymbol{x})}$ of the generated samples at $\boldsymbol{x}$, namely, by taking samples to the proportion of those under the shaded area, the correct distribution $\pi(\boldsymbol{x})$ can be sampled up to a constant.

3.3 MARKOV CHAINS AND METROPOLIS MONTE CARLO SAMPLING

3.3.1 Properties of Markov Chains

A more general framework to obtain samples from a specific target distribution is that of the Markov chain Monte Carlo (MCMC) method. Monte Carlo integration of Eq. (3.2) can be achieved by running a cleverly constructed Markov chain. This technique can be used to study many complex problems.

Conditions on Markov Model. We first discuss Markov chains informally. Given a set of states, a Markov process or Markov chain moves successively from one state $\boldsymbol{x}_i$ to another state $\boldsymbol{x}_j$, with a transition probability $\pi(\boldsymbol{x}_j|\boldsymbol{x}_i)$. It is assumed that the

Markov property holds, namely, $\pi(\boldsymbol{x}_j|\boldsymbol{x}_i)$ does not depend on any state the process visited before $\boldsymbol{x}_i$.

To apply the MCMC method, the Markov process needs to satisfy certain conditions. We assume that the Markov process obeys the irreducibility condition, the recurrent state condition, and the aperiodicity condition. The irreducibility condition dictates that it is possible to travel from any state to any other state, albeit through perhaps many time steps. With the recurrent state condition, all states have the property that the expected number of visits to this state is infinite. With the aperiodicity condition, there is always a nonzero probability to be at any state after a certain number of steps of Markov chain.

Steady State of Markov Chain. If a Markov chain satisfies the above three conditions, it will converges to a steady-state probability distribution as time increases. Denoting the probability distribution at the ith step of a Markov chain as $\pi_i(\boldsymbol{x})$, we have

$$\lim_{n\to\infty} \pi_n(\boldsymbol{x}) = \lim_{n\to\infty} \pi_0(\boldsymbol{x})\boldsymbol{P}^n = \pi_\infty(\boldsymbol{x}),$$

where $\boldsymbol{P}$ is the Markov transition probability matrix, with the transition probability $p_{i,j}$ as its (i, j)th element, and $\pi_\infty(\boldsymbol{x}_i) > 0$ as every state is recurrent. Hence, this Markov chain will converge to a steady-state distribution $\pi_\infty(\boldsymbol{x})$, which is unique and will not depend on the initial distribution $\pi_0(\boldsymbol{x})$. $\pi_\infty(\boldsymbol{x})$ is the unique solution of $\pi(\boldsymbol{x})\boldsymbol{P} = \pi(\boldsymbol{x})$, and for each state $\boldsymbol{x}_i$ we have

$$\lim_{n\to\infty} p_{i,j} = \pi(\boldsymbol{x}_j) \qquad \text{for any } i.$$

This suggests that once we are able to generate a random sample $\boldsymbol{x}$ at time t from the stationary distribution $\pi_\infty(\boldsymbol{x})$, all subsequent (correlated) samples are also generated from $\pi_\infty(\boldsymbol{x})$. Solution to $\pi(\boldsymbol{x})\boldsymbol{P} = \pi(\boldsymbol{x})$ further implies that after a long duration, the probability of finding the process in state $\boldsymbol{x}_j$ is π_j, regardless of the starting state. That is, $\pi_\infty(\boldsymbol{x}_j)$ is the long run mean fraction of time that the process is in state $\boldsymbol{x}_j$.

Time Reversibility. Time reversibility plays a central role in constructing a Markov chain for Monte Carlo simulation. When a movie is played backward, we immediately take notice. However, there are processes which are impossible to distinguish by observing their trajectories if it is played forward or backward in time. These are time reversible processes.

Specifically, if the joint probability a Markov process takes the sequence of states $(\boldsymbol{x}_0, \boldsymbol{x}_1, \ldots, \boldsymbol{x}_n)$ at time $(0, 1, \ldots, n)$ is the same as it takes the sequence of states $(\boldsymbol{x}_n, \boldsymbol{x}_{n-1}, \ldots, \boldsymbol{x}_1)$ at time $(0, 1, \ldots, n)$ for arbitrary n, namely,

$$\pi(\boldsymbol{x}_0, \boldsymbol{x}_1, \ldots, \boldsymbol{x}_n) = \pi(\boldsymbol{x}_n, \boldsymbol{x}_{n-1}, \ldots, \boldsymbol{x}_1, \boldsymbol{x}_0),$$

this Markov process is *time reversible*.

For a time reversible process, if $\boldsymbol{x}_1 = \boldsymbol{x}_0$, we will have $\pi_1(\boldsymbol{x}) = \pi_0(\boldsymbol{x})$. Since $\pi_1(\boldsymbol{x}) = \pi_o(\boldsymbol{x})P$, we have $\pi_0(\boldsymbol{x}) = \pi_0(\boldsymbol{x})\boldsymbol{P}$, that is, the initial distribution is stationary. In addition, since the joint probability $\pi(\boldsymbol{x}_0 = i, \boldsymbol{x}_1 = j)$ is $\pi(\boldsymbol{x}_1 = i, \boldsymbol{x}_0 = j)$ for a time-reversible Markov chain, we have

$$\pi(i)p_{i,j} = \pi(j)p_{j,i} \qquad \text{for all } i, j.$$

This is called the *detailed balance* condition.

Overall, one can show that the Markov chain is time-reversible if and only if the detailed balance condition is satisfied and the initial state is already in the steady state.

3.3.2 Markov Chain Monte Carlo Sampling

The basis of Markov chain Monte Carlo simulation, also called Metropolis Monte Carlo, is that if a Markov chain satisfies the conditions discussed above, it will converge to the stationary distribution $\pi(\boldsymbol{x})$. Therefore, we can obtain dependent samples following $\pi(\boldsymbol{x})$. Assuming it takes m steps for the Markov chain to reach the stationary state, we can discard the first m samples and obtain our estimates of Eq. (3.2) as

$$\mathbb{E}[f(\boldsymbol{x})] = \sum_{m+1}^{n} f(\boldsymbol{x}_i).$$

The key issue is then, how can we construct a Markov chain such that its stationary distribution is the same as the distribution $\pi(\boldsymbol{x})$ of our interest, namely, the target distribution.

The basic idea was laid out by Metropolis and colleagues in the 1950s, which was further developed by Hastings in the 1970s. Briefly, we sample a candidate state $\boldsymbol{y}$ from a proposal distribution $q(\cdot|\boldsymbol{x}_t)$. $\boldsymbol{y}$ is accepted as $\boldsymbol{x}_{t+1}$, with the probability $\alpha(\boldsymbol{x}, \boldsymbol{y})$:

$$\alpha(\boldsymbol{x}, \boldsymbol{y}) = \min\left[1, \frac{\pi(\boldsymbol{y})q(\boldsymbol{x}|\boldsymbol{y})}{\pi(\boldsymbol{x})q(\boldsymbol{y}|\boldsymbol{x})}\right]. \tag{3.3}$$

We can summarize the Markov chain Monte Carlo method as Algorithm 3.3.2

Here the proposal distribution $q(\boldsymbol{y}|\boldsymbol{x})$ can have any form, but the stationary distribution will always be $\pi(\boldsymbol{x})$. It is remarkable that this seemingly simple approach works.

We now discuss the rationale behind MCMC. The transition probability for the Markov chain to travel from $\boldsymbol{x}_t$ to a different state $\boldsymbol{x}_{t+1}$ can be decomposed into two events. First, we generated $\boldsymbol{x}_{t+1}$ with the proposal probability $q(\boldsymbol{y}|\boldsymbol{x}_t)$. Second, $\boldsymbol{y}$ is then accepted as $\boldsymbol{x}_{t+1}$ with the probability $\alpha(\boldsymbol{x}, \boldsymbol{y})$. When $\boldsymbol{x}_{t+1} \neq \boldsymbol{x}$, the transition probability $p(\boldsymbol{x}_{t+1}|\boldsymbol{x}_t)$ is the product $q(\boldsymbol{y}|\boldsymbol{x}_t) \cdot \alpha(\boldsymbol{x}, \boldsymbol{y})$. When $\boldsymbol{x}_{t+1} = \boldsymbol{x}$, the transition probability $p(\boldsymbol{x}_{t+1}|\boldsymbol{x}_t)$ is $\mathbb{I}(\boldsymbol{x}_{t+1} = \boldsymbol{x}_t)\left[1 - \int_{\boldsymbol{y}\neq\boldsymbol{x}} q(\boldsymbol{y}|\boldsymbol{x}_t)\alpha(\boldsymbol{x}, \boldsymbol{y})\right]$. From Eq. (3.3), we know

$$\pi(\boldsymbol{x}_t)q(\boldsymbol{x}_{t+1}|\boldsymbol{x}_t)\alpha(\boldsymbol{x}_t, \boldsymbol{x}_{t+1}) = \min[\pi(\boldsymbol{x}_t)q(\boldsymbol{x}_{t+1}|\boldsymbol{x}_t), \qquad \pi(\boldsymbol{x}_{t+1})q(\boldsymbol{x}_t|\boldsymbol{x}_{t+1})].$$

Algorithm 3.3.2 Markov chain Monte Carlo

Set $t = 0$ and initialize state $\boldsymbol{x}_0$;
repeat
  Sample a candidate new state $\boldsymbol{y}$ from a proposal distribution function $q(\cdot|\boldsymbol{x}_t)$;
  $\alpha(\boldsymbol{x}, \boldsymbol{y}) \leftarrow \min\left[1, \frac{\pi(\boldsymbol{y})q(\boldsymbol{x}|\boldsymbol{y})}{\pi(\boldsymbol{x})q(\boldsymbol{y}|\boldsymbol{x})}\right]$;
  Draw u from $\mathcal{U}[0, 1]$
  if $(u \leq \alpha(\boldsymbol{x}, \boldsymbol{y}))$ **then**
    $\boldsymbol{x}_{t+1} \leftarrow \boldsymbol{y}$
  else
    $\boldsymbol{x}_{t+1} \leftarrow \boldsymbol{x}_t$
  end if
  Increment t
until convergency

Similarly, we know

$$\pi(\boldsymbol{x}_{t+1})q(\boldsymbol{x}_t|\boldsymbol{x}_{t+1})\alpha(\boldsymbol{x}_{t+1}, \boldsymbol{x}_t) = \min[\pi(\boldsymbol{x}_{t+1})q(\boldsymbol{x}_t|\boldsymbol{x}_{t+1}), \qquad \pi(\boldsymbol{x}_t)q(\boldsymbol{x}_{t+1}|\boldsymbol{x}_t)]).$$

That is, we have

$$\pi(\boldsymbol{x}_t)q(\boldsymbol{x}_{t+1}\boldsymbol{x}_t)\alpha(\boldsymbol{x}_t, \boldsymbol{x}_{t+1}) = \pi(\boldsymbol{x}_{t+1})q(\boldsymbol{x}_t|\boldsymbol{x}_{t+1})\alpha(\boldsymbol{x}_{t+1}, \boldsymbol{x}_t),$$

which is the same as the detailed balance condition

$$\pi(\boldsymbol{x}_t)p(\boldsymbol{x}_{t+1}|\boldsymbol{x}_t) = \pi(\boldsymbol{x}_{t+1})p(\boldsymbol{x}_t|\boldsymbol{x}_{t+1}). \tag{3.4}$$

With the detailed balance condition satisfied, if we integrate both sides of Eq. (3.4) with regard to $\boldsymbol{x}_t$, we have

$$\int \pi(\boldsymbol{x}_t)p(\boldsymbol{x}_{t+1}|\boldsymbol{x}_t)\, d\boldsymbol{x}_t = \pi(\boldsymbol{x}_{t+1}).$$

That is, if $\boldsymbol{x}_t$ is drawn from the stationary distribution $\pi(\boldsymbol{x})$, then $\boldsymbol{x}_{t+1}$ is also drawn from the stationary distribution. We have now shown that once a sample is generated from the stationary distribution $\pi(\boldsymbol{x})$, all subsequent samples will be from that distribution.

Remark. Although Markov chain Monte Carlo sampling has found wide applications, there are a number of issues: Sometimes it is difficult to assess whether the Markov chain has reached the stationary distribution or whether adequate sampling is achieved. It is also often difficult to ensure that samples drawn from the steady-state distribution have small variance. A critical component of an effective sampling strategy is the design of the trial function or proposal function $q(\cdot|bx_t)$, often also called the move set, which generates candidate $\boldsymbol{x}_{t+1}$. A well-designed move set will increase the convergence rate significantly. However, this often requires expertise and specific consideration of the problem and is a challenging task.

3.4 SEQUENTIAL MONTE CARLO SAMPLING

The sequential Monte Carlo (SMC) method offers another general framework to obtain samples from a target distribution. It has its origin in early studies of chain polymers, where a growth strategy was used to generate self-avoiding polymer chains, with chains grown one monomer at a time until the desired length is reached [25]. By generating many independently grown chains, estimation such as Eq. (3.1) can be made. This approach was subsequently extensively studied and extended with wide applications [1,3,7,13–15,18,19,21,22,32,33].

3.4.1 Importance Sampling

Because of the high dimensionality, sampling is more effective if we focus on important regions instead of uniformly sampling everywhere. If our goal is to generate samples from the target distribution $\pi(\boldsymbol{x})$, it would be more productive to sample more frequently in regions where $\pi(\boldsymbol{x})$ has larger values.

It is often easier to sample from a trial distribution $g(\boldsymbol{x})$ than from the desired target distribution $\pi(\boldsymbol{x})$. The distribution $g(\boldsymbol{x})$ should be designed such that it has the same support as $\pi(\boldsymbol{x})$ and is as close in shape to $f(\boldsymbol{x})\pi(\boldsymbol{x})$ as possible. In fact, it is possible that a well-designed $g(\boldsymbol{x})$ is a better distribution than $\pi(\boldsymbol{x})$ to sample from for estimating $\mathbb{E}[f(\boldsymbol{x})]$.

As $g(\boldsymbol{x})$ is usually different from $\pi(\boldsymbol{x})$, correcting its bias is necessary. In the rejection sampling method discussed earlier, this was achieved by accepting only a fraction of generated samples. A more general approach is to assign weights to generated samples. When drawing samples $\boldsymbol{x}^{(1)}, \boldsymbol{x}^{(2)}, \ldots, \boldsymbol{x}^{(m)}$ from a trial distribution $g(\boldsymbol{x})$, we calculate the importance weight associated with each sample:

$$w^{(j)} = \frac{\pi(\boldsymbol{x}^{(j)})}{g(\boldsymbol{x}^{(j)})}.$$

The estimation of Eq. (3.1) can then be calculated as

$$\mathbb{E}[f(\boldsymbol{x})] \approx \left[\frac{w^{(1)} \cdot f(\boldsymbol{x}^{(1)}) + \cdots + w^{(m)} \cdot f(\boldsymbol{x}^{(m)})}{w^{(1)} + \cdots + w^{(m)}}\right]. \tag{3.5}$$

3.4.2 Sequential Importance Sampling

Designing an effective trial distribution or sampling distribution $g(\boldsymbol{x})$ can be very challenging. A problem is the high dimension of $\boldsymbol{x}$. For example, to sample conformations of a chain polymer such as a model protein molecule, we need to consider the configurations of many monomers (residues). One approach is to decompose the chain into individual monomers or residues and adopt the chain growth strategy. By growing the chain one monomer at a time, we can adaptively build up a trial function, one monomer at a time.

Specifically, we can decompose a high-dimensional random variable $\boldsymbol{x}$ as $\boldsymbol{x} = (x_1, \ldots, x_d)$. In the case of a chain polymer, $\boldsymbol{x}$ is the configuration of the full-length chain, and $x_i \in \mathbb{R}^3$ is the coordinates of the ith monomer. We can build up a trial distribution as

$$g(\boldsymbol{x}) = g_1(x_1)g_2(x_2|x_1)\ldots g_d(x_d|x1,\ldots,x_{d-1}),$$

where $g_1(x_1)$ is the trial distribution to generate x_1, and $g_i(x_i|x_1,\ldots,x_{i-1})$ is the trial distribution to generate x_i condition on already generated $(x_1,\ldots,x_{i-1})$. The target distribution can also be written analogously as

$$\pi(\boldsymbol{x}) = \pi_1(x_1)\pi_2(x_2|x_1)\cdots\pi_d(x_d|x1,\ldots,x_{d-1}), \tag{3.6}$$

The weight for a sample can be written as

$$w^{(j)} = \frac{\pi(\boldsymbol{x}^{(j)})}{g(\boldsymbol{x}^{(j)})} = \frac{\pi_1(x_1)\pi_2(x_2|x_1)\cdots\pi_d(x_d|x1,\ldots,x_{d-1})}{g_1(x_1)g_2(x_2|x_1)\cdots g_d(x_d|x1,\ldots,x_{d-1})}.$$

It is more convenient to calculate the weight by incrementally updating the weight as more components are added. At an intermediate step t, we can have the weight as

$$w_t = w_{t-1}\cdot\frac{\pi(x_t|(x_1,\ldots,x_{t-1})}{g(x_t|(x_1,\ldots,x_{t-1})}.$$

As we sequentially add x_i, the ideal decomposition of the target distribution shown in Eq. (3.6) is difficult to compute, as the target distribution at any intermediate step t, namely, $\pi(\boldsymbol{x}_t) = \pi_1(x_1)\pi_2(x_2|x_1)\cdots\pi_t(x_t|x1,\ldots,x_{t-1}) = \int \pi(x_1,\ldots,x_d)dx_{t+1}\cdots dx_d$, can only be computed through integrating out all other components, an often no less challenging task than the problem we set out to solve itself.

We can introduce instead an intermediate distribution $\pi_t(x_1,\ldots,x_t)$ when adding the tth component/monomer, which can be viewed as an approximation to the marginal distribution of the partial sample $\boldsymbol{x}_t = (x_1,\ldots,x_t)$:

$$\pi_t(x_1,\ldots,x_t) \approx \int \pi(\boldsymbol{x})\,dx_{t+1}\cdots dx_d.$$

When all components are added or the chain is grown to its full length, the intermediate distribution coincides with our target distribution:

$$\pi_d(x_1,\ldots,x_d) = \pi(x_1,\ldots,x_d).$$

It is natural to use $\pi_t(x_1,\ldots,x_t)$ to design the trial sampling distribution. For example, we can have our trial distribution as

$$g(x_t|x_1,\ldots,x_{t-1}) = \pi_t(x_t|x_1,\ldots,x_{t-1}).$$

To correct the bias in sampling using these intermediate distributions, we can calculate the weight incrementally as we add x_t:

$$w_t = w_{t-1}\frac{\pi_t(x_1,\ldots,x_t)}{\pi_{t-1}(x_1,\ldots,x_t)}\cdot\frac{1}{g_t(x_t|x_1,\ldots,x_{t-1})}.$$

Using the intermediate distributions $\pi_t(x_1, \dots, x_t)$ has a number of advantages. As $\pi_t(x_1, \dots, x_t)$ more or less tracks the target distribution $\pi(x_1, \dots, x_d)$, we can judge the quality of samples before they are completed. For example, we can improve computing efficiency by eliminating further simulation if a sample has very small w_t before finishing adding all components to x_d.

Overall, the sampling process is carried out sequentially: We first draw x_1 from $g_1(x_1)$, and then we draw the x_2 condition on the existing x_1 from g_2. This is repeated until x_d is drawn. In the example of generating a chain polymer, we first draw x_1 for the placement of the first monomer, and then we add the second monomer to the location x_2, which is drawn from g_2, until the chain reaches the full length d. To sample $\boldsymbol{x}$ from $\pi(\boldsymbol{x})$, the individual $g_i()$s are to be designed so the joint trial distribution $g(\boldsymbol{x})$ resembles $\pi(\boldsymbol{x})$ or $f(\boldsymbol{x})\pi(\boldsymbol{x})$ as much as possible.

The sequential Monte Carlo algorithm can be summarized as Algorithm 3.4.2.

Algorithm 3.4.2 Sequential Monte Carlo

Draw $x_1^{(j)}, j = 1, \dots, m$ from $g_1(x_1)$
for $t = 1$ to $n - 1$ **do**
  for $j = 1$ to m **do**
    Sampling the $(t + 1)$th monomer for the jth sample
    Draw position $x_{t+1}^{(j)}$ from $g_{t+1}(x_{t+1}|x_1^{(j)} \dots x_t^{(j)})$
    Compute the incremental weight

$$u_{t+1}^{(j)} \leftarrow \frac{\pi_{t+1}(x_1^{(j)} \dots x_{t+1}^{(j)})}{\pi_t(x_1^{(j)} \dots x_t^{(j)}) \cdot g_{t+1}(x_{t+1}^{(j)}|x_1^{(j)} \dots x_t^{(j)})}$$

$$w_{t+1}^{(j)} \leftarrow u_{t+1}^{(j)} \cdot w_t^{(j)}$$

  end for
  (Resampling)
end for

Self-Avoiding Walk in Two-Dimensional Lattice. As an illustration, we examine the problem of estimating the average end-to-end distance of a chain polymer of length N, which is modeled as a self-avoiding walk on a planar lattice. The configuration of the molecule is denoted as $\boldsymbol{x} = (x_1, \dots, x_N)$, where the ith monomer is located at $x_i = (a, b)$, where a, b are the coordinates of the monomer. As a chain polymer, distances between neighboring monomers x_i and x_{i+1} are exactly 1. As this molecule has excluded volume, none of the lattice sites can be occupied by more than one monomer.

Our goal is to estimate the average end-to-end distance $\mathbb{E}(||x_N - x_1||^2)$ of the self-avoiding walks under the uniform distribution $\pi(\boldsymbol{x}) = 1/Z_N$, where Z_N is the total number of SAWs. If we use the approach of Metropolis Monte Carlo, we would start with a particular configuration, for example, an extended chain, and apply various moves, and run the simulation for a long period to ensure the stationary distribution is reached, and then calculate $\mathbb{E}(||x_N - x_1||^2)$ from collected correlated samples.

Here we discuss how to apply the chain growth-based sequential Monte Carlo technique to solve this problem. This is a widely used approach to study chain polymers [7,13–15,25,32,33]. Naively, we can start at $(0, 0)$ and repeatedly choose with equal probability one of the three neighboring sites for placing the next monomer. If a site is already occupied, we go back to the origin and start with a new chain, until the surviving chain reaches the full length. However, the success rate of this approach is very small, as most attempts will end up with running into an occupied site prematurely.

This approach, due to Rosenbluth et al. , is to look one step ahead when placing a monomer. At step t, we examine all neighboring sites of $x_t = (i, j)$ at $(i \pm 1, j)$ and $(i, j \pm 1)$. If all neighbors have been visited, this chain is terminated, with an assigned weight of 0. Otherwise, we select one of the available sites with equal probability to place the next monomer. Specifically, we draw the position of x_{t+1} condition on current configuration of the chain $(x_1, \ldots, x_t)$ according to the probability distribution

$$\pi(x_{t+1}) = p[(k, l)|x_1, \ldots, x_t] = \frac{1}{n_t},$$

where (k, l) is one of the unoccupied neighbors, and n_t is the total number of such unoccupied neighbors. Figure 3.3 illustrates this approach, where the process of generating all 5-mers starting from a 2-mer is shown.

However, samples generated are not uniformly distributed. When growing a 2-mer to all possible 5-mers (Fig. 3.3), as there are a total of 25 5-mers that can be generated from the initial dimer, each chain should be generated with an equal probability of 1/25 as the uniform distribution is our target distribution. However, these 5-mers will be generated with unequal probability. For the uniform target distribution, we should have $\pi(\boldsymbol{x}) \propto 1$, but the chains are sampled differently, with $\boldsymbol{x} \propto (n_1 \times n_2 \times \cdots \times n_{N-1})^{-1}$, with n_i being the number of empty sites neighboring the last added monomer at step i. Such bias can be corrected by assigning a weight to each chain generated, which is

$$w(\boldsymbol{x}) = n_1 \times n_2 \times \cdots \times n_{N-1}.$$

Within the sequential importance sampling framework, our sampling distribution is

$$g_t(x_t|\boldsymbol{x}_{t-1}) = \begin{cases} \frac{1}{n_{t-1}} & \text{when } n_{t-1} > 0\,, \\ 0 & \text{when } n_{t-1} = 0. \end{cases} \quad (3.7)$$

That is, if there are empty sites neighboring x_{t-1}, x_t then occupies one of the n_{t-1} available sites. The sequence of intermediate distribution functions are $\pi_t(\boldsymbol{x}_t)$, $t = 1, \ldots, N-1$, which is a sequence of uniform distribution of SAWs with t-monomers:

$$\pi_t(\boldsymbol{x}_t) = \frac{1}{Z_t},$$

where Z_t is the total number of SAWs with t monomers.

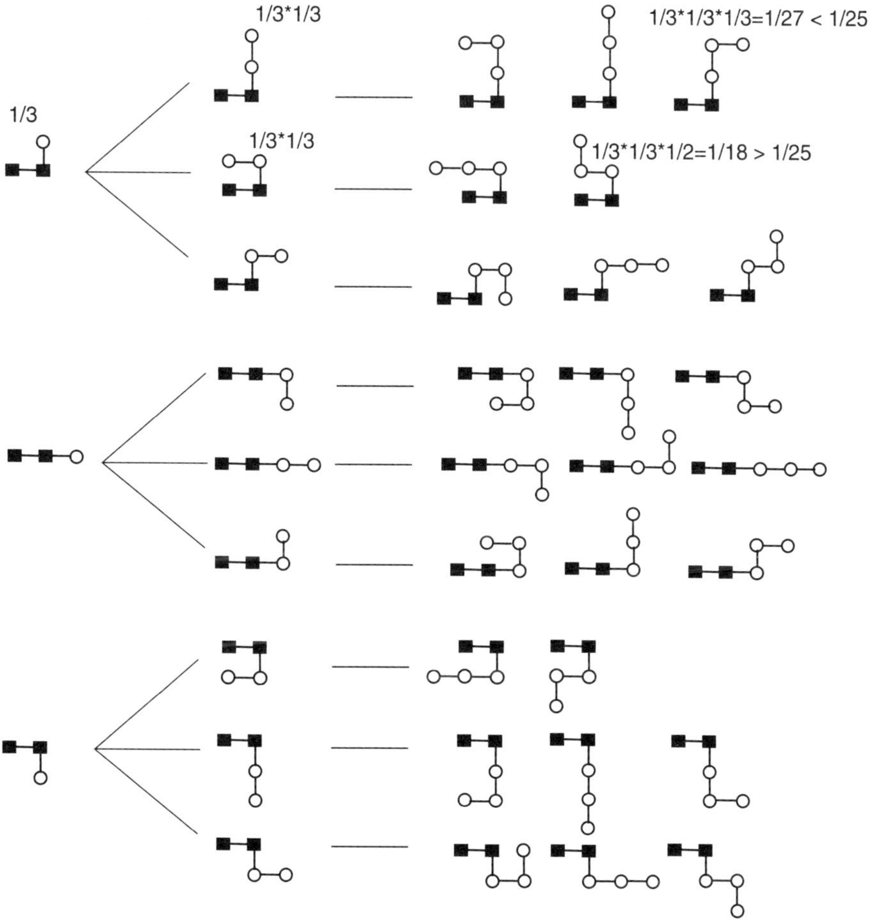

FIGURE 3.3 Generating a self-avoiding chain by sequential importance sampling. Following Rosenbluth [25], samples of self-avoiding walks are grown from 2-mer to 5-mer by adding a monomer at each of the unoccupied neighboring site with equal probability. To draw samples from the uniform distribution of 5-mers, all of the 25 conformations should have equal probability of $1/25$ to be generated. However, the conformations are generated with unequal probability. For example, some are generated with the probability of $1/3 \cdot 1/3 \cdot 1/2 = 1/18$, and others are generated with the probability of $1/3 \cdot 1/3 \cdot 1/3 = 1/27$. Such bias can be corrected by assigning a proper weight to each of the conformations.

3.4.3 Resampling

As more components or monomers are added, these unfinished partial samples may have very diverse weights. Many samples may have very small weights, while a few may have very large weights. Samples with small weights will contribute little to our estimation of Eq. (3.1). Samples with very small values of $f(\boldsymbol{x})$ will also have little contributions if our task is to estimate $\mathbb{E}_\pi[f(\boldsymbol{x})]$. In the case of growing polymer chain, we may find that the chain has reached a dead end and can no longer be grown further. It may be more profitable to replace these samples with other more promising samples.

The technique of resampling can be used to reduce sample variance. We can first assign a resampling probability to each of the current sample. The assigned values reflect our preference to either encourage or discourage this sample to be taken again. Let the set of m samples be $\{\boldsymbol{x}_t^{(j)}\}, j = 1, \ldots, m$, and let the associated resampling probabilities be $\{\alpha^{(j)}\}, j = 1, \ldots, m$. We draw n samples $\boldsymbol{x}_t^{(j')}$ from the existing set of m samples according to probabilities $(\alpha^{(1)}, \cdots, a^{(m)})$. A new weight $w_t^{(j')} = \frac{w_t^{(j)}}{\alpha^{(j)}}$ is then assigned to the resampled $\boldsymbol{x}_t^{(j')}$. The resulting samples will be properly weighted according to the target distribution $\pi(\boldsymbol{x})$. The resampling algorithm can be summarized as shown in Algorithm 3.4.3.

Algorithm 3.4.3 Resampling

m: number of original samples.
$\{(x_1^{(j)}, \ldots, x_t^{(j)}), w^{(j)}\}_{j=1}^m$: original properly weighter samples
for $*j = 1$ to m **do**
Draw $*j$th sample from original samples $\{(x_1^{(j)}, \ldots, x_t^{(j)})\}_{j=1}^m$ with probabilities $\propto \{\alpha^{(j)}\}_{j=1}^m$
Each sample in the newly formed sample is assigned a new weight
$*j$th chain in new sample is a copy of kth chain in original sample
$w^{(*j)} \leftarrow w^{(k)}/\alpha^{(k)}$
end for

As our goal is to prune away poor samples, it is important to design effective resampling probabilities $\{\alpha^{(j)}\}$. One generic approach is to set $\alpha^{(j)}$ as a monotonic function of $w^{(j)}$. For targeted resampling, we will choose $\alpha^{(j)}$ based on the objective of interest. Physical considerations and future information often can help to design $\alpha^{(j)}$. We will also need to balance the desire to have diverse samples and to have many samples with large weights.

3.5 APPLICATIONS

3.5.1 Markov Chain Monte Carlo for Evolutionary Rate Estimation

We now discuss how Markov chain Monte Carlo can be applied using the example of estimating substitution rates of amino acid residues, an important task in analysis of the pattern of protein evolution.

Protein Function Prediction. When a protein is found to be evolutionarily related to another protein—for example, through sequence alignment—one can often make

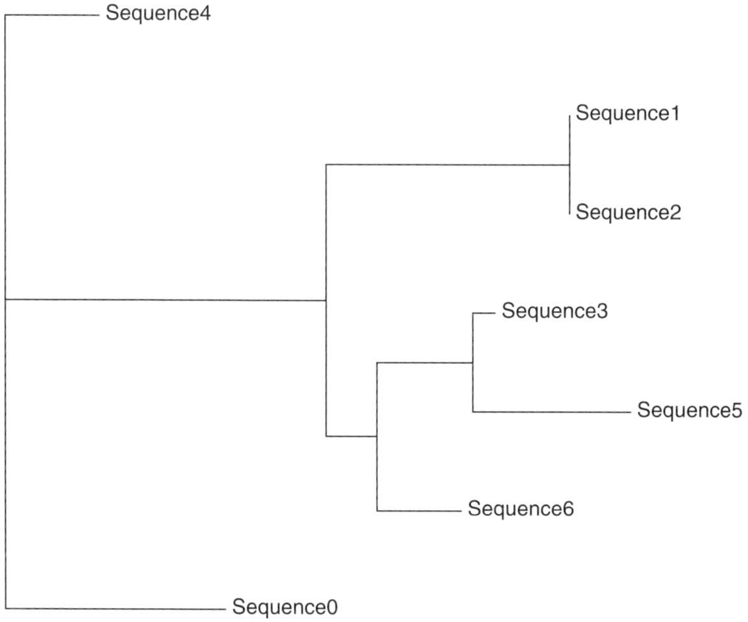

FIGURE 3.4 An example of a phylogenetic tree. The branching topology of the tree represents the ancestor and descendent relationship. The length of the edge represents the evolutionary time in relative units.

inference on its biochemical functions. The success in detecting such evolutionary relationship between two proteins depends on the use of a scoring matrix to quantify the similarity between two aligned sequences.

Scoring matrices can be derived from analysis of substitution rates of amino acid residues. The widely used Pam and Blosum scoring matrices are based on empirical models of amino acid residue substitutions [2,11]. A more recent approach is to employ an explicit continuous-time Markov process to model the history of evolution of the specific protein of interests [28,29,31]. We discuss below how Markov chain Monte Carlo can be used to estimate substitution rates of amino acid residues.

Continuous-Time Markov Process for Residue Substitution. Assuming that a phylogenetic tree is given, which captures the evolutionary relationship between protein sequences, we can use a reversible continuous-time Markov process to model substitutions of amino acid residues [5,30]. The model parameters are the 20×20 rate matrix $\boldsymbol{Q}$, in which the entries q_{ij}s are instantaneous substitution rates of amino acid residues for the set $\mathcal{A}$ of 20 amino acid residues, with the diagonal element taken as $q_{i,i} = -\sum_{i,j \neq i} q_{i,j}$. This matrix of instantaneous rates can be used to calculate the transition probabilities after time t [16]:

$$\boldsymbol{P}(t) = \{p_{ij}(t)\} = \boldsymbol{P}(0) \exp(\boldsymbol{Q} \cdot t),$$

where $\boldsymbol{P}(0) = \boldsymbol{I}$. Here $p_{ij}(t)$ represents the probability that a residue of type i will mutate into a residue of type j after time t.

Likelihood Function of a Fixed Phylogeny. For sequence k and sequence l separated by divergence time t_{kl}, the time-reversible probability of observing residue x_k at a position h in k and residue x_l at the same position in l is

$$\pi(x_k)p_{x_k x_l}(t_{kl}) = \pi(x_l)p_{x_l x_k}(t_{kl}).$$

For a set $\mathcal{S}$ of s multiple-aligned sequences $(\boldsymbol{x}_1, \boldsymbol{x}_2, \dots, \boldsymbol{x}_s)$ of length n amino acid residues, we assume that a reasonably accurate phylogenetic tree is known. We denote the tree as $\boldsymbol{T} = (\mathcal{V}, \mathcal{E})$. Here $\mathcal{V}$ is the set of sequences, namely, the union of the set of observed s sequences $\mathcal{L}$ (leaf nodes), and the set of $s - 1$ ancestral sequences $\mathcal{I}$ (internal nodes). $\mathcal{E}$ is the set of edges of the tree, where each edge represents a ancestor–descendent relationship, with edge length representing the evolutionary time. Let the vector $\boldsymbol{x}_h = (x_1, \dots, x_s)^T$ be the observed residues at position h, with $h \in \{1, \dots, n\}$ for the s sequences. The probability of observing s number of residues $\boldsymbol{x}_h$ at position h according to our model is

$$p(\boldsymbol{x}_h|\boldsymbol{T}, \boldsymbol{Q}) = \pi_{x_k} \sum_{\substack{i \in \mathcal{I} \\ x_i \in \mathcal{A}}} \prod_{(i,j) \in \mathcal{E}} p_{x_i x_j}(t_{ij})$$

after summing over the set $\mathcal{A}$ of all possible residue types for the internal nodes $\mathcal{I}$. The probability $P(\mathcal{S}|\boldsymbol{T}, \boldsymbol{Q})$ of observing all residues in the aligned region is

$$P(\mathcal{S}|\boldsymbol{T}, \boldsymbol{Q}) = P(\boldsymbol{x}_1, \cdots, \boldsymbol{x}_s|\boldsymbol{T}, \boldsymbol{Q}) = \prod_{h=1}^{n} p(\boldsymbol{x}_h|\boldsymbol{T}, \boldsymbol{Q}).$$

Bayesian Estimation of Instantaneous Rates. We adopt a Bayesian approach to estimated $\boldsymbol{Q}$. We describe the instantaneous substitution rate $\boldsymbol{Q} = \{q_{ij}\}$ by a posterior distribution $\pi(\boldsymbol{Q}|\mathcal{S}, \boldsymbol{T})$. We use a prior distribution $\pi(\boldsymbol{Q})$ to encode our past knowledge of amino acid substitution rates for proteins. $\pi(\boldsymbol{Q}|\mathcal{S}, \boldsymbol{T})$ summarizes prior information available on the rates $\boldsymbol{Q} = \{q_{ij}\}$ and the information contained in the observations $\mathcal{S}$ and $\boldsymbol{T}$. It can be estimated up to a constant as

$$\pi(\boldsymbol{Q}|\mathcal{S}, \boldsymbol{T}) \propto \int P(\mathcal{S}|\boldsymbol{T}, \boldsymbol{Q}) \cdot \pi(\boldsymbol{Q})\, d\boldsymbol{Q}.$$

Markov Chain Monte Carlo. We can run a Markov chain to generate samples drawn from the target distribution $\pi(\boldsymbol{Q}|\mathcal{S}, \boldsymbol{T})$. Starting from a rate matrix $\boldsymbol{Q}_t$ at time t, we generate a new rate matrix $\boldsymbol{Q}_{t+1}$ using the proposal function $T(\boldsymbol{Q}_t, \boldsymbol{Q}_{t+1})$. The proposed new matrix $\boldsymbol{Q}_{t+1}$ will be either accepted or rejected by the acceptance ratio $r(\boldsymbol{Q}_t, \boldsymbol{Q}_{t+1})$. Specifically, we have

$$\boldsymbol{Q}_{t+1} = A(\boldsymbol{Q}_t, \boldsymbol{Q}_{t+1}) = T(\boldsymbol{Q}_t, \boldsymbol{Q}_{t+1}) \cdot r(\boldsymbol{Q}_t, \boldsymbol{Q}_{t+1}).$$

To ensure that the Markov chain will reach the stationary state, we need to satisfy the requirement of detailed balance, that is,

$$\pi(\boldsymbol{Q}_t|S,\boldsymbol{T}) \cdot A(\boldsymbol{Q}_t,\boldsymbol{Q}_{t+1}) = \pi(\boldsymbol{Q}_{t+1}|S,\boldsymbol{T}) \cdot A(\boldsymbol{Q}_{t+1},\boldsymbol{Q}_t).$$

This is achieved by using the Metropolis–Hastings acceptance ratio $r(\boldsymbol{Q}_t,\boldsymbol{Q}_{t+1})$ discussed earlier to either accept or reject $\boldsymbol{Q}_{t+1}$, depending on whether the following inequality holds:

$$u \le r(\boldsymbol{Q}_t,\boldsymbol{Q}_{t+1}) = \min\left\{1, \frac{\pi(\boldsymbol{Q}_{t+1}|S,\boldsymbol{T}) \cdot T(\boldsymbol{Q}_{t+1},\boldsymbol{Q}_t)}{\pi(\boldsymbol{Q}_t|S,\boldsymbol{T}) \cdot T(\boldsymbol{Q}_t,\boldsymbol{Q}_{t+1})}\right\},$$

where u is a random number drawn from the uniform distribution $\mathcal{U}[0,1]$. With the assumption that the underlying Markov process satisfy the conditions outlined earlier, a Markov chain generated following these rules will reach the stationary state [8,24].

Once the stationary state is reached, we can collect m correlated samples of the $\boldsymbol{Q}$ matrix. The posterior means of the rate matrix can then estimated as

$$\mathbb{E}_\pi(\boldsymbol{Q}) \approx \sum_{i=1}^{m} \boldsymbol{Q}_i \cdot \pi(\boldsymbol{Q}_i|S,\boldsymbol{T}).$$

Rate Matrix and Scoring Matrix. With the estimated $\boldsymbol{Q}$ matrix, we can derive scoring matrices of different evolutionary time intervals [12]. The (i,j)th entry $b_{ij}(t)$ of a scoring matrix between residues i and j at different evolutionary time t can be calculated as

$$b_{ij}(t) = \frac{1}{\lambda} \log \frac{p_{ij}(t)}{\pi_j},$$

where λ is a scaling factor.

3.5.2 Sequential Chain Growth Monte Carlo for Estimating Conformational Entropy of RNA Loops

Conformational entropy makes an important contribution to the stability and folding of a biomolecule, but it is challenging to compute conformational entropy. Here we study the problem of computing the RNA loop entropy. Using a discrete k-state model for each nucleotide of the RNA molecule, we can model loops as self-avoiding walks in three-dimensional space and then calculate the loop entropy using sequential Monte Carlo.

For a loop of length n, where n is the number of unpaired nucleotides, its entropy change can be defined as [33]:

$$-\frac{\Delta S(n)}{k_B} = \ln\left(\frac{\Omega_{\text{coil}}}{\Omega_{\text{loop}}}\right), \tag{3.8}$$

where Ω_{coil} is the number of all possible conformations of a random coil of length n, and Ω_{loop} is the number of loop conformations that are compatible with the stem that closes the loop.

We use the sequential Monte Carlo algorithm to calculate the RNA loop entropy. During the process of chain growth, we generates a set of properly weighted conformations with respect to the target distribution of uniformly distributed RNA molecules, along with correct weights of the conformations. We use the following scheme:

1. **Initialization.** We set the initial sampling size to $m_1 = 1$, with weight $w_1^{(1)} = 1$. At step $t - 1$, we have m_{t-1} partial conformations with corresponding weights, denoted as $\{(S_{t-1}^{(j)}, w_{t-1}^{(j)}),\ j = 1, \ldots, m_{t-1}\}$.
2. **Chain Growth.** For each partially grown conformation $S_{t-1}^{(j)}$, we exhaustively test all possible attachments of the next nucleotide, with a total of $k_t^{(j)}$ different possibilities. This will generate no greater than k different partial conformations of length t, $\bar{S}_t^{(j,l)} = (S_{t-1}^{(j)}, s_t)$, with temporary weights $\bar{w}_t^{(j,l)} = w_{t-1}^{(j)}$. We denote all such samples generated as $\{(\bar{S}_t^{(l)}, \bar{w}_t^{(l)}), l = 1, \ldots, L\}$, where $L = \sum_{j=1}^{m_{t-1}} k_t^{(j)}$.
3. **Resampling.** If $L \leq m$, which is the upper bound of Monte Carlo sample size, we keep all of the samples and their corresponding weights and set $m_t = L$. If $L > m$, we choose $m_t = m$ distinct samples with marginal probabilities proportional to a set of priority scores $\beta_t^{(l)}$. Intuitively, the priority score $\beta_t(S_t)$ reflects the chain's ''growth perspective'' and is used to encourage the growth of chain S_t to specific directions.
4. **Estimation.** When the target loop length n is reached, Ω_{coil} is estimated as $\sum_{j=1}^{m_n} w_n^{(j)} \mathbb{I}(S_n^{(j)})$, where m_n is the number of samples at length n, $w_n^{(j)}$ is the importance weight of samples $S_n^{(j)}$, and $\mathbb{I}()$ is the identity function of 1.

Details of the resampling strategy, including the design of the priority scores, can be found in reference 33.

The calculated loop entropy for hairpin loops of length 3–50 has excellent agreement with values extrapolated from the Jackson–Stockmayer model. However, calculations reveal that loop entropies of more complex RNA secondary structures are significantly different from the extrapolated values for long internal loops. Overall, conformational entropy of different RNA secondary structures with loops can be calculated with accuracy beyond extrapolation of simplified theoretical models.

3.6 DISCUSSION AND SUMMARY

In this chapter, we discussed the general problem of characterizing ensemble properties of biological systems through integration by sampling, along with the difficulties of sampling in high-dimensional space. We briefly examined the approach of rejection sampling, and were discussed in more detail two general frameworks in Monte Carlo sampling, namely, the Markov chain Monte Carlo (MCMC) or Metropolis Monte Carlo method and the sequential Monte Carlo method. We discussed basic concepts such as sampling from a desired target distribution, properties of Markov chains, time reversibility, detailed balance, and the stationary state. This was followed by the example of estimating evolutionary substitution rates of amino acids. For sequential Monte Carlo, we discussed the general principle of importance sampling, the approach of sequentially building up the target distribution, and the technique of resampling for variance reduction. The applications in generating self-avoiding walks for studying chain polymers and calculating RNA loop entropy were then presented.

Remark. Generating samples from a target distribution for tasks such as Eq. (3.1) is a fundamental problem in science and engineering. Among the two general frameworks of Monte Carlo sampling, the Metropolis Monte Carlo or the Markov chain Monte Carlo (MCMC) method can generate correlated samples from a target distribution, and the sequential Monte Carlo or the sequential importance sampling method can generate samples from a trial distribution different from the target distribution. Samples are then adjusted according their importance weights so they follow the target distribution.

The MCMC method has its origin in the 1950s [23], where the idea of an evolving Markov chain was first introduced for sampling from a target distribution. The extension to allow nonsymmetric transition rules was made by Hastings [10]. Multilevel sampling methods were subsequently developed, including the umbrella sampling method [26] and parallel tempering or replica exchange sampling methods [4,6]. The application of MCMC for studying molecular evolution can be found in reference 28.

The sequential importance sampling method was first described in the work of Rosenbluth et al. in generating chain polymers using a chain growth strategy [25]. Further development of the look-ahead strategy was subsequently developed in reference 22. The theory of sequential importance sampling with resampling was developed by Liu and Chen [17] and in simulating chain polymers by Grassberger [7]. The topic of proper rejection control can be found in reference 21. The general theoretical framework of sequential Monte Carlo can be found in [18,20]. Further studies of chain polymers, including those under severe constraints, including void formation, protein packing, generating conformations from contact maps, and generating transition state ensemble of protein folding, can be found in references 13–15 and 32. Loop entropy calculation for various RNA secondary structures using sequential Monte Carlo can be found in reference 34. A study on the effects of spatial confinement of cell nucleus in determining the folding landscape, including scaling behavior of long range chromatin interactions, is described in reference 9.

REFERENCES

1. R. Chen and J. Liu. Sequential monte carlo methods for dynamic systems. *J. Am. Statist. Assoc.*, **93**:1032–1043, 1998.

2. M. O. Dayhoff, R. M Schwartz, and B. C Orcutt. *A model of evolutionary change in proteins. Atlas of Protein Sequence and Structure, Vol. 5, Suppl. 3.*, In pp. 345–352. National Biomedical Research Foundation, Washington, D.C., 1978.
3. A. Doucet, N. De Freitas, N. Gordon, and A. Smith. *Sequential Monte Carlo Methods in Practice*. Springer Verlag, New York, 2001.
4. David J. Earl and Michael W. Deem. Parallel tempering: Theory, applications, and new perspectives. *Phys. Chem. Chem. Phys.*, **7**:3910–3916, 2005.
5. J. Felsenstein. Evolutionary trees from DNA sequences: A maximum likelihood approach. *J. Mol. Evol.*, **17**:368–376, 1981.
6. C. J. Geyer. Markov chain Monte Carlo maximum likelihood. In *Proceedings of the 23rd Symposium on the Interface*, p. 156. American Statistical Association, New York, 1991.
7. P. Grassberger. Pruned-enriched Rosenbluth method: Simulation of θ polymers of chain length up to 1,000,000. *Phys. Rev. E.*, **56**:3682–3693, 1997.
8. G. R. Grimmett and D. R. Stizaker. *Probability and Random Processes*. Oxford University Press, New York, 2001.
9. G. Güsoy, Y. Xu, A. L. Kenter, and J. Liang. Spatial confinement is a major determinant of the folding landscape of human chromosomes. *Nucleic Acid Res.*, **42**:8223–8230, 2014.
10. W. K. Hastings. Monte Carlo sampling methods using Markov chains and their applications. *Biometrika*, **57**:97–109, 1970.
11. S. Henikoff and J. G. Henikoff. Amino acid substitution matrices from protein blocks. *Proc. Natl. Acad. Sci. USA*, **89**:10915–10919, 1992.
12. S. Karlin and S. F. Altschul. Methods for assessing the statistical significance of molecular sequence features by using general scoring schemes. *Proc. Natl. Acad. Sci. USA*, **87**:2264–2268, 1990.
13. J. Liang, J. Zhang, and R. Chen. Statistical geometry of packing defects of lattice chain polymer from enumeration and sequential Monte Carlo method. *J. Chem. Phys.*, **117**:3511–3521, 2002.
14. M. Lin, H. M. Lu, R. Chen, and J. Liang. Generating properly weighted ensemble of conformations of proteins from sparse or indirect distance constraints. *J. Chem. Phys.*, **129**(9):094101, 2008.
15. M. Lin, J. Zhang, H. M. Lu, R. Chen, and J. Liang. Constrained proper sampling of conformations of transition state ensemble of protein folding. *J. Chem. Phys.*, **134**(7):075103, 2011.
16. P. Liò and N. Goldman. Models of molecular evolution and phylogeny. *Genome Res.*, **8**:1233–1244, 1998.
17. J. S. Liu and R. Chen. Blind deconvolution via sequential imputations. *J. Am. Statist. Assoc.*, **90**:567–576, 1995.
18. J. S. Liu and R. Chen. Sequential Monte Carlo methods for dynamic systems. *J. Am. Statist. Assoc.*, **93**:1032–1044, 1998.
19. J. S. Liu, R. Chen, and T. Logvinenko. A theoretical framework for sequential importance sampling and resampling. In J. F. G. de Freitas A. Doucet and N. Gordon, editors, *Sequential Monte Carlo Methods in Practice*. Cambridge University Press, New York, 2000.

20. J. S. Liu, R. Chen, and T. Logvinenko. A theoretical framework for sequential importance sampling and resampling. In A. Doucet, J. F. G. de Freitas, and N. J. Gordon, editors, *Sequential Monte Carlo in Practice*. Springer-Verlag, New York, 2001.
21. J. S. Liu, R. Chen, and W. H. Wong. Rejection control and importance sampling. *J. Am. Statist. Assoc.*, **93**:1022–1031, 1998.
22. H. Meirovitch. A new method for simulation of real chains: Scanning future steps. *J. Phys.A: Math. Gen.*, **15**:L735–L741, 1982.
23. N. Metropolis, A. W. Rosenbluth, M. N. Rosenbluth, A. H. Teller, and E. Teller. Equations of state calculations by fast computing machines. *J. Chem. Phys*, **21**:1087–1091, 1953.
24. C. P. Robert and G. Casella. *Monte Carlo Statistical Methods*. Springer-Verlag, New York., 2004.
25. M. N. Rosenbluth and A. W. Rosenbluth. Monte Carlo calculation of the average extension of molecular chains. *J. Chem. Phys.*, **23**:356–359, 1955.
26. G. Torrie and J. Valleau. Nonphysical sampling distributions in Monte Carlo free-energy estimation: Umbrella sampling. *J. Comput. Phys.*, **23**(2):187–199, 1977.
27. Y. Y. Tseng and J. Liang. Are residues in a protein folding nucleus evolutionarily conserved? *J. Mol. Biol.*, **335**:869–880, 2004.
28. Y. Y. Tseng and J. Liang. Estimation of amino acid residue substitution rates at local spatial regions and application in protein function inference: A Bayesian Monte Carlo approach. *Mol. Biol. Evol.*, **23**(2):421–436, 2006.
29. S. Whelan and N. Goldman. A general empirical model of protein evolution derived from multiple protein families using a maximum-likelihood approach. *Mol. Biol. Evol.*, **18**:691–699, 2001.
30. Z. Yang. Estimating the pattern of nucleotide substitution. *J. Mol. Evol.*, **39**:105–111, 1994.
31. Z. Yang, R. Nielsen, and M. Hasegawa. Models of amino acid substitution and applications to mitochondrial protein evolution. *Mol. Biol. Evol.*, **15**:1600–1611, 1998.
32. J. Zhang, R. Chen, C. Tang, and J. Liang. Origin of scaling behavior of protein packing density: A sequential Monte Carlo study of compact long chain polymers. *J. Chem. Phys.*, **118**:6102–6109, 2003.
33. J. Zhang, M. Lin, R. Chen, and J. Liang. Discrete state model and accurate estimation of loop entropy of rna secondary structures. *J. Chem. Phys.*, **128**(125107):1–10, 2008.

EXERCISES

3.1 In the Markov chain Monte Carlo method, the final stationary distribution reached after the chain convergency is the desired target contribution:

$$\int \pi(\boldsymbol{x})A(\boldsymbol{x}, \boldsymbol{y})\, d\boldsymbol{x} = \pi(\boldsymbol{y}),$$

where $\boldsymbol{x}$ is the state variable and $A(\boldsymbol{x}, \boldsymbol{y}) = T(\boldsymbol{x}, \boldsymbol{y}) \cdot r(\boldsymbol{x}, \boldsymbol{y})$ is the actual transition function, that is, the product of the proposal function $T(\boldsymbol{x}, \boldsymbol{y})$ and an acceptance–rejection rule $r(\boldsymbol{x}, \boldsymbol{y})$. The proposal function $T(\boldsymbol{x}, \boldsymbol{y})$ suggests a possible move from $\boldsymbol{x}$ to $\boldsymbol{y}$. The acceptance-rejection rule decides whether the proposed move to $\boldsymbol{y}$ will be accepted: Draw a random number u from the uniform distribution $\mathcal{U}[0,1]$. If $u \leq r(\boldsymbol{x},\boldsymbol{y})$, the move is accepted and $\boldsymbol{y}$ is taken as the new position. Otherwise stay with $\boldsymbol{x}$.

In the original Metropolis Monte Carlo method, the proposal function is symmetric: $T(\boldsymbol{x},\boldsymbol{y}) = T(\boldsymbol{y},\boldsymbol{x})$, and the acceptance–rejection rule is simply

$$r(\boldsymbol{x},\boldsymbol{y}) = \min\{1, \pi(\boldsymbol{y})/\pi(\boldsymbol{x})\}$$

Since the target distribution is the Boltzmann distribution $\pi(\boldsymbol{x}) \sim \exp(h(\boldsymbol{x}))$, where $h(\boldsymbol{x})$ is an energy function, the acceptance rule is often written as $u \leq r(\boldsymbol{x},\boldsymbol{y}) = \exp(-[h(\boldsymbol{y}) - h(\boldsymbol{x})])$. This strategy will work, for example, if the proposal function gives equal probability $1/n(\boldsymbol{x})$ to each of the $n(\boldsymbol{x})$ conformations that can be reached from conformation $\boldsymbol{x}$:

$$T(\boldsymbol{x},\boldsymbol{y}) = 1/n(\boldsymbol{x}),$$

and if $n(\boldsymbol{x}) = n(\boldsymbol{y})$ for $\boldsymbol{x}$ and $\boldsymbol{y}$ that are connected by a move.

However, the number of possible moves for a conformation $\boldsymbol{x}$ frequently depends on the local geometry. For example, it is more difficult in protein simulation to move an amino acid residue that is buried in the interior than moving a residue located in a loop region. In other words, the number of allowed moves is different: $n(\boldsymbol{x}) \neq n(\boldsymbol{y})$, although each can be computed exactly.

When $n(\boldsymbol{x}) \neq n(\boldsymbol{y})$, what rules can you devise to generate a Markov chain such that its stationary distribution is the same Boltzman distribution. First, write your answer in pseudocode; second, show that indeed your strategy works.

3.2 Hastings first realized that the proposal distribution does not need to be symmetric, but can be arbitrarily chosen so long as the condition of detailed balance is satisfied. His generalization leads to the modified acceptance rule:

$$u \leq r(\boldsymbol{x},\boldsymbol{y}) = \min\left\{1, \frac{\pi(\boldsymbol{y})T(\boldsymbol{y},\boldsymbol{x})}{\pi(\boldsymbol{x})T(\boldsymbol{x},\boldsymbol{y})}\right\},$$

so more flexible and efficient sampling strategy can be developed, which still generates samples following the desired target distribution. Answer the following questions and show your proofs:

(**a**) Show that Hasting's rule satisfies the detailed balance condition $\pi(\boldsymbol{x})A(\boldsymbol{x},\boldsymbol{y}) = \pi(\boldsymbol{y})A(\boldsymbol{y},\boldsymbol{x})$.

(**b**) Why does Hasting's rule work? That is, why is the equilibrium distribution the same as the desired target distribution?

(**c**) According to the same principle, will the following trial acceptance rule work?

$$u \leq r(\boldsymbol{x},\boldsymbol{y}) = \min\left\{1, \frac{\pi(\boldsymbol{x})T(\boldsymbol{y},\boldsymbol{x})}{\pi(\boldsymbol{y})T(\boldsymbol{y},\boldsymbol{x}) + \pi(\boldsymbol{y})T(\boldsymbol{x},\boldsymbol{y})}\right\}.$$

(d) How about the next rule below?

$$u \leq r(\boldsymbol{x}, \boldsymbol{y}) = \min\left\{1, \frac{\pi(\boldsymbol{y})T(\boldsymbol{y}, \boldsymbol{x})}{\pi(\boldsymbol{y})T(\boldsymbol{y}, \boldsymbol{x}) + \pi(\boldsymbol{x})T(\boldsymbol{x}, \boldsymbol{y})}\right\}.$$

3.3 Here we use the Monte Carlo method to fold a sequence of a simplified protein model on a two-dimensional lattice. The conformation of a protein of length n is denoted as $\mathbf{x} = (\mathbf{x}_1, \ldots, \mathbf{x}_n) = ((x_1, y_1), \ldots, (x_n, y_n))$, where (x_i, y_i) is the coordinates of the ith residue in the two-dimensional lattice. The energy function of this HP model is $H(\mathbf{x}) = -\sum_{i+1<j} \Delta(\mathbf{x}_i, \mathbf{x}_j)$, where $\Delta(\mathbf{x}_i, \mathbf{x}_j) = 1$ if $\mathbf{x}_i$ and $\mathbf{x}_j$ are non-bonded spatial neighbors and if both are H residues. Otherwise, $\Delta(x_i, x_j) = 0$. We can use the move sets of end move, corner move, crank-shift move, and pivot move. For the end move, the ends of the chain move to an empty adjacent site. For the corner move, a single monomer is flipped. For the crankshaft move, two monomers are simultaneously moved. For pivot move, we choose a node along the chain as a pivot and apply a symmetry operation to the rest of the chain subsequent to the pivot. On a two-dimensional lattice, symmetry operation include rotation and reflection.

Our goal is to search for this sequence the conformation of the lowest energy, that is, we want to fold this model protein. The conformations follow the Boltzmann distribution $\pi \propto \exp\{-U(\mathbf{x})/T\}$, where T is temperature. Start from an extended conformation and write a program implementing the Metropolis–Hastings algorithm (or the Markov chain Monte Carlo method) with simulated annealing for this problem. Use different temperatures as needed–for example, a temperature ladder of $T_0 = 2, T_1 = 0.9 \cdot T_0, \ldots$. Be careful to allow enough burn-in period.

- Specify the actual move sets you used.
- Write down your transition rules.
- Write down your acceptance criterion for a move.
- Justification for the choice of burning-period.

You should run the simulation as many times as you can afford, and keep the conformation and energy value of your best result for some of the chains you run.

- Your output for a chain should include (a) drawing of the folded conformation with the lowest energy found and (b) printed energy value.
- Select four temperature values that are most interesting to you, and plot the trajectory of the energy values of the protein starting at the end of the burning period.

3.4 In importance sampling, it is essential to keep samples properly weighted, as this enables one to calculate many macroscopic properties of the target population. In model studies of protein folding, one approach to estimate thermodynamic properties of HP model proteins is to sample from the uniform distribution $u(\boldsymbol{x})$ of all SAWs on a lattice, using a sampling distribution function $g(\boldsymbol{x})$.

(1) What would be the proper weight w_i of each sampled conformation $\boldsymbol{x}_i$?

(2) If the goal is to estimate properties of the Boltzman distribution $\pi(\boldsymbol{x})$ which HP molecules follow, write down how you would re-adjust the weight of samples properly weighted for the uniform distribution $\{(\boldsymbol{x}_1, w_1), \cdots, (\boldsymbol{x}_m, w_m)\}$?

(3) Now you decide to sample conformations of HP sequence directly from the Boltzmann distribution $\pi(\boldsymbol{x})$. Write down the incremental weight one has to keep at each step of the growth where dimension increases by one. Here sampling is based on the trial function $g_i(\boldsymbol{x}_i)$ using the sequence of auxiliary functions $\pi_i(\boldsymbol{x}_i)$.

4

STOCHASTIC MOLECULAR NETWORKS

4.1 INTRODUCTION

Biomolecular networks formed by interacting biomolecules form the basis of regulatory machineries of many cellular processes. Stochasticity plays important roles in many networks. These include networks responsible for gene regulation, protein synthesis, and signal transduction [3,21,32,38,44]. The intrinsic stochasticity in these cellular processes originates from reactions involving small copy numbers of molecules. It frequently occurs in a cell when molecular concentrations are in the range of μM to nM [3,40]. For example, the regulation of transcriptions depends on the binding of often a few proteins to a promoter site. The synthesis of protein peptides on a ribosome involves a small number of molecules. Patterns of cell differentiation also depend on events with initially a small number of molecules. In these biological processes, fluctuations intrinsic in low copy number events play important roles.

With the importance of stochasticity in cellular functions well recognized [39,44,46,61,62], it is important to understand the stochastic nature and its consequences in cellular processes. In this chapter, we first discuss the basic theoretical framework of the probability landscape of a stochastic network and the underlying discrete chemical master equation (dCME). We then discuss a computational method that optimally enumerate the state space essential for solving the dCME, as well as methods for calculating the steady state and the dynamically evolving probability landscape. We will then describe approaches to simplify the state space. We further discuss the formulation of the continuous chemical master equation (cCME),

Models and Algorithms for Biomolecules and Molecular Networks, First Edition. Bhaskar DasGupta and Jie Liang.

which approximates the dCME, as well as its further simplifications in the form of Fokker–Planck and Langevin models. This is followed by a discussion of the approach of Monte Carlo simulations to study stochastic network, with the Gillespie algorithm discussed in some detail.

4.2 REACTION SYSTEM AND DISCRETE CHEMICAL MASTER EQUATION

Molecular Species and Reactions. We assume a well-stirred system with a constant volume at a constant temperature. It contains n molecular species $\mathcal{X} = \{X_1, \ldots, X_n\}$, with X_i denoting the label of the ith molecular species. There are m chemical reactions $\mathcal{R} = \{R_1, \ldots, R_m\}$ in the network. We denote the copy number of the ith molecular species as x_i. The combination of the copy numbers at time t is a vector of integers $\boldsymbol{x}(t) = (x_1(t), \ldots, x_n(t)) \in \mathbb{N}^n$. We call $\boldsymbol{x}(t)$ the *microstate* of the system at time t.

The probability for the system to be in state $\boldsymbol{x}(t)$ is denoted as $p(\boldsymbol{x}, t)$. The set Ω of all possible combinations of copy numbers, $\Omega = \{\boldsymbol{x}(t) | t \in (0, \infty)\}$, is the *state space* of the system. The collection of probabilities associated with each of the microstate in Ω at time t is the *probability landscape* $\boldsymbol{p}(t)$. The time-evolving probability landscape $\boldsymbol{p}(t)$ provides a full description of the properties of a stochastic molecular network [2,8,11,28,52].

Stoichiometry. A chemical reaction k can be written as

$$c_1(k)X_1 + c_2(k)X_2 + \cdots + c_n(k)X_n \to c_1'(k)X_1 + c_2'(k)X_2 + \cdots + c_n'(k)X_n.$$

It brings the system from a microstate $\boldsymbol{x}_i$ to another microstate $\boldsymbol{x}_j$. The difference between $\boldsymbol{x}_i$ and $\boldsymbol{x}_j$ is the stoichiometry vector $\boldsymbol{s}_k$ of reaction k:

$$\boldsymbol{s}_k = \boldsymbol{x}_i - \boldsymbol{x}_j = (c_1(k) - c_1'(k), \ldots, c_n(k) - c_n'(k)). \tag{4.1}$$

Here $\boldsymbol{s}_k$ has the same dimension as the microstate, and $\boldsymbol{s}_k$ can admit 0 entries if a molecular species does not participate in the reaction.

As an example, the reaction

$$A + 2B \to C$$

reduces the number of A and B by 1 and 2, respectively, and increases the number of C by 1. Its stoichiometry vector has $c_A = +1, c_B = +2, c_C = -1$, and $\boldsymbol{s} = (+1, +2, -2)$. If there are other molecular species, their coefficients are all 0 for this reaction. By treating microscopic states of reactants explicitly, linear and nonlinear reactions, such as synthesis, degradation, dimeric binding, and multimerization, can all be modeled as transitions between microstates.

Reaction Rate. The reaction rate $A_k(\boldsymbol{x}_i, \boldsymbol{x}_j)$, namely, the transition probability per unit time from $\boldsymbol{x}_i$ to $\boldsymbol{x}_j$ due to the kth reaction that connects state $\boldsymbol{x}_i$ to state $\boldsymbol{x}_j$, is determined by the intrinsic reaction rate constant r_k and the copy numbers of relevant reactants at the beginning of the reaction, which is given by the state $\boldsymbol{x}_i$:

$$A_k(\boldsymbol{x}_i, \boldsymbol{x}_j) = A_k(\boldsymbol{x}_i) = r_k \prod_{l=1}^{n} \binom{x_l}{c_l(k)}, \tag{4.2}$$

assuming the convention $\binom{0}{0} = 1$. The intrinsic transition rate r_k is determined by the physical properties of the molecules and the cell environment [18], and the reaction rate $A(\boldsymbol{x}_i, \boldsymbol{x}_j)$ only depends on the starting state $\boldsymbol{x}_i$.

If the kth reaction can lead the system from state $\boldsymbol{x}_i$ to state $\boldsymbol{x}_j$, we have $A_k(\boldsymbol{x}_i, \boldsymbol{x}_j) > 0$, otherwise $A_k(\boldsymbol{x}_i, \boldsymbol{x}_j) = 0$. Although often only one reaction occurs to connect two microstates, in principle it is possible to have more than one reaction connecting $\boldsymbol{x}_i$ to $\boldsymbol{x}_j$. Therefore, we have the overall reaction rate that brings the system from $\boldsymbol{x}_i$ to $\boldsymbol{x}_j$ as

$$A(\boldsymbol{x}_i, \boldsymbol{x}_j) = \sum_{R_k \in \mathcal{R}} A_k(\boldsymbol{x}_i, \boldsymbol{x}_j).$$

Overall, we have the transition rate matrix

$$\boldsymbol{A} = \{A(\boldsymbol{x}_i, \boldsymbol{x}_j)\}, \tag{4.3}$$

where the diagonal elements are defined as

$$A(\boldsymbol{x}_i, \boldsymbol{x}_i) = -\sum_{i \neq j} A(\boldsymbol{x}_i, \boldsymbol{x}_j). \tag{4.4}$$

Discrete Chemical Master Equation. The chemical master equation that governs the change of the probability landscape can be written as

$$\frac{dp(\boldsymbol{x}, t)}{dt} = \sum_{\boldsymbol{x}'} \Big[A(\boldsymbol{x}', \boldsymbol{x}) p(\boldsymbol{x}', t) - A(\boldsymbol{x}, \boldsymbol{x}') p(\boldsymbol{x}, t) \Big]. \tag{4.5}$$

Here the probability $p(\boldsymbol{x}, t)$ is continuous in time, but the states are discrete. We call this the *discrete chemical master equation* (dCME). In matrix form, it can be written as

$$\frac{d\boldsymbol{p}(t)}{dt} = \boldsymbol{A}\boldsymbol{p}(t). \tag{4.6}$$

The dCME describes the gain and loss in probability associated with each microstate due to chemical reactions. These chemical reactions can be regarded as jump processes upon firings of reactions, which bring the system from one combination of copy number of molecular species to a different combination of copy number of molecular species. The dCME fully accounts for the stochastic jumps between states, regardless whether the copy numbers $\boldsymbol{x}_i$ and $\boldsymbol{x}_j$ are small or large. The overall stochasticity due to small copy number events is therefore fully described. It provides a fundamental framework to study stochastic molecular networks [18,58].

However, it is challenging to study a realistic system using the dCME. Analytical solutions exists only for very simple cases, such as self-regulating genes [23,60], or for small problems with strong assumptions of separation of reaction rates [28,52]. Exact numerical solution of the dCME is also difficult, as a nontrivial number of species of small copy numbers may be involved. A major hurdle is the expected exponential increase in the size of the state space when the number of molecular species and their copy numbers increase and when the network becomes complex.

4.3 DIRECT SOLUTION OF CHEMICAL MASTER EQUATION

4.3.1 State Enumeration with Finite Buffer

The technique of optimally enumerating microstates for a given initial condition now allows certain realistic systems to be studied using dCME, under the condition of finite buffer [11]. Below we describe how microstates can be enumerated optimally.

For a network with n molecular species and m reactions, we calculate all microstates that the network can reach starting from a given initial condition, under the *finite buffer* constraint. We use a buffer of finite capacity to represent a reservoir of molecules, from which synthesis reactions generate new molecules, and to which degradation reactions deposit molecules removed from the network. Synthesis reaction is allowed to occur only if the buffer capacity is not exhausted. This is necessary due to the limitation of computing resources. As the microstate of a specific combination of copy numbers is $\boldsymbol{x} = (x_1, \ldots, x_n)$, we add x_{n+1} to denote the current buffer capacity, namely, the number of net new molecules that can still be synthesized at this microstate. A synthesis reaction occurs only if $x_{n+1} > 0$ when using the state enumeration algorithm.

Under these conditions, the set of all possible microstates that can be reached from an initial condition constitute the *state space* Ω of the system. The set of allowed transitions is $\boldsymbol{T} = \{t_{ij}\}$, in which t_{ij} maps the microstate $\boldsymbol{x}_i$ before the reaction to the microstate $\boldsymbol{x}_j$ after the reaction. The initial condition of the reaction system is now given as $\boldsymbol{x}(0) = (x_1(0), x_2(0), \ldots, x_n(0), x_{n+1}(0))$, where $x_i(0)$ is the initial copy number of the ith molecular species at time $t = 0$, and $x_{n+1}(0) = B$ is the predefined buffer capacity.

The algorithm for enumerating the state space is summarized as Algorithm 4.3.1. After initialization, it starts with the given initial microstate $\boldsymbol{x}(0)$. Each reaction is then examined in turn to determine if this reaction can occur for the current microstate. If so, and if the buffer is not used up, the state that this reaction leads to is generated. If the newly generated state was never encountered before, we declare it as a new state and add it to our collection of states for the state space. We repeat this process for all new states, with the aid of a stack data structure. This process terminates when all new states are exhausted [11].

Under the finite buffer constraint, the time complexity of this algorithm is optimal. Since only the unseen state will be pushed onto the stack, every state is pushed and popped at most once, and each state will be generated/visited at most twice before

Algorithm 4.3.1 Finite Buffer State Enumerator($\mathcal{X}, \mathcal{R}, B$)

Network model: $\boldsymbol{N} \leftarrow \{\mathcal{X}, \mathcal{R}\}$
Initial condition: $\boldsymbol{x}(0) \leftarrow \{x_1(0), x_2(0), \ldots, x_n(0)\}$;
Set the value of buffer capacity: $x_{(n+1)}(0) \leftarrow B$;
Initialize the state space and the set of transitions: $\Omega \leftarrow \emptyset$; $\boldsymbol{T} \leftarrow \emptyset$;
Stack $ST \leftarrow \emptyset$; `Push`(ST, $\boldsymbol{x}(0)$); *StateGenerated* ←`FALSE`
while $ST \neq \emptyset$ **do**
  $\boldsymbol{x}_i \leftarrow$ `Pop` (ST);
  for $k = 1$ to m **do**
    if reaction R_k occurs under condition $\boldsymbol{x}_i$ **then**
      if reaction R_k is a synthetic reaction and generates u_k new molecules **then**
        $x_{n+1} \leftarrow x_{n+1} - u_k$
        if $x_{n+1} \geq 0$ **then**
          Generate state $\boldsymbol{x}_j$ that is reached by following reaction R_k from $\boldsymbol{x}_i$;
          StateGenerated ←`TRUE`
        end if
      else
        if reaction R_k is a degradation reaction and breaks down u_k molecules **then**
          $x_{n+1} \leftarrow x_{n+1} + u_k$
        end if
        Generate state $\boldsymbol{x}_j$ that is reached by following reaction R_k from $\boldsymbol{x}_i$;
        StateGenerated ←`TRUE`
      end if
      if (*StateGenerated* = `TRUE`) and ($\boldsymbol{x}_j \notin \Omega$) **then**
        $\Omega \leftarrow \Omega \cup \{\boldsymbol{x}_j\}$;
        `Push`(ST, $\boldsymbol{x}_j$);
        $\boldsymbol{T} \leftarrow \boldsymbol{T} \cup \{t_{i,j}\}$;
        $a_{i,j} \leftarrow$ Transition Coefficient($\boldsymbol{x}_i$, R_k)
      end if
    end if
  end for
end while
Assign all diagonal elements of $\boldsymbol{A}$ using Eq. (4.2).
Output Ω, $\boldsymbol{T}$ and $\boldsymbol{A} = \{a_{i,j}\}$.

it is popped from the stack. As access to each state and to push/pop operations take $O(1)$ time, the total time required for the stack operations is $O(|\Omega|)$. As the algorithm examines each of the reactions for each reached state, the complexity of total time required is $O(m|\Omega|)$, where m is usually a modest constant (*e.g.*, < 50). Based on the same argument, it is also easy to see that the algorithm is optimal in storage, as only valid states and valid transitions are recorded. Using this algorithm, all states reachable from an initial condition within the finite buffer constraint will be accounted for, and no irrelevant states will be included. Furthermore, all possible transitions will be recorded, and no infeasible transitions will be attempted [11].

With this optimal method for enumerating the microstates of a finite system, numerical methods for solving large linear systems can be applied to solve the dCME equation. Very realistic systems can now be directly studied, such as the decision network of phage lambda [8,11].

4.3.2 Generalization and Multi-Buffer dCME Method

Reaction rates in a network can vary greatly: Many steps of fast reactions can occur within a given time period, while only a few steps of slow reactions can occur in the same time period. The efficiency of state enumeration can be greatly improved when memory allocation is optimized based on different behavior of these reactions.

The finite buffer method can be further extended for accurate solution of the chemical master equation [9]. A reaction network can be decomposed into components of birth and death processes, which are the only reactions that can add or remove molecules from the system. If we regard reactions as a set of vertices $\boldsymbol{V}$, and a pair of reactions $\boldsymbol{R}_i$ and $\boldsymbol{R}_j$ are connected by an edge e_{ij} if they share either reactant(s) or product(s), we can then construct an undirected reaction graph $\boldsymbol{G}_R$. It can be decomposed into u number of disjoint *independent reaction components* $\{\boldsymbol{H}_i\}$: $\boldsymbol{G}_R = \bigcup_{i=1}^{u} \boldsymbol{H}_i$, with $\boldsymbol{E}(\boldsymbol{H}_i) \cap \boldsymbol{E}(\boldsymbol{H}_j) = \emptyset$ for $i \neq j$. We focus on those independent reaction components $\boldsymbol{H}_j$s, called *independent Birth–Death* (iBD) components $\{\boldsymbol{H}_j^{iBD}\}$, which contain at least one synthesis reaction. The multi-buffer algorithm for state enumeration is a generalization of the finite-buffer algorithm, in which each iBD is equipped with its own buffer queue. This leads to improved efficiency and increases size reduction in enumerated state space. Details can be found in reference discussing the ACME method [9,10].

4.3.3 Calculation of Steady-State Probability Landscape

We can obtain a Markovian state transition matrix $\boldsymbol{M}$ from the reaction rate matrix $\boldsymbol{A}$: $\boldsymbol{M} = \boldsymbol{I} + \boldsymbol{A} \cdot \Delta t$ [26], where $\boldsymbol{I}$ is the identity matrix, and Δt is the time increment that satisfies $\Delta t < \min\{1/A_0(\boldsymbol{x}_i)\}$. The steady state probability landscape over the microstates, namely, the probability distribution function $\boldsymbol{p}(t = \infty)$ of the microstates at time $t = \infty$, can be obtained by solving the system of equations:

$$p(\infty) = \boldsymbol{M} p(\infty)$$

These linear equations can be solved using iterative solvers such as the simple Jacobi algorithm or more advanced algorithms. A GPU-based algorithm solving such large linear systems can lead to speed up of ≈ 30 times [37].

4.3.4 Calculation of Dynamically Evolving Probability Landscape

The solution to the dCME

$$\frac{d\boldsymbol{p}(t)}{dt} = \boldsymbol{A}\boldsymbol{p}(t)$$

can be written in the form of matrix exponential:

$$p(t) = e^{\boldsymbol{A}t}\boldsymbol{p}(0), \tag{4.7}$$

where $e^{\boldsymbol{A}t}$ is defined as an infinite series [22]: $e^{\boldsymbol{A}t} = \sum_0^\infty \boldsymbol{A}^k/k!$. We discuss the computation of the dynamically evolving probability landscape $\boldsymbol{p}(t)$ below.

4.3.5 Methods for State Space Truncation for Simplification

For large systems, the dCME can be solved numerically if the dimension of the state space can be reduced. This can be achieved by projecting the high-dimensional state space to a lower-dimensional finite space.

Krylov Subspace Method. The rate matrix $\boldsymbol{A}$ has a very large dimension but is sparse. This is because a microstate will have $\leq m$ reactions leading to $\leq m$ different microstates. One can convert the costly problem of exponentiating a large sparse matrix to that of exponentiating a small dense matrix. This can be achieved by projecting the original matrix A to the Krylov subspace $\mathcal{K}_k$, which is easy to compute [36]:

$$\mathcal{K}_m(\boldsymbol{A}t, \pi(0)) \equiv \text{Span}\{\pi(0), \ldots, (\boldsymbol{A}t)^{d-1}\pi(0)\}. \tag{4.8}$$

The Krylov subspace used is of a very small dimension of d =30–60, although the resulting matrix is dense. Denoting $||\cdot||_2$ as the 2-norm of a vector or matrix, the approximation then becomes $\boldsymbol{p}(t) \approx ||\boldsymbol{p}(0)||_2 \boldsymbol{V}_{d+1} \exp\left(\bar{\boldsymbol{H}}_{d+1}t\right) \boldsymbol{e}_1$, where $\boldsymbol{e}_1$ is the first unit basis vector, $\boldsymbol{V}_{d+1}$ is an $(n+1) \times (d+1)$ matrix formed by the orthonormal basis of the Krylov subspace, and $\bar{\boldsymbol{H}}_{d+1}$ is the upper Heisenberg matrix; both can be computed from an Arnoldi algorithm [15]. The error is bounded by

$$\mathcal{O}(e^{d-t||\boldsymbol{A}||_2}(t||\boldsymbol{A}||_2/d)^d).$$

One only needs to compute explicitly $\exp\left(\bar{\boldsymbol{H}}_{d+1}t\right)$. This is a much simpler problem as d is much smaller. A special form of the Padé rational of polynomials instead of Taylor expansion can be used to avoid numerical instability, which arises when summing terms with alternating signs [19,54]:

$$e^{tbH_{d+1}} \approx N_{pp}(tbH_{d+1})/N_{pp}(-tbH_{d+1}).$$

Here $N_{pp}(t\bar{\boldsymbol{H}}_{d+1}) = \sum_{l=0}^{p} c_l (t\bar{\boldsymbol{H}}_{k+1})^l$ and $c_l = c_{l-1} \cdot \frac{p+1-l}{(2p+1-l)l}$. The EXPOKIT software by Sidje provides an excellent implementation of the Krylov subspace method for computing matrix exponential [54]. This approach has been shown to be very effective in studying large dynamic system ($n = 8.0 \times 10^5$) such as protein folding [26], signaling transmission in macromolecular assembly of GroEL-GroES [35], and in the stochastic network of phage lambda [8].

The Krylov subspace method concurrently evaluates the matrix exponential. The overall scheme can be expressed as

$$p(t) \approx \exp(\tau_T \boldsymbol{A}) \ldots \exp(\tau_0 \boldsymbol{A}) \boldsymbol{p}(0),$$

with $t = \sum_{i=0}^{K} \tau_i$, in which the evaluation is from right to left. Here $\{\tau_i\}$ are the sizes of time steps, and T is the total number of time steps [36].

MacNamara et al. further extended the Krylov subspace method by splitting the rate matrix $\boldsymbol{A}$. In some case, one can divide the states into the ''fast partition'' and the ''slow partition'' [5]. Here the condition is that two states belong to the same subset of the fast partition if and only if one can be reached from the other via a sequence of finite fast reactions [5]. Correspondingly, the matrix can be split into two:

$$\boldsymbol{A} = \boldsymbol{A}_f + \boldsymbol{A}_s,$$

where $\boldsymbol{A}_f$ corresponds to the fast CME, and $\boldsymbol{A}_s$ corresponds to the slow CME. We have

$$\frac{d\boldsymbol{p}_f(t)}{dt} = \boldsymbol{A}_f \boldsymbol{p}_f(t)$$

and

$$\frac{d\boldsymbol{p}_s(dt)}{t} = \boldsymbol{A}_s \boldsymbol{p}_s(t).$$

With this deliberate separation, both $\boldsymbol{A}_f$ and $\boldsymbol{A}_s$ should maintain the important property of being infinitesimal generators of continuous time Markov processes by themselves [5]. With more elaborated splitting scheme for aggregation of Markov processes, the Krylov subspace projection method have been shown to be computationally very efficient [36].

Finite State Projection. When the state space is too large and enumeration is no longer feasible, another approach is simply including only a subset of the original microstates [43]. Munsky and Khammash made two insightful observations. Denote two sets of indices of the microstates being chosen as J_1 and J_2, and assume $J_1 \subseteq J_2$. The reduced rate matrix obtained by selecting states in J_1 and J_2 are $\boldsymbol{A}_{J_1}$ and $\boldsymbol{A}_{J_2}$, respectively. The first observation is

$$[e^{\boldsymbol{A}_{J_2}}]_{J_1} \geq e^{\boldsymbol{A}_{J_1}} \geq \boldsymbol{0}. \tag{4.9}$$

This assures that

$$[e^{\boldsymbol{A}_{J_2} t}]_{J_1} \boldsymbol{p}(\boldsymbol{x}_{J_1}, 0) \geq (e^{\boldsymbol{A}_{J_1} t}) \boldsymbol{p}(\boldsymbol{x}_{J_1}, 0).$$

This inequality implies that by increasing the size of the selected subset of states, the approximation improves monotonically. The second observation is, if one obtains a reduced state space by selecting states contained in the index set J, and if

$\mathbf{1}^T e^{t\boldsymbol{A}_J} \boldsymbol{p}(\boldsymbol{x}_J, 0) \geq 1 - \epsilon$ for $\epsilon > 0$, then

$$e^{t\boldsymbol{A}_J} \boldsymbol{p}(\boldsymbol{x}_J, 0) \leq \boldsymbol{p}(\boldsymbol{x}_J, t) \leq e^{t\boldsymbol{A}_J} \boldsymbol{p}(\boldsymbol{x}_J, 0) + \epsilon \mathbf{1}. \tag{4.10}$$

That is, starting with the initial probability of the reduced vector $\boldsymbol{p}(\boldsymbol{x}_J, 0)$, we can compute the probability vector in the reduced space $e^{tA_J} \boldsymbol{p}(\boldsymbol{x}_J, 0)$ at time t using the reduced rate matrix A_J. If the inner product of this vector with $\mathbf{1}$ is no less than $1 - \epsilon$, the difference of this vector from the projected true vector $\boldsymbol{p}(\boldsymbol{x}_J, t)$ of the true probability $\boldsymbol{p}(\boldsymbol{x}, t)$ is also no more than $\epsilon \mathbf{1}$. This inequality guarantees that the approximation obtained with reduced state space will never exceed the actual solution, and its error is bounded by ϵ [43].

These key observations led to the Finite State Project algorithm, which iteratively adds new states to an initial reduced state space, until the approximation error is within a prescribed bound [43]. The original Finite State Projection method was further extended [42], and it was recommended that the initial non-sparse probability vector $\boldsymbol{p}(\boldsymbol{x}, 0)$ should be determined by running a few steps of stochastic simulation discussed in a later section.

However, there is no known general strategy as to what states to add to a finite projection to most effectively improve the approximation accuracy. Furthermore, as an absorption state is introduced to account for all microstates not included in the reduced state space in calculation, the finite state projection method is not appropriate for computing the steady-state probabilistic landscape, as the approximation of the absorption state will lead to errors that increases rapidly with time.

4.4 QUANTIFYING AND CONTROLLING ERRORS FROM STATE SPACE TRUNCATION

Analysis based on the multi-buffer algorithm for state enumeration enables the establishment of upper bounds of errors resulting from state space truncation, which is inevitable when solving the discrete CME of a complex network. For ease of discussion, we examine networks with only one iBD component and therefore one buffer queue. We can factor the enumerated states Ω into $N + 1$ groups of subsets by the net number of tokens of the buffer queue in use: $\Omega \equiv \{\mathcal{G}_0, \mathcal{G}_1, \ldots, \mathcal{G}_N\}$. We can further construct a permuted transition rate matrix $\tilde{\boldsymbol{A}}$ from the original dCME matrix $\boldsymbol{A}$ in Eq. (4.3):

$$\tilde{\boldsymbol{A}} = \begin{pmatrix} \boldsymbol{A}_{0,0} & \boldsymbol{A}_{0,1} & \cdots & \boldsymbol{A}_{0,N} \\ \boldsymbol{A}_{1,0} & \boldsymbol{A}_{1,1} & \cdots & \boldsymbol{A}_{1,N} \\ \cdots & \cdots & \cdots & \cdots \\ \boldsymbol{A}_{N,0} & \boldsymbol{A}_{N,1} & \cdots & \boldsymbol{A}_{N,N} \end{pmatrix}, \tag{4.11}$$

where each block submatrix $\boldsymbol{A}_{i,j}$ includes all transitions from states in factored group $\mathcal{G}_i$ to states in factored group $\mathcal{G}_j$.

Furthermore, we can construct an aggregated continuous-time Markov process with a rate matrix $\boldsymbol{B}_{(N+1)\times(N+1)}$ on the partition $\{\mathcal{G}_0, \mathcal{G}_1, \ldots, \mathcal{G}_N\}$. The steady-state probability of the aggregated Markov process gives the same steady-state probability distribution as that given by the original matrix $\boldsymbol{A}$ on the partitioned groups $\{\mathcal{G}_s\}$. The $(N+1) \times (N+1)$ transition rate matrix $\boldsymbol{B}$ can be constructed as

$$\boldsymbol{B} = \begin{pmatrix} -\alpha_0^{(N)} & \alpha_0^{(N)} & 0 & \cdots & \cdots & \cdots & \cdots & \cdots \\ \beta_1^{(N)} & -\alpha_1^{(N)} - \beta_1^{(N)} & \alpha_1^{(N)} & 0 & \cdots & \cdots & \cdots & \cdots \\ & & & \cdots & & \cdots & & \\ \cdots & \cdots & 0 & \beta_i^{(N)} & -\alpha_i^{(N)} - \beta_i^{(N)} & \alpha_i^{(N)} & 0 & \cdots \\ & & & \cdots & & \cdots & & \\ \cdots & \cdots & \cdots & \cdots & \cdots & 0 & \beta_N^{(N)} & -\beta_N^{(N)} \end{pmatrix}, \tag{4.12}$$

where the aggregated synthesis rate $\alpha_i^{(N)}$ for the group $\mathcal{G}_i$ and the aggregated degradation rate $\beta_{i+1}^{(N)}$ for the group $\mathcal{G}_{i+1}$ at the steady state are two constants:

$$\alpha_i^{(N)} \equiv \frac{\mathbb{1}^T \boldsymbol{A}_{i,i+1}^T \tilde{\pi}(\mathcal{G}_i)}{\mathbb{1}^T \tilde{\pi}(\mathcal{G}_i)} \quad \text{and} \quad \beta_{i+1}^{(N)} \equiv \frac{\mathbb{1}^T \boldsymbol{A}_{i+1,i}^T \tilde{\pi}(\mathcal{G}_{i+1})}{\mathbb{1}^T \tilde{\pi}(\mathcal{G}_{i+1})}, \tag{4.13}$$

in which vector $\tilde{\pi}(\mathcal{G}_i)$ and $\tilde{\pi}(\mathcal{G}_{i+1})$ are the steady state probability vector over the permuted microstates in the group $\mathcal{G}_i$ and $\mathcal{G}_{i+1}$, respectively [9].

This is equivalent to transforming the transition rate matrix $\tilde{\boldsymbol{A}}$ in Eq. (4.11) to $\boldsymbol{B}$ by substituting each block submatrix $\boldsymbol{A}_{i,i+1}$ of synthesis reactions with the corresponding aggregated synthesis rate $\alpha_i^{(N)}$ and each block $\boldsymbol{A}_{i+1,i}$ of degradation reactions with the aggregated degradation rate $\beta_{i+1}^{(N)}$, respectively. The steady-state probability $\tilde{\pi}_i^{(N)}$ of the aggregated state $\mathcal{G}_i$ then can be written as

$$\tilde{\pi}_N^{(N)} = \frac{\prod_{k=0}^{N-1} \frac{\alpha_k^{(N)}}{\beta_{k+1}^{(N)}}}{1 + \sum_{j=1}^{N} \prod_{k=0}^{j-1} \frac{\alpha_k^{(N)}}{\beta_{k+1}^{(N)}}}, \tag{4.14}$$

based on well-known analytical solution to the steady state probability distribution of simple birth–death processes [56]. The error due to state truncation asymptotically obey the following inequality when the buffer capacity N increases to infinity:

$$\mathrm{Err}^{(N)} \leq \frac{\dfrac{\alpha_N^{(\infty)}}{\beta_{N+1}^{(\infty)}}}{1 - \dfrac{\alpha_N^{(\infty)}}{\beta_{N+1}^{(\infty)}}} \cdot \tilde{\pi}_N^{(N)}. \tag{4.15}$$

We can construct bounds to the right-hand side of Eq. (4.15). The maximum aggregated synthesis rates from the block submatrix $\boldsymbol{A}_{i,i+1}$ can be computed as

$$\bar{\alpha}_i^{(N)} = \max\{\mathbb{1}^T \boldsymbol{A}_{i,i+1}^T\} \tag{4.16}$$

and the minimum aggregated degradation rates from the block submatrix $\boldsymbol{A}_{i+1,i}$ can be computed as

$$\underline{\beta}_{i+1}^{(N)} = \min\{\mathbb{1}^T \boldsymbol{A}_{i+1,i}^T\},. \tag{4.17}$$

Both $\bar{\alpha}_i^{(N)}$ and $\underline{\beta}_{i+1}^{(N)}$ can be calculated once the permuted transition rate matrix $\tilde{\boldsymbol{A}}$ is defined. An upperbound $\bar{\tilde{\pi}}_N^{(N)}$ to $\tilde{\pi}_N^{(N)}$ can then be computed as

$$\bar{\tilde{\pi}}_N^{(N)} = \frac{\prod_{k=0}^{N-1} \dfrac{\bar{\alpha}_k^{(N)}}{\underline{\beta}_{k+1}^{(N)}}}{1 + \sum_{j=1}^{N} \prod_{k=0}^{j-1} \dfrac{\bar{\alpha}_k^{(N)}}{\underline{\beta}_{k+1}^{(N)}}}, \tag{4.18}$$

With Inequality (4.15) and Eqs. (4.16) and (4.17), an asymptotic error bound to the truncation error is:

$$\mathrm{Err}^{(N)} \leq \frac{\dfrac{\bar{\alpha}_{N-1}^{(N)}}{\underline{\beta}_N^{(N)}}}{1 - \dfrac{\bar{\alpha}_{N-1}^{(N)}}{\underline{\beta}_N^{(N)}}} \cdot \frac{\prod_{k=0}^{N-1} \dfrac{\bar{\alpha}_k^{(N)}}{\underline{\beta}_{k+1}^{(N)}}}{1 + \sum_{j=1}^{N} \prod_{k=0}^{j-1} \dfrac{\bar{\alpha}_k^{(N)}}{\underline{\beta}_{k+1}^{(N)}}}, \tag{4.19}$$

This can be estimated *a priori*, without costly trial solutions to the dCME. Generalization to truncating state space to all buffer queues can be found in references 9 and 10.

Remark. Studying the behavior of a stochastic network is challenging. Even with a correctly constructed stochastic network, it is generally not known if an accurate solution to the dCME has been found. For example, it is difficult to know if all major probabilistic peaks have been identified or important ones with significant probability mass in the usually high-dimensional space are undetected. It is also difficult to know if the locations of identified probabilistic peak are correctly mapped. One also does not know if a computed probabilistic landscapes is overall erroneous and how such errors can be quantified. Furthermore, the best possible accuracy one can achieve with finite computing resources is generally unknown. We also do not know what computing resource is required, so solutions with errors within a predefined tolerance can be obtained. The development of theory and methods for error estimations such as those described here can help to resolve these important issues.

4.5 APPROXIMATING DISCRETE CHEMICAL MASTER EQUATION

There exists a large body of work in formulating stochastic differential equations to study reaction networks. Below we discuss several well-known approaches, which can be viewed as approximations of varying degrees to the dCME.

4.5.1 Continuous Chemical Master Equation

If we treat the state space as continuous, that is, if we assume the amount of a molecular species x_i is measured by a real value (such as concentration) instead of an integer (copy numbers), the vector $\boldsymbol{x}(t)$ becomes a real-valued vector $\boldsymbol{x}(t) \in \mathbb{R}^n$. We then have the continuous chemical master equation, which is equivalent to the dCME of Eq. (4.5):

$$\frac{\partial p(\boldsymbol{x}, t)}{\partial t} = \int_{\boldsymbol{x}'} [A(\boldsymbol{x}', \boldsymbol{x})p(\boldsymbol{x}', t) - A(\boldsymbol{x}, \boldsymbol{x}')p(\boldsymbol{x}, t)]\, d\boldsymbol{x}', \tag{4.20}$$

where the kernel $A(\boldsymbol{x}', \boldsymbol{x})$ represents the transition probability function per unit time from $\boldsymbol{x}'$ to $\boldsymbol{x}$. We call this the continuous Chemical Master Equation (cCME). The cCME in this form is equivalent to the Chapman–Kolmogorov equation frequently used to describe continuous Markov processes [27].

The continuous state space version of the CME requires strong assumptions. It is only appropriate if one can assume that the difference in the amount of molecules in neighboring states is infinitesimally small, which is valid only if the copy number of the molecular species in the system is much larger than 1, and also much larger than the changes in the numbers of molecules when a reaction occurs. cCME therefore cannot be used when the total amount of molecules involved is very small—for example, in systems of a single or a handful of particles. In these cases, dCME should be used instead.

4.5.2 Stochastic Differential Equation: Fokker–Planck Approach

When cCME is appropriate, one can further approximate the cCME with various formulations of Stochastic Differential Equations (SDEs). One such formulation is the Fokker–Planck equation. Similar to the cCME, it describes the evolution of probability landscape of the system, but with the transition kernel in the CME replaced by a differential operator of second order.

We follow the disposition of van Kampen [59] and briefly describe how the Fokker–Planck equation is related to the cCME, with discussion on additional assumptions and approximations involved beyond those necessary for the cCME.

Assumptions of the Fokker–Planck Equation. To approximate the cCME, the first assumption we make is that the jumps between states described in Eq. (4.5) must be small, that is, the "before" and the "after" states are in close neighborhood:

$$||\boldsymbol{x}_j - \boldsymbol{x}_i|| < \epsilon,$$

where $\epsilon \in \mathbb{R}$ is infinitesimally small. Second, the transition probability varies slowly:

$$A(\boldsymbol{x}, \boldsymbol{y}) \approx A(\boldsymbol{x}', \boldsymbol{y}'), \quad \text{if } ||\boldsymbol{x} - \boldsymbol{x}'|| \leq \epsilon \text{ and } ||\boldsymbol{y} - \boldsymbol{y}'|| \leq \epsilon.$$

Third, the probability $\boldsymbol{p}(\boldsymbol{x}, t)$ must also vary slowly:

$$\boldsymbol{p}(\boldsymbol{x}, t) \approx \boldsymbol{p}(\boldsymbol{x}', t) \qquad \text{if } |\boldsymbol{x} - \boldsymbol{x}'| \leq \epsilon.$$

As a consequence of these assumptions, the transition kernel $A(\boldsymbol{x}, \boldsymbol{y})$ is differentiable to a high order. It is clear that the cCME cannot be used to study discrete jump processes.

With these assumptions, the first term in Eq. (4.20), where the full detail of the transition kernel $A(\boldsymbol{x}', \boldsymbol{x})$ is needed, can be approximated. The goal is to replace $A(\boldsymbol{x}', \boldsymbol{x})$ with its Taylor expansion centered around $\boldsymbol{x}$. For ease of illustration, we express transitions as a function of the starting point and the jump. We first reparameterize $A(\boldsymbol{x}', \boldsymbol{x})$ as $A(\boldsymbol{x}'; \boldsymbol{s})$, where $\boldsymbol{s} = \boldsymbol{x} - \boldsymbol{x}'$. Similarly, for $A(\boldsymbol{x}, \boldsymbol{x}')$ in the second term, we have $A(\boldsymbol{x}, \boldsymbol{x}') = A(\boldsymbol{x}; -\boldsymbol{s})$. Equation (4.20) then can be rewritten as

$$\frac{\partial p(\boldsymbol{x}, t)}{\partial t} = \int_{\boldsymbol{S}} A(\boldsymbol{x} - \boldsymbol{s}; \boldsymbol{s}) p(\boldsymbol{x} - \boldsymbol{s}, t)\, d\boldsymbol{s} - p(\boldsymbol{x}, t) \int_{\boldsymbol{S}} A(\boldsymbol{x}; -\boldsymbol{s})\, d\boldsymbol{s}. \tag{4.21}$$

Approximation Through Taylor Expansion. The first term of Eq. (4.21) is then expanded around $\boldsymbol{x}$ using Taylor expansion as

$$\begin{aligned} &\int A(\boldsymbol{x} - \boldsymbol{s}; \boldsymbol{s}) \boldsymbol{p}(\boldsymbol{x} - \boldsymbol{s}, t)\, d\boldsymbol{s} \\ &\quad = \int A(\boldsymbol{x}; \boldsymbol{s}) \boldsymbol{p}(\boldsymbol{x}, t)\, d\boldsymbol{s} - \sum_i \frac{\partial}{\partial \boldsymbol{x}_i} [\int \boldsymbol{s} \cdot A(\boldsymbol{x}; \boldsymbol{s})\, d\boldsymbol{s} \cdot \boldsymbol{p}(\boldsymbol{x}, t)] \\ &\quad + \frac{1}{2} \sum_{i,j} \frac{\partial^2}{\partial \boldsymbol{x}_i \partial \boldsymbol{x}_j} \left[\int \boldsymbol{s}^2 \cdot A(\boldsymbol{x}; \boldsymbol{s})\, d\boldsymbol{s} \cdot \boldsymbol{p}(\boldsymbol{x}, t) \right] \\ &\quad - \frac{1}{3!} \sum_{i,j,k} \frac{\partial^3}{\partial \boldsymbol{x}_i \partial \boldsymbol{x}_j \partial \boldsymbol{x}_k} \left[\int \boldsymbol{s}^3 \cdot A(\boldsymbol{x}; \boldsymbol{s})\, d\boldsymbol{s} \cdot \boldsymbol{p}(\boldsymbol{x}, t) \right] + \cdots . \end{aligned} \tag{4.22}$$

Putting it back to Eq. (4.21), and approximating by dropping terms higher than second order, we obtain the Fokker–Planck equation:

$$\frac{\partial p(\boldsymbol{x}, t)}{\partial t} = -\sum_i \frac{\partial}{\partial \boldsymbol{x}_i}\left[\int \boldsymbol{s} \cdot A(\boldsymbol{x}; \boldsymbol{s})\, d\boldsymbol{s} \cdot p(\boldsymbol{x}, t)\right] + \frac{1}{2}\sum_{i,j} \frac{\partial^2}{\partial \boldsymbol{x}_i \partial \boldsymbol{x}_j}\left[\int \boldsymbol{s}^2 \cdot A(\boldsymbol{x}; \boldsymbol{s})\, d\boldsymbol{s} \cdot p(\boldsymbol{x}, t)\right]. \quad (4.23)$$

Using a simpler notation, we have

$$\frac{\partial p(\boldsymbol{x}, t)}{\partial t} = -\sum_i \frac{\partial}{\partial \boldsymbol{x}_i}[\mathcal{F}(\boldsymbol{x}) p(\boldsymbol{x}, t)] + \frac{1}{2}\sum_{i,j} \frac{\partial^2}{\partial \boldsymbol{x}_i \partial \boldsymbol{x}_j} \cdot [\mathcal{G}(\boldsymbol{x}) p(\boldsymbol{x}, t)], \quad (4.24)$$

where $\mathcal{F}(\boldsymbol{x}) = \int \boldsymbol{s} \cdot A(\boldsymbol{x}; \boldsymbol{s})\, d\boldsymbol{s}$ and $\mathcal{G}(\boldsymbol{x}) = \int \boldsymbol{s}^2 \cdot A(\boldsymbol{x}; \boldsymbol{s})\, d\boldsymbol{s}$.

4.5.3 Stochastic Differential Equation: Langevin Approach

Another formulation is that of the Langevin equation. When the macroscopic behavior of a reaction system can be determined, a general approach to study its stochastic behavior is to combine a diffusion term describing the macroscopic behavior with a separate noise term describing the stochastic fluctuations of the system.

Random fluctuations in the copy numbers of molecules occur because of the random jumps due to spontaneous firings of reactions. Such reactions will introduce changes in the copy numbers of molecular species—for example, by the amount of $-\boldsymbol{s}_k$ for each firing of the kth reaction. Assume that the jump is small, namely, $\boldsymbol{x}(t + \Delta t) = \boldsymbol{x}(t) - \boldsymbol{s}_k \approx \boldsymbol{x}(t)$, and reaction R_k occurs during a small time interval dt at $t + \Delta t$. These assumptions would result in unchanged reaction rate:

$$A_k(\boldsymbol{x}(t + \Delta t)) \approx A_k(\boldsymbol{x}(t)) = r_k \prod_{l=1}^{n} \binom{x_l}{c_{lk}}. \quad (4.25)$$

With these assumptions, the vector of the amount of molecular species $\boldsymbol{x}(t + \Delta t)$ at time $t + \Delta t$ can be written as

$$\boldsymbol{x}(t + \Delta t) = \boldsymbol{x}(t) - \sum_{R_k \in \mathcal{R}} n_k(\boldsymbol{x}, \Delta t) \cdot \boldsymbol{s}_k,$$

assuming that several reactions may occur. Here $n_k(\boldsymbol{x}, \Delta t)$ is the number of reaction R_k occurs during the period Δt. Under the assumption $\boldsymbol{x}(t + \Delta t) \approx \boldsymbol{x}(t)$, the copy numbers of molecular species for calculating reaction rate using Eq. (4.25) do not change during Δt; therefore, the reaction rates also do not change during Δt. With this assumption, all reactions occurring during Δt can be considered independent of each other.

This assumption is valid only if the copy numbers in $\boldsymbol{x}(t)$ are large, so the stoichiometry coefficients c_i forming the jump vector $\boldsymbol{s}_k$ are all comparatively small. Such an assumption breaks down when the copy number of molecular species is not significantly larger than the stoichiometry coefficients, and therefore this approximation cannot be employed to describe systems of a handful of particles.

We now introduce further approximations. A reasonable model for the random variable $n_k(\boldsymbol{x}, \Delta t)$ is that of a Poisson process: $n_k \sim \mathcal{P}(\lambda_k)$, where $\lambda_k = A_k(\boldsymbol{x}(t)) \cdot \Delta t$. With the additional assumption that Δt is sufficiently long such that a large number ($\gg 1$) of reactions occur during Δt, the Poisson distribution for the number of spontaneous reactions can be approximated by a Gaussian distribution [17]. Note that this assumption is contradictory to the earlier assumption of small jumps (changes in $\boldsymbol{x}$ is small). Approximating the Poisson distribution by a Gaussian distribution will be accurate when λ_k is large—for example, $\lambda_k > 1000$. With this, we now have $n_k \sim \mathcal{N}(\mu, \sigma^2)$, with $\mu = \sigma^2 = \lambda_k$, or alternatively, $n_k \sim \lambda_k + \lambda_k^{1/2}\mathcal{N}(0, 1) = A_k(\boldsymbol{x}(t)) \cdot \Delta t + [A_k(\boldsymbol{x}(t) \cdot \Delta t]^{1/2}\mathcal{N}(0, 1)$.

Under these assumptions, the fluctuations of the amount of molecules follow m independent Gaussian processes, one for each reaction:

$$\boldsymbol{x}(t + \Delta t) = \boldsymbol{x}(t) - \sum_{R_k \in \mathcal{R}} A_k(\boldsymbol{x}(t)) \cdot \Delta t \cdot \boldsymbol{s}_k - \sum_{R_k \in \mathcal{R}} [A_k(\boldsymbol{x}(t)) \cdot \Delta t]^{1/2} \cdot \boldsymbol{s}_k \cdot \mathcal{N}(0, 1). \tag{4.26}$$

This leads to the following equation:

$$\frac{\partial \boldsymbol{x}(t)}{\partial t} = - \sum_{R_k \in \mathcal{R}} A_k(\boldsymbol{x}(t)) \cdot \boldsymbol{s}_k - \sum_{R_k \in \mathcal{R}} [A_k(\boldsymbol{x}(t))]^{1/2} \cdot \boldsymbol{s}_k \cdot [\frac{1}{\partial t^{1/2}} \cdot \mathcal{N}(0, 1)].$$

Denote $\mathcal{G}(t) = \frac{1}{\partial t^{1/2}} \cdot \mathcal{N}(0, 1) = \mathcal{N}(0, 1/\partial t)$, we have the Langevin equation:

$$\frac{\partial \boldsymbol{x}(t)}{\partial t} = \sum_{R_k \in \mathcal{R}} A_k(\boldsymbol{x}(t)) \cdot \boldsymbol{s}_k + \sum_{R_k \in \mathcal{R}} [A_k(\boldsymbol{x}(t))]^{1/2} \cdot \boldsymbol{s}_k \cdot \mathcal{G}(t). \tag{4.27}$$

This is the chemical Langevin equation described in reference 17.

4.5.4 Other Approximations

There are alternatives to the Fokker–Planck and Langevin approaches to account for the stochasticity differently. One can replace the diffusion term with a term for the variance–covariance between pairs of the molecular reactions [20], or between concentrations of different molecular species [57], without the explicit inclusion of a random process. Here the magnitude of the covariance is determined by the Hessian matrix of the second-order partial derivative of the propensity functions of the reactions [20,57]. This inclusion of the second moments to account for the stochasticity is the basis of the stochastic kinetic model [20] and the mass fluctuation kinetic model (MFK) [57]. These approaches can model reactions involving one or two molecules well [20,57]. They are similar in spirit to the Fokker–Planck equation model by including a second moment term for better approximation, but are fundamentally different as they are macroscopic in nature and do not involve any random processes. Yet another approach is to directly model explicitly the stochastic coupling of the macroscopic concentrations of molecular species, in addition to the Gaussian noise of the original Langevin model [12].

4.6 STOCHASTIC SIMULATION

A widely used method to study stochastic networks is to carry out Monte Carlo simulations. By following the trajectories of reactions, one can gather statistics of reaction events at different time to gain understanding of the network behavior [18,41]. We discuss the underlying algorithm, called the *stochastic simulation algorithm (SSA)*, which is also known as the *Gillespie algorithm*.

4.6.1 Reaction Probability

We denote the probability that after the last reaction at t, the first reaction, which happens to be the kth reaction, occurs during dt at an infinitesimally small time interval dt after $t + \Delta t$ to be

$$p[\boldsymbol{x}(t + \Delta t), \Delta t, k|\boldsymbol{x}(t)]\, dt.$$

If we divide the time interval Δt into H subintervals and assume that the occurrence of reactions following a Poisson process, the probability that none of the m reactions have occurred during the time prior to the end of a small time interval $\epsilon = \Delta t/H$ is

$$\prod_{k=1}^{m}[1 - A_k(\boldsymbol{x}(t))\epsilon] \approx \sum_{k=1}^{m}[1 - A_k(\boldsymbol{x}(t))\epsilon].$$

As the probability of no reactions for each of the H intervals is the same, no reactions have occurred during Δt is

$$\lim_{H\to\infty}\sum_{k=1}^{m}[1 - A_k(\boldsymbol{x}(t))\epsilon]^H = e^{-A(\boldsymbol{x}(t))\Delta t}.$$

As the instantaneous state transition probability for reaction k at $t + \Delta t$ is $A_k[\boldsymbol{x}(t + \Delta t)]dt$, we have

$$p[\boldsymbol{x}(t + \Delta t), \Delta t, k|\boldsymbol{x}(t)] = A_k[\boldsymbol{x}(t + \Delta t)]\, e^{-A(\boldsymbol{x}(t))\Delta t}dt.$$

4.6.2 Reaction Trajectory

Let the state the system is in at time t to be $\boldsymbol{x}(t)$. After a time interval Δt, reaction k occurs at an infinitesimally small time interval dt at $t + \Delta t$, and the system is brought to the state $\boldsymbol{x}(t + \Delta t)$. We can observe the trajectory of a sequence of such reactions. Starting from state $\boldsymbol{x}(t_0)$ at time t_0, after a series of time intervals $(\Delta t_0, \Delta t_1, \ldots, \Delta t_{T-1})$, the system reaches the state $\boldsymbol{x}(t_T)$, after traversing the sequence of states $(\boldsymbol{x}(t_0), \boldsymbol{x}(t_1), \ldots, \boldsymbol{x}(t_{T-1}))$, with reactions $k_0, k_1, \ldots, k_{T-1}$ occurring along the way. Let $(t_1, t_2, \ldots, t_T)$ be the sequence of time points when a reaction occurs. The trajectory of reactions can be denoted as

$$[\boldsymbol{x}(t_0);\ \boldsymbol{x}(t_1), k_0;\ \ldots;\ \boldsymbol{x}(t_{T-1}), k_{T-2};\ \boldsymbol{x}(t_T), k_{T-1}].$$

Alternatively, we can denote the time intervals by its increments $(\Delta t_0, \ldots, \Delta t_{T-1})$.

4.6.3 Probability of Reaction Trajectory

Assuming a Markovian process—namely, that future reactions depend only on current state but not on any past state—the probability associated with a time trajectory is

$$\begin{aligned}\pi[&\boldsymbol{x}(t_0); \boldsymbol{x}(t_1), k_1;\ \cdots;\ \boldsymbol{x}(t_{T-1}), k_{T-1};\ \boldsymbol{x}(t_T), k_T] \\ &= \pi[\boldsymbol{x}(t_1), \Delta t_0, k_0 | \boldsymbol{x}(t_0)].\pi[\boldsymbol{x}(t_2), \Delta t_1, k_1 | \boldsymbol{x}(t_1)] \cdots \pi[\boldsymbol{x}(t_T), \Delta t_{T-1}, k_{T-1} | \boldsymbol{x}(t_{T-1})].\end{aligned} \tag{4.28}$$

In principle, the probability of starting from state $\boldsymbol{x}(t_0)$ and reaching state $\boldsymbol{x}(t_T)$ can then be obtained by integrating over all possible paths:

$$\pi[\boldsymbol{x}(t_T)|\boldsymbol{x}(t_0)] = \sum_{(t_1,\cdots,t_{T-1}),(k_1,\ldots,k_T)} \pi[\boldsymbol{x}(t_0);\ \boldsymbol{x}(t_1), k_1;\ \ldots;\ \boldsymbol{x}(t_{T-1}), k_{T-1};\ \boldsymbol{x}(t_T), k_T].$$

4.6.4 Stochastic Simulation Algorithm

If we can generate many independent samples of reaction trajectories that follow a proper probabilistic model starting from the same initial condition (e.g., Eq. (4.28)), we can collect the reaction trajectories at the same sampling time intervals. These properly sampled trajectories can then be used to study the behavior of the stochastic network.

The Stochastic Simulation Algorithm (SSA) or the Gillespie algorithm was designed to perform such simulations [18]. It is summarized in Algorithm 4.6.4.

Generating Random Variables $(\Delta t, k)$.

A key component of the Gillespie algorithm is to generate a pair of random variables $(\Delta t, k)$, the time interval Δt until the next reaction occurs, and the specific kth reaction as the reaction that actually occurred next. We have

$$\pi(\Delta t, k) = \pi_1(\Delta t) \cdot \pi_2(k|\Delta t),$$

where $\pi_1(\Delta t)$ is the probability that the next reaction, regardless which specific one, will occur at $t + \Delta t + dt$, and $\pi_2(k|\Delta t)$ is the probability that the next reaction will be the kth reaction. As $\pi_1(\Delta t) = \sum_i \pi(i, \Delta t)$, where $\pi(i, \Delta t)$ is the probability that reaction i occurs at time $t + \Delta t + dt$, we have $\pi_2(k|\Delta t) = \frac{\pi(k,\Delta t)}{\sum_i \pi(i,\Delta t)}$. As we assume a

Algorithm 4.6.4 Stochastic Simulation Algorithm

Set initial condition: $\boldsymbol{x}(0) \leftarrow \{x_1(0), x_2(0), \ldots, x_n(0)\}$
Initialize $\boldsymbol{r} = (r_1, \cdots, r_k)$, and set $t = 0$.
Generate a series of sampling time $(t_1, t_2, \cdots, t_T)$
$t \leftarrow 0$
while $t < t_T$ or $A(\boldsymbol{x}(t)) \neq \boldsymbol{0}$ **do**
  Generate a pair of random variables $(\Delta t, k)$ following
    $\pi(\Delta t, k) \sim A_k(\boldsymbol{x}(t))e^{-A(\boldsymbol{x}(t))\Delta t}$
  Update $\boldsymbol{x}(t + \Delta t)$ for $\boldsymbol{x}(t)$ with the occurred reaction k
  Update $A(\boldsymbol{x}(t + \Delta t))$
  if $t < t_i$ and $t + \Delta t > t_i$ **then**
    Record $\boldsymbol{x}(t)$ as $\boldsymbol{x}(t_i)$
    Update t_i to t_{i+1}
  end if
  $t \leftarrow t + \Delta t$
end while

model of Poisson process, we have

$$\pi_1(\Delta t) = A(\boldsymbol{x}(t))e^{-A(\boldsymbol{x}(t))\Delta t} \quad \text{and} \quad \pi_2(k|\Delta t) = \frac{A_k(\boldsymbol{x}(t))}{A(\boldsymbol{x}(t))}.$$

That is, if we can generate a random variable Δt following $\pi_1(\Delta t)$, and another random integer k according to $\pi_2(k|\Delta t)$, the resulting pair $(\Delta t, k)$ will follow the desired distribution $\pi(\Delta t, k)$.

Assume we can generate a random number r following the uniform distribution $r \sim \mathcal{U}_{[0,1]}$. A general approach to obtain a random variable x that follows a distribution $\mathcal{F}$ is to calculate the transformation of r through the inverse function $\mathcal{F}^{-}$: $x = \mathcal{F}^{-}(r)$. Since $\pi_1(\Delta t) = e^{-A(\boldsymbol{x}(t))\Delta t}$, we can have

$$\Delta t = \frac{1}{A(\boldsymbol{x}(t))} \ln \frac{1}{r_1}, \qquad \text{where} \quad r_1 \sim \mathcal{U}_{[0,1]}.$$

To sample the next reaction k, we can generate again first a uniformly distributed random variable $r_2 \sim \mathcal{U}_{[0,1]}$. We can take the kth reaction such that

$$\sum_{i=1}^{k-1} A_i(\boldsymbol{x}(t_i)) < r_2 A(\boldsymbol{x}(t)) \leq \sum_{i=1}^{k} A_i(\boldsymbol{x}(t_i)).$$

Another approach to generate a pair of random variable $(\Delta t, k)$ is to first calculate the probability at time $t + \Delta t$ for a reaction i to occur during an infinitesimally small time interval at $t + \Delta t + dt$, assuming that there were no changes between $\boldsymbol{x}(t)$ and $\boldsymbol{x}(t + \Delta t)$; that is, there is no occurrence of other reactions. We can generate a tentative reaction time Δt_l for reaction l as

$$\Delta t_l = \frac{1}{A_l(\boldsymbol{x}(t))} \ln \frac{1}{r_l}, \qquad \text{where} \quad r_l \sim \mathcal{U}_{[0,1]}.$$

From this set of pairs of random variables $(l, \Delta t_l)$, we select the pair of random variables of the shortest time Δt, at which the next reaction k would occur:

$$\Delta t = \min\{\Delta t_l\}$$

and

$$k = \arg_l \min\{\Delta t_l\}$$

as the next reaction k.

Remark. There are a number of issues in carrying out studies using stochastic simulation, as adequate sampling is challenging when the network becomes complex. There is no general guarantee that simulation can provide a full account of the network stochasticity, as it is difficult to determine whether simulations are extensive enough for accurate statistics. It is also difficult to determine whether adequate sampling has been achieved for individual trajectory. In addition, it is often difficult to characterize rare events that may be biologically important, as simulations follow high-probability paths. Much recent work has been focused on improving SSA, for example, by introducing data structure so the generation of the two random variables of τ and reaction k is more efficient [16]. In addition, an approach to speed up SSA is to find the best time step τ such that the copy numbers of the molecular species, hence the reaction rates, do not change much, so the simulation can leap forward with large time step [6]. Recent interest in introducing bias in selection of the next reaction and in altering the reaction rate showed promise in improved sampling of rare events [14,31,51]. Adaptively adjusted bias of reactions based on look-ahead strategy showed that barrier crossing can be engineered for efficient and accurate sampling of rare events [7].

4.7 APPLICATIONS

We now discuss how stochastic networks can be modeled by directly solving the underlying chemical master equation using two biological examples.

4.7.1 Probability Landscape of a Stochastic Toggle Switch

Toggle switch is one of the smallest genetic networks that can present bistability [52]. It is a small network consisting of two genes, say, A and B. Single copies of gene A and gene B in the chromosome each encode a protein product. The protein product of each gene represses the other gene: When two protein monomers associate, they bind to the appropriate operator site and repress the transcription of the other gene.

(a)

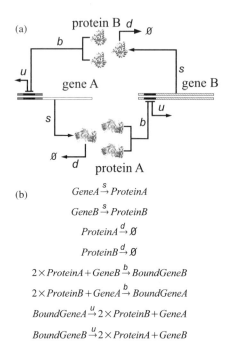

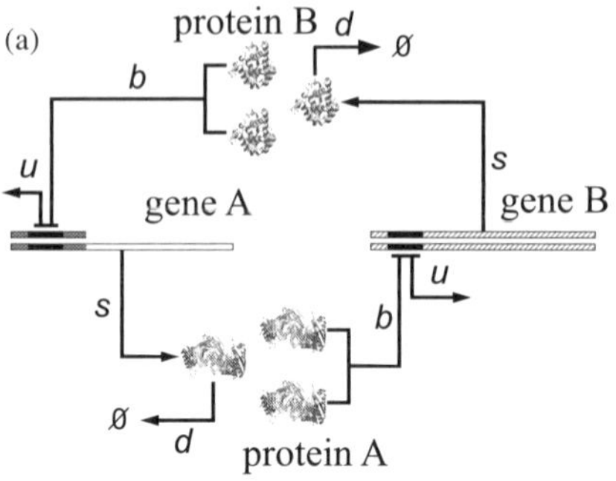

(b)

$$GeneA \xrightarrow{s} ProteinA$$

$$GeneB \xrightarrow{s} ProteinB$$

$$ProteinA \xrightarrow{d} \emptyset$$

$$ProteinB \xrightarrow{d} \emptyset$$

$$2 \times ProteinA + GeneB \xrightarrow{b} BoundGeneB$$

$$2 \times ProteinB + GeneA \xrightarrow{b} BoundGeneA$$

$$BoundGeneA \xrightarrow{u} 2 \times ProteinB + GeneA$$

$$BoundGeneB \xrightarrow{u} 2 \times ProteinA + GeneB$$

FIGURE 4.1 The stochastic network of a toggle switch. (**a**) The topology of the network and the reaction rates. (**b**) The corresponding chemical reactions of the 8 stochastic processes.

The molecular species and the network topology of a toggle switch model are shown in Fig. 4.1a. The stochastic processes include: the synthesis and degradation of proteins A and B, with reaction constants denoted as s and d, respectively and the binding and unbinding of the operator site of one gene by the protein products of the other gene at rate b and u, respectively (Fig. 4.1b). The binding states of the two operator sites are ''on-on/unbound-unbound'', ''on-off/unbound-bound,'' ''off-on/bound-unbound,'' and ''off-off/bound-bound.'' The synthesis rates of both proteins A and B depend on the binding state of the operator sites [11,52]. Even for this simple network, no general exact solutions are known.

The exact probably landscape of the toggle switch model at steady state can be computed numerically. We can choose the parameter values as $s = 100d, u = d/10$, and $b = d/100{,}000$ in units of degradation rate d, and set the initial condition to: 1 copy of unbound gene A, 1 copy of unbound gene B, 0 copies of bound gene A and bound gene B, 0 copies of their protein products, and the buffer size for the total protein A and protein B combined that can be synthesized of 300. We then enumerate the state space of the toggle switch using the finite buffer algorithm. The steady state probability landscape of the network can then be computed (Fig. 4.2). It is clear that a toggle switch has four different states, corresponding to the ''on/on,'' ''on/off,'' ''off/on,'' and ''off/off'' states. With these chosen parameters, the toggle/switch exhibits clear bistability; that is, it has high probabilities for the ''on/off'' and ''off/on''

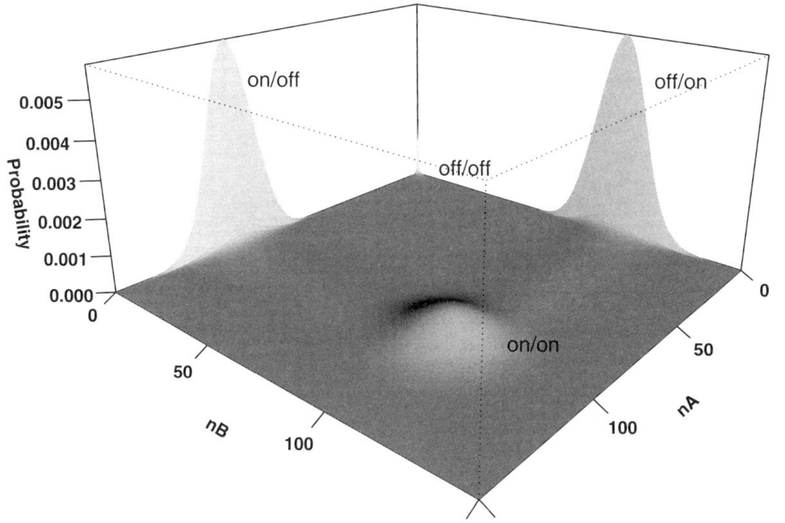

FIGURE 4.2 The steady-state probability landscape of a toggle switch. A toggle switch has four different states, corresponding to different binding state of genes A and B. At the condition of small value of u/b, the off/off state is strongly suppressed for any value of u/d, and the system exhibits bistability.

states, but has a low probability for the ''on/on'' state. The ''off/off'' state is severely suppressed [11].

4.7.2 Epigenetic Decision Network of Cellular Fate in Phage Lambda

Bacteriophage lambda is a virus that infects *E. coli* cells (Fig. 4.3). Of central importance is the molecular circuitry that controls phage lambda to choose between two productive modes of development, namely, the lysogenic phase and the lytic phase (Fig. 4.3A). In the lysogenic phase, phage lambda represses its developmental function, integrates its DNA into the chromosome of the host *E. coli* bacterium, and is replicated in cell cycles for potentially many generations. When threatening DNA damage occurs—for example, when UV irradiation increases—phage lambda switches from the epigenetic state of lysogeny to the lytic phase and undergoes massive replications in a single cell cycle and releases 50–100 progeny phages upon lysis of the *E. coli* cell. This switching process is called *prophage induction* [49].

The molecular network that controls the choice between these two different physiological states has been studied extensively [1,3,4,24,25,34,48,49,53]. All of the major molecular components of the network have been identified, binding constants and reaction rates have been characterized, and there is a good experimental understanding of the general mechanism of the molecular switch [49]. Theoretical studies have also contributed to the illumination of the central role of stochasticity [3] and the stability of lysogen against spontaneous switching [4,63].

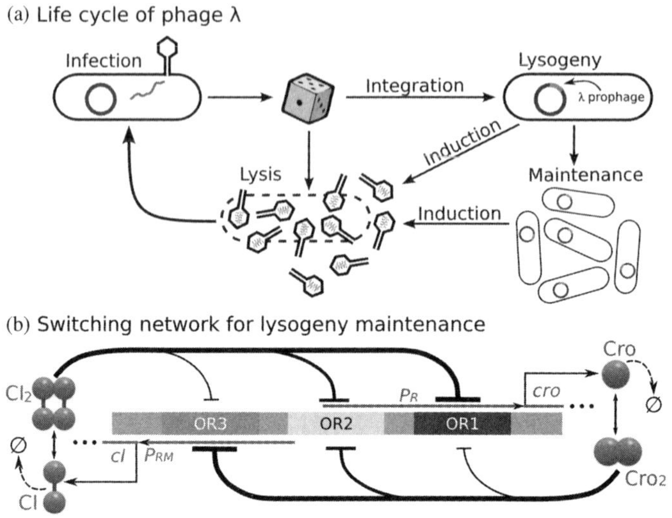

FIGURE 4.3 Different selection of cell fate of *E. coli* infected by phage lambda and a model of the epigenetic circuit for lysogeny maintenance. (A) The lysogenic and lytic phases of phage lambda. (B) A simplified model of the epigenetic switch for lysogeny maintenance.

To study how lysogeny is maintained and how it transitions to the lytic state, we can use a simplified stochastic model for the molecular regulatory network that controls the epigenetic switch in phage lambda (Fig. 4.3b) [8]. Using a total of 54 biochemical reactions involving 13 molecular species, this model explicitly includes key components, essential reactions, and cooperativities of the phage lambda decision circuitry. The effects of UV irradiation can be modeled by increasing the CI degradation rates k_d due to the response of the SOS system. This epigenetic network model can reach around 1.7 million microstates. The steady-state probability associated with each of these microstates can be computed from dCME after the microstates are enumerated by the finite buffer algorithm [8].

Figure 4.4 (row 1) shows the probability landscape of the phage lamed at five different UV irradiation conditions, each modeled with a different CI degradation rate k_d. Although there are 13 molecular species, we can project the landscape to the two-dimensional subspace and record the total copy numbers of CI_2 dimer and Cro_2 dimer molecules. With a high copy number of CI_2 repressors, the lysogenic phase of the phage lambda is maintained, whereas a high copy number of Cro_2 protein signifies the lytic phase [25]. A clear picture of the landscape in lysogeny, at the start of transition, during mid-transition, at the end of transition, and in lysis can be seen.

The stochastic network models can also be used to aid in understanding of the mechanism of how the decision network works. It is well known that cooperativity among proteins play important roles. After removing all cooperativities between neighboring proteins in the model, phage lambda cannot enter lysogeny regardless the dosage of the UV irradiation (Fig. 4.4, row 2). However, when the cooperativity

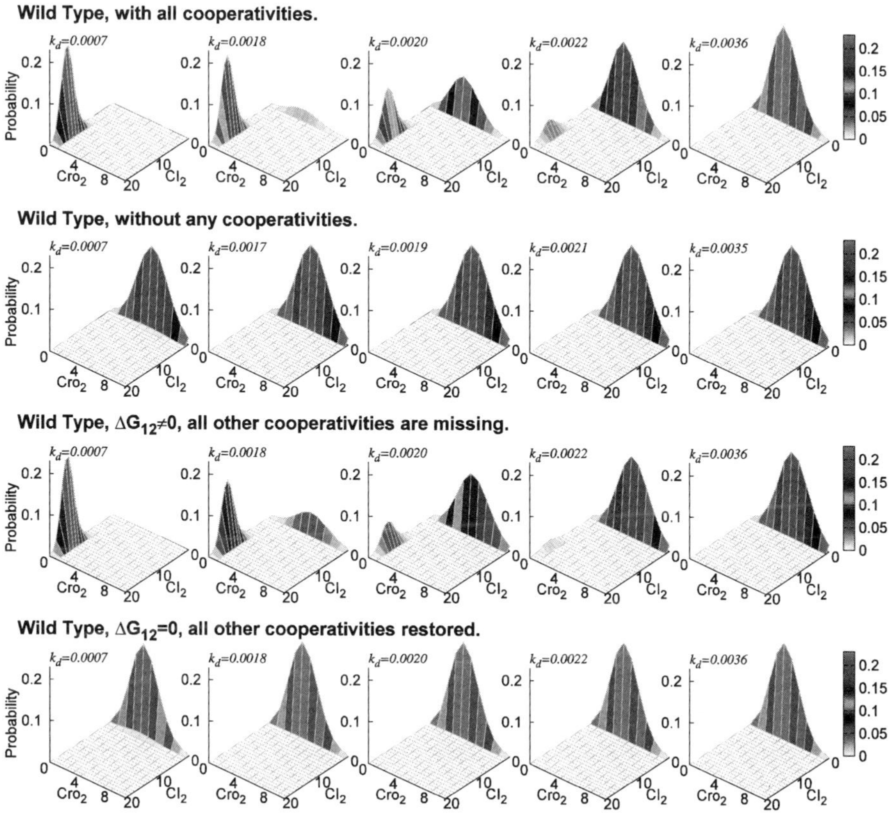

FIGURE 4.4 The probability landscape of the epigenetic circuits of lysogeny maintenance in phage lambda. (Row 1) For wild-type phage lambda, at the CI degradation rate of $k_d = 7.0 \times 10^{-4}$/s, probability landscape centers at locations with high copy numbers of CI_2 and close to 0 copy of Cro_2. This corresponds to the lysogenic phase of phage lambda. When k_d increases from $k_d = 1.8 \times 10^{-3}$/s to 2.2×10^{-3}/s, the peak located at the lysogenic phase gradually diminishes, whereas the peak located at lytic phase gradually increases. At about $k_d = 2.0 \times 10^{-3}$/s, phage lambda has about equal probability to be in either lysogenic or lytic phase. When CI is degraded at a faster rate of $k_d = 3.6 \times 10^{-3}$/s, the probability landscape centers at locations where there are higher copy numbers of Cro dimer and close to 0 copy of CI. This corresponds to the lytic phase of phage lambda. (Row 2) When all cooperativities are removed from the model, lysogeny cannot be achieved. (Row 3) When only the cooperativity of ΔG_{12} is restored, wild-type behavior is largely restored. (Row 4) When all other cooperativities except ΔG_{12} are restored, lysogeny still cannot be achieved.

ΔG_{12} between two CI dimer proteins when binding to operator sites are restored, the lysogeny is largely restored (Fig. 4.4, row 3). In contrast, if all other cooperativities are restored except ΔG_{12}, phage lambda still lacks the ability to enter the lysogeny phase (Fig. 4.4, row 4). These calculations suggest that the cooperativity ΔG_{12} plays key roles in maintaining the properties of the network.

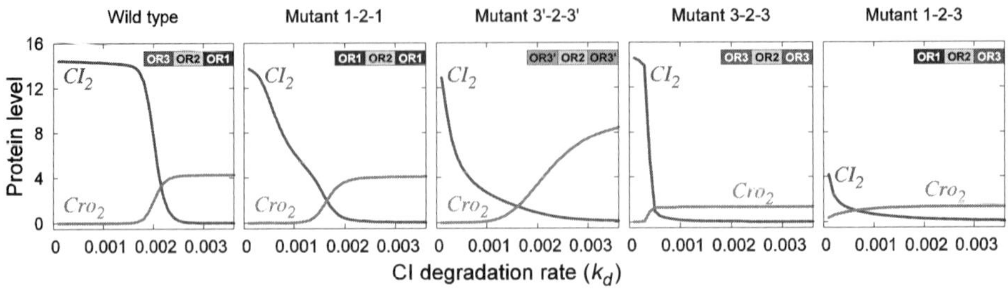

FIGURE 4.5 Instability, shallow threshold, and switching inefficiency of the network against fluctuation in UV irradiation in mutant phage lambda, in which the wild type operators are mutated (3'-2-3'), or their locations permuted (1-2-1, 3-2-3, and 1-2-3) [34]. Wild-type phage with operator site OR3-OR2-OR1 (3-2-1) maintains a stable level of CI_2, but respond to further UV irradiation after a set point and switches to lytic phage efficiently. In contrast, mutant 1-2-1, where an OR3 is replaced by an OR1, and mutant 3′-2-3′ (slighted mutated OR3 replacing original OR1 and OR3) do not maintain a stable level of CI_2. They are leaky and responds gradually to graded changes in k_d. Their thresholds and that of mutant 3-2-3 for lytic transition are much shallower. Mutant 1-2-3 does not maintain a sufficient amount of CI_2 and therefore cannot maintain lysogeny.

An important property of biological stochastic network is its robustness against changes in the molecular components of the epigenetic network. Experimental studies showed that when the ordering of operator sites are changed, mutants of phage lambda all have functional epigenetic circuits, but have markedly different tolerance to UV irradiation. Calculations from solving the dCME model showed that the wild-type lysogen has a high threshold toward lysis and is overall insensitive to fluctuation of UV dosage, if it is below certain threshold (Fig. 4.5). That is, the switching network of phage lambda is very stable and is strongly buffered with a high threshold against fluctuations in CI degradation rate due to environmental changes in UV irradiation. This high threshold against environmental fluctuations is important for the self-perpetuating nature of the epigenetic state of *E. coli* cells, allowing lysogeny to be passed on to its offspring. Once the degradation rate of CI reaches a threshold, phage lambda switches very efficiently to the lytic phase, and this efficiency is not built at the expense of stability against random fluctuation. Wild-type phage lambda therefore can integrate signaling in the form of different CI degradation rates and can distinguish a true signal above the high threshold from random noise fluctuating below this threshold.

In contrast, all mutant variants exhibit the behavior of a hair trigger and require much less UV irradiation for the onset of lysis induction (Fig. 4.5). In addition, they are ''leaky'' and respond in a graded fashion toward increased UV irradiation, instead of the well-behaved threshold behavior observed in wild-type phage lambda. In the case of mutant 1-2-3, the mutant phage lambda cannot enter the lysogenic state. These results are in full agreement with experimental findings [8,34].

4.8 DISCUSSIONS AND SUMMARY

In this chapter, we have discussed the significance of the chemical master equation (CME) as a theoretical framework for modeling nonlinear, biochemical reaction networks. This formulation provides a foundation to study stochastic phenomena in biological networks. Its role is analogous to that of the Schrödinger equation in quantum mechanics [50]. Developing computational solutions to the CME has important implications, just as the development of computational techniques for solving the Schrödinger equation for systems with many atoms is [29,30]. By computing the time-evolving probability landscape of cellular stochastic networks, we may gain understanding of the possible mechanisms of cellular states, as well as the inheritable phenotypes with a distributive epigenetic code, in which the network architecture and its landscape dictate the physiological metastases of the cell under different conditions [47,63].

Overall, studying the behavior of a stochastic network is challenging. and solving a given CME is a computationally challenging task. We have outlined several key difficulties, as well as some of the progresses that have been made so far. The finite buffer algorithm allows direct numerical solution to the discrete CME and can be applied to study stochasticity of systems with a handful of particles, as well as larger networks arising from very realistic biological problem, such as that of the lysogeny-lysis control circuit of the phage lambda [8]. As an exact method, it can also be used to study model systems of finite size to gain insight into stochastic behavior of networks. The ability to compute error due to state truncation *a priori* enables ability to ensure the correctness of the computational solution to the dCME, as well as knowledge of its level of accuracy. Furthermore, it provide means to develop optimized strategies to minimize truncation error when finite computing resources is given. The stochastic simulation algorithm offers the approach of studying the stochastic network through simulations. The formulation of stochastic differential equation such as the Langevin equation allows exploration of more complex stochastic systems, at the expense of less rigorous assumptions and perhaps more errors.

An important task is to integrate different stochastic methods for efficient computational solution of complex stochastic networks at large scale, with controlled accuracy. For example, one may apply the finite buffer algorithm to solve dCME directly for certain critical parts of the network, where rare events need to be assessed very accurately. One may use the Langevin stochastic equation to study other parts of the network where general stochastic behavior needs to be determined. In addition, one may also wish to apply the stochastic simulation algorithm to certain parts of the network to probe their behavior. Furthermore, one may wish to apply ordinary differential equation (ODE) models to study parts of the system where copy numbers of molecules are large and there are little stochastic effects.

A great challenge is to develop a general strategy so the best methods can be applied to specific parts of the network and the results integrated to provide an overall picture of the stochastic dynamic behavior of the network. It would be desirable that the resulting errors due to approximations of varying degree are bounded within a tolerance level, while maintaining necessary computing speed and resource requirement. A further challenge is to develop such hybrid methods to compute the overall spatiotemporal stochastic dynamic properties of different cellular compartments, multicellular tissue, with consideration of different spatial distribution or gradient of molecules such as oxygen, nutrient, morphogenes, and other signaling factors, all with stochasticity appropriately considered.

The complex nature of the stochastic dynamics arising from biochemical networks bears some resemblance to another complex system, namely, that of protein folding. Both have very large space of microstates, and both can be modeled by transitions between microstates using master equations [13,26,45]. However, these two systems differ in several important aspects. First, while protein folding can be modeled as a relaxation process toward the equilibrium state, biochemical networks are intrinsically open, with synthesis and degradation of molecules an integral part of the system, hence there are no equilibrium states. Instead, one frequently seeks to study the non-equilibrium steady state. Second, once the energy of a protein conformation is known, the relative probability of its sequence adopting this conformation in the equilibrium state can be calculated from the Boltzmann distribution, without the need of knowing all other possible conformations and their associated probabilities. In contrast, it is not possible to calculated the relative probability of a specific microstate of copy numbers *a priori* without solving the entire CME, as the probability distribution of network states do not generally follow any specific analytical forms, and there are no detailed balance and there exists cyclic probability fluxes [33].

REFERENCES

1. L. M. Anderson and H. Yang. DNA looping can enhance lysogenic CI transcription in phage lambda. *Proc. Natl. Acad. Sci. USA*, **105**(15):5827–5832, 2008.
2. P. Ao, C. Kown, and H. Qian. On the existence of potential landscape in the evolution of complex systems. *Complexity*, **12**:19–27, 2007.
3. A. Arkin, J. Ross, and H. H. McAdams. Stochastic kinetic analysis of developmental pathway bifurcation in phage lambda-infected *Escherichia coli* cells. *Genetics*, **149**(4):1633–1648, 1998.
4. E. Aurell, S. Brown, J. Johanson, and K. Sneppen. Stability puzzles in phage λ. *Phys. Rev. E*, **65**(5):051914, 2002.
5. Y. Cao, D. T. Gillespie, and L. R. Petzold. The slow-scale stochastic simulation algorithm. *J. Chem. Phys.*, **122**(1):14116, 2005.
6. Y. Cao, D. T. Gillespie, and L. R. Petzold. Efficient step size selection for the tau-leaping simulation method. *J. Chem. Phys.*, **124**(4):044109, 2006.
7. Y. Cao and J. Liang. Adaptively biased sequential importance sampling for rare events in reaction networks with comparison with exact solutions from finite buffer dCME method. *J. Chem. Phys.*, 2013.
8. Y. Cao, H. M. Lu, and J. Liang. Probability landscape of heritable and robust epigenetic state of lysogeny in phage lambda. *Proc. Natl. Acad. Sci. USA*, **107**(43):18445–18450, 2010.

9. Y. Cao, A. Terebus, and J. Liang. State space truncation with quantified effects for accurate solution of discrete chemical master equation. *Bull. Math. Biol.*, 2015, in review.
10. Y. Cao, A. Terebus, and J. Liang. Accurate chemical master equation solution method with multi-finite buffers for time-evolving and steady state probability landscapes and first passage times. *SIAM Multiscale Model. Simul.*, 2015, in review.
11. Y. Cao and J. Liang. Optimal enumeration of state space of finitely buffered stochastic molecular networks and exact computation of steady state landscape probability. *BMC Systems Biology*, **2**:30, 2008.
12. Y. Cao and J. Liang. Nonlinear coupling for improved stochastic network model: A study of Schnakenberg model. In *Proceedings of 3rd Symposium on Optimization and Systems Biology (OSB)*, 2009.
13. M. Cieplak, M. Henkel, J. Karbowski, and J. R. Banavar. Master equation approach to protein folding and kinetic traps. *Phys. Rev. Lett.*, **80**:3654–3657, 1998.
14. B. J. Daigle, Jr., M. K. Roh, D. T. Gillespie, and L. R. Petzold. Automated estimation of rare event probabilities in biochemical systems. *J. Chem. Phys.*, **134**(4):044110, 2011.
15. B. N. Datta. *Numerical Linear Algebra and Applications*. Brooks/Cole Publishing Company, Independence, KY, 1995.
16. M.A. Gibson and J. Bruck. Exact stochastic simulation of chemical systems with many species and many channels. *J. Phys. Chem.*, **105**:1876–1889, 2000.
17. D.. T. Gillespie. The chemical langevin equation. *J. Chem. Phys.*, **113**:297–306, 2000.
18. D. T. Gillespie. Exact stochastic simulation of coupled chemical reactions. *J. Phys. Chem.*, **81**:2340–2361, 1977.
19. G. H Golub and C. F. Van Loan. *Matrix Computations*. Johns Hopkins University Press, Baltimore, 1996.
20. J. Goutsias. Classical versus stochastic kinetics modeling of biochemical reaction systems. *Biophys. J.*, **92**(7):2350–2365, 2007.
21. J. Hasty, J. Pradines, M. Dolnik, and J. J. Collins. Noise-based switches and amplifiers for gene expression. *Proc. Natl. Acad. Sci. USA*, **97**(5):2075–2080, 2000.
22. R. A. Horn and C. R. Johnson. *Topics in Matrix Analysis*. Cambridge University Press, New York, 1991.
23. J. E. Hornos, D. Schultz, G. C. Innocentini, J. Wang, A. M. Walczak, J. N. Onuchic, and P.G. Wolynes. Self-regulating gene: an exact solution. *Phys. Rev. E. Stat. Nonlin. Soft Matter. Phys.*, **72**(5 Pt 1):051907, 2005.
24. F. Jacob and J. Monod. Genetic regulatory mechanisms in the synthesis of proteins. *J. Mol. Biol.*, **3**:318–356, 1961.
25. A. D. Johnson, A. R. Poteete, G. Lauer, R. T. Sauer, G. K. Ackers, and M. Ptashne. lambda repressor and cro–components of an efficient molecular switch. *Nature*, **294**(5838):217–223, 1981.
26. S. Kachalo, H. M. Lu, and J. Liang. Protein folding dynamics via quantification of kinematic energy landscape. *Phys. Rev. Lett.*, **96**(5):058106, 2006.
27. S. Karlin and H. M. Taylor. *A First Course in Stochastic Processes*, 2nd edition, Academic Press, 1975.
28. K. Y. Kim and J. Wang. Potential energy landscape and robustness of a gene regulatory network: Toggle switch. *PLoS Comput. Biol.*, **3**(3):e60, 2007.

29. J. Kohanoff. *Electronic Structure Calculations for Solids and Molecules: Theory and Computational Methods*. Cambridge University Press, New York, 2006.
30. W. Kohn and L. J. Sham. Self-consistent equations including exchange and correlation effects. *Phys. Rev.*, **140**:A1133–A1138, 1965.
31. H. Kuwahara and I. Mura. An efficient and exact stochastic simulation method to analyze rare events in biochemical systems. *J. Chem. Phys.*, **129**(16):165101, 2008.
32. M. D. Levin. Noise in gene expression as the source of non-genetic individuality in the chemotactic response of *Escherichia coli*. *FEBS Lett.*, **550**(1–3):135–138, 2003.
33. J. Liang and H. Qian. Computational cellular dynamics based on the chemical master equation: A challenge for understanding complexity. *J. Computer Sci. Tech.*, **25**(1):154–168, 2010.
34. J. W. Little, D. P. Shepley, and D. W. Wert. Robustness of a gene regulatory circuit. *The EMBO J.*, **18**(15):4299–4307, 1999.
35. H.-M. Lu and J. Liang. Perturbation-based markovian transmission model for probing allosteric dynamics of large macromolecular assembling: A study of groel-groes. *PLoS Comput. Biol.*, **5**(10):e1000526, 10 2009.
36. S. Macnamara, A.M. Bersani, K. Burrage, and R.B. Sidje. Stochastic chemical kinetics and the total quasi-steady-state assumption: Application to the stochastic simulation algorithm and chemical master equation. *J. Chem. Phys.*, **129**(9):095105, 2008.
37. M. Maggioni, T. Berger-Wolf, and J. Liang. GPU-based steady-state solution of the chemical master equation. In *12th IEEE International Workshop on High Performance Computational Biology (HiCOMB)*, 2013.
38. H. H. McAdams and A. Arkin. Stochastic mechanisms in gene expression. *Proc. Natl. Acad. Sci. USA*, **94**(3):814–819, 1997.
39. J. T. Mettetal, D. Muzzey, J. M. Pedraza, E. M. Ozbudak, and A. van Oudenaarden. Predicting stochastic gene expression dynamics in single cells. *Proc. Natl. Acad. Sci. USA*, **103**(19):7304–7309, 2006.
40. Y. Morishita and K. Aihara. Noise-reduction through interaction in gene expression and biochemical reaction processes. *J. Theor. Biol.*, **228**(3):315–325, 2004.
41. Y. Morishita, T.J. Kobayashi, and K. Aihara. An optimal number of molecules for signal amplification and discrimination in a chemical cascade. *Biophys. J.*, **91**(6):2072–2081, 2006.
42. B. Munsky and M. Khammash. A multiple time interval finite state projection algorithm for the solution to the chemical master equation. *J. Comput. Phys.*, **226**:818–835, 2007.
43. B. Munsky and M. Khammash. The finite state projection algorithm for the solution of the chemical master equation. *J. Chem. Phys.*, **124**(4):044104, 2006.
44. E. M. Ozbudak, M. Thattai, I. Kurtser, A. D. Grossman, and A. van Oudenaarden. Regulation of noise in the expression of a single gene. *Nat. Genet.*, **31**(1):69–73, 2002.
45. S. B. Ozkan, I. Bahar, and K. A. Dill. Transition states and the meaning of ϕ-values in protein folding kinetics. *Folding & Design*, **3**:R45–R58, 1998.
46. J. Paulsson and M. Ehrenberg. Random signal fluctuations can reduce random fluctuations in regulated components of chemical regulatory networks. *Phys. Rev. Lett.*, **84**(23):5447–5450, 2000.
47. M. Ptashne. On the use of the word 'epigenetic'. *Curr. Biol.*, **17**(7):R233–6, 2007.

48. M. Ptashne, K. Backman, M.Z. Humayun, A. Jeffrey, R. Maurer, B. Meyer, and R.T. Sauer. Autoregulation and function of a repressor in bacteriophage lambda. *Science*, **194**(4261):156–161, 1976.

49. M. Ptashne. *A Genetic Switch: Phage Lambda Revisited*, 3rd edition. Cold Spring Harbor Laboratory Press, Cold Spring Harbor, NY, 2004.

50. H. Qian and D. Beard. *Chemical Biophysics: Quantitative Analysis of Cellular Systems*. Cambridge University Press, New York, 2010.

51. M. K. Roh, B. J. Daigle, Jr., D. T. Gillespie, and L. R. Petzold. State-dependent doubly weighted stochastic simulation algorithm for automatic characterization of stochastic biochemical rare events. *J. Chem. Phys.*, **135**(23):234108, 2011.

52. D. Schultz, J. N. Onuchic, and P.G. Wolynes. Understanding stochastic simulations of the smallest genetic networks. *J. Chem. Phys.*, **126**(24):245102, 2007.

53. M. A. Shea and G. K. Ackers. The OR control system of bacteriophage lambda a physical–chemical model for gene regulation. *J. Mol. Biol.*, **181**(2):211–230, 1985.

54. R. B. Sidje. Expokit: A software package for computing matrix exponentials. *ACM Trans. Math. Softw.*, **24**(1):130–156, 1998.

55. E. Sontag. Lecture notes on mathematical systems biology, 2013.

56. H. M. Taylor and S. Karlin. *An Introduction to Stochastic Modeling*, 3rd edition. Academic Press, New York, 1998.

57. C. A. Uribe and G. C. Verghese. Mass fluctuation kinetics: Capturing stochastic effects in systems of chemical reactions through coupled mean-variance computations. *J. Chem. Phys.*, **126**(2):024109, 2007.

58. N. G. Van Kampen. *Stochastic Processes in Physics and Chemistry*. North Holland, Amsterdam, 1992.

59. N. G. Van Kampen. *Stochastic Processes in Physics and Chemistry*, 3rd edition. Elsevier Science and Technology Books, 2007.

60. M. Vellela and H. Qian. A quasistationary analysis of a stochastic chemical reaction: Keizer's paradox. *Bull. Math. Biol.*, **69**:1727–1746, 2007.

61. D. Volfson, J. Marciniak, W. J. Blake, N. Ostroff, L. S. Tsimring, and J. Hasty. Origins of extrinsic variability in eukaryotic gene expression. *Nature*, **439**(7078):861–864, 2006.

62. T. Zhou, L. Chen, and K. Aihara. Molecular communication through stochastic synchronization induced by extracellular fluctuations. *Phys. Rev. Lett.*, **95**(17):178103, 2005.

63. X. M. Zhu, L. Yin, L. Hood, and P. Ao. Robustness, stability and efficiency of phage lambda genetic switch: Dynamical structure analysis. *J. Bioinform. Comput. Biol.*, **2**(4):785–817, 2004.

EXERCISES

4.1 If the microstates of a stochastic network can be enumerated, one can solve the underlying dCME directly. For a network with m molecular species with r reactions, assume that each molecular species can have at most n copies of molecules.

(**a**) Without knowing the details of the reactions if one ignores all dependency between molecules and allows the possibility that all molecular species may simultaneously have the maximum of n copies of molecules, provide an upper bound on the size of the state space.

(**b**) As different molecular species are coupled through chemical reactions, they are not independent. Because of these couplings, the effective number of independent species is less than m. Let the stoichiometry matrix of the network be $\boldsymbol{C}$, which is an $m \times r$ matrix, with its kth column representing the kth stoichiometry vector $\boldsymbol{s}_k$ as defined in Eq. (4.1). $\boldsymbol{C}$ describes the coupling between different molecular species of the network. The degree of reduction in independent molecular species due to coupled reactions is specified by the rank of $\boldsymbol{C}$, denoted as rank $\boldsymbol{C}$ [55]. How can you make a better estimation of the size of the state space?

4.2 We examine a model of stochastic reactions in some detail.

(**a**) Suppose the kth reaction can be written as

$$c_A A + c_B B + c_C C \rightarrow c_D D + c_E E.$$

It has an intrinsic rate of r_k. Please write down the rate of the reaction $A_k(\boldsymbol{x})$ that depends on the state of the system—for example, the copy numbers $(\boldsymbol{x}_A, \boldsymbol{x}_B, \boldsymbol{x}_C)$ of the system.

(**b**) Assuming a Poisson process, show that the probability of no reaction occurring during Δt is $e^{-A(\boldsymbol{x})\cdot\Delta t}$, where $A(\boldsymbol{x}) = \sum_k A_k(\boldsymbol{x})$. We first divide Δt into H intervals of small durations $\epsilon = \Delta/H$.

(**1**) What is the approximate probability that none of the m reactions have occurred during the time prior to the end of the first time interval ϵ? You can ignore the higher-order terms.

(**2**) What is the approximate probability that none of the reactions have occurred during Δt?

(**3**) Show that when taking the limit with $H \rightarrow \infty$, this probability is $e^{-A(\boldsymbol{x})\cdot\Delta t}$.

4.3 In the stochastic simulation algorithm, one needs to update the rates of all reactions after each time step. Namely, one needs to recalculate $A(x(t + \Delta t)$, as the vector of copy numbers of molecules is altered after each sampled reaction. This is computationally expensive. To speed up the calculation, one can choose to update the reaction rates only after a time interval $\Delta\tau$, when the accumulated error would otherwise exceed some tolerance. That is, we can leap the system forward by $\Delta\tau$ without updating the copy numbers of the molecular species [6]. Suppose a tolerance threshold θ_i is specified to indicate the acceptable level of error $\theta_i \boldsymbol{x}_i$ in the copy number $\boldsymbol{x}_i$ of the ith molecular species, such that during the interval of $\Delta\tau$, we have $\Delta \boldsymbol{x}_i \leq \max\{\theta_i \boldsymbol{x}_i, 1\}$. What is overall the best $\Delta\tau$ one can use so errors in all species are within tolerance?

4.4 The Langevin equation of Eq. (4.26) assumes that the number $n_k(\boldsymbol{x}, \Delta t)$ of spontaneously fired reactions R_k occurring during Δt follows a Gaussian distribution. This is valid only when many reactions occur during Δt. A more realistic model is to assume that $n_k(\boldsymbol{x}, \Delta t)$ follows a Poisson distribution, if the copy numbers of molecules do not change significantly, and hence the reaction rates also do not change significantly during Δt, as discussed above.

(**a**) Describe in pseudocode a method for generating a random variable $n_k(\boldsymbol{x}, \Delta t)$ that follows the appropriate Poisson distribution. You can use the inverse function method or any other method.

(**b**) During Δt, all of the r reactions can occur. They are fired randomly following r independent Poisson processes. Modify the standard Langevin formula in Eq. (4.26) so the random fluctuation follows Poisson processes.

4.5 In the stochastic simulation algorithm, one needs to generate two random variables: the time interval Δt until the next reaction occurs and the specific reaction k that would occur. Although the stochastic simulation algorithm by Gillespie is accurate, it samples reaction trajectories according to their overall probabilities and therefore can be very inefficient in sampling rare events.

(**a**) One can bias towards a reaction by artificially accelerating its reaction rate. Instead of sampling Δt_k for reaction k following $A_k(\boldsymbol{x}(t))e^{-A(\boldsymbol{x}(t))\Delta t}$, one can bias to accelerate the reaction by introducing an inflation constant α_k, so the time interval Δt_k will be selected following the modified probability

$$\alpha_k \cdot A_k(\boldsymbol{x}(t))e^{-A(\boldsymbol{x}(t))\Delta t}.$$

However, this bias needs to be corrected. Modify Algorithm 2 in pseudocode so appropriate bias α_k (as input) is introduced, and the final results are appropriately corrected.

(**b**) One can also improve the sampling efficiency by introducing desirable bias towards specific reactions that would otherwise occur rarely. Instead of selecting the kth reaction as the next reaction according to the probability $\frac{A_k(\boldsymbol{x}(t))}{A(\boldsymbol{x}(t))}$, one can bias towards/away from this reaction by introducing an inflation/deflation factor β_k, so the kth reaction will be selected following the modified probability $\beta_k \cdot \frac{A_k(\boldsymbol{x}(t))}{A(\boldsymbol{x}(t))}$. Again, this bias needs to be corrected. Modify Algorithm 2 in pseudocode so that the appropriate bias β_k (as input) is introduced, and the final results are appropriately corrected.

(**c**) Write down in pseudocode a modified stochastic simulation algorithm incorporating both bias factors α_k and β_k, and describe how to ensure that proper corrections are included.

5

CELLULAR INTERACTION NETWORKS

Cells are the fundamental and smallest units of life that are capable of independent functioning. A living organism may be uni- or multicellular and is made up of one of two basic types of cells, the *prokaryotic* cells and the *eukaryotic* cells, that evolved from a common ancestor cell but still share many common features. The biological functioning and life of a cell is controlled by signaling and energy transfer interactions among its numerous constituents such as proteins, RNAs, DNAs, and other small molecules, allowing them to adapt to changing environments [1,25]. Such interactions may involve a *cascade* of biochemical reactions. Systematic approaches to understanding cellular interaction networks involve several steps such as data collection and integration of available information, adopting an appropriate model for the system, experimenting on a global level, and generation of new hypotheses about the interaction patterns. With the advancement in digital and computing technologies in the last few decades, it is now possible to perform genome-wide experimental studies to identify interactions among thousands of proteins and genes via DNA micro-arrays, florescent proteins, and Western blots. This in turn has already generated massive amounts of interaction data.

As stated above, after data collection and integration, an investigation of a molecular interaction network is continued by selecting, implicitly or explicitly, a model to characterize the interactions between components of the cellular environment. Naturally, the selection of the model depends on several factors such as the level of details desired, the characteristics of the particular interaction data studied, and the

Models and Algorithms for Biomolecules and Molecular Networks, First Edition. Bhaskar DasGupta and Jie Liang.

overall goal of the investigation. In very broad terms, there are two types of models for interaction, which we will discuss in the next two sections.

5.1 BASIC DEFINITIONS AND GRAPH-THEORETIC NOTIONS

We briefly review some basic graph-theoretic concepts before proceeding any further; see standard textbooks such as reference 17 for a more comprehensive discussion of these and related concepts. A directed or undirected graph $G = (V, E)$ consists of a set V of nodes and a set E of directed edges (also called arcs) or undirected edges (which we will simply refer to as edges). An edge $e = \{u, v\} \in E$ or an arc $e = (u, v) \in E$ denotes an edge between the nodes u and v or an arc directed from node u to node v, respectively. We will also denote an edge $\{u, v\}$ or an arc (u, v) by $u - v$ or $u \to v$, respectively. A *path* of length k is an ordered sequence of edges (for undirected graphs) $\left(u_1 - u_2, u_2 - u_3, \ldots, u_{k-1} - u_k, u_k - u_{k+1}\right)$, or an ordered sequence of arcs (for directed graphs) $\left(u_1 \to u_2, u_2 \to u_3, \ldots, u_{k-1} \to u_k, u_k \to u_{k+1}\right)$; a cycle is a path for which $u_{k+1} = u_1$. A directed graph $G = (V, E)$ is *strongly connected* provided that, for *any* pairs of nodes u and v, there exists a path from u to v and also a path from v to u.

5.1.1 Topological Representation

In this type of representation, the physical, chemical, or statistical dependencies among various components in the molecular network is represented by ignoring the kinetic (i.e., time-varying) components of the interactions. Typically, such a model is represented by a directed or undirected graph $G = (V, E)$ in which the cellular components are the set of nodes V, and the arcs or edges in E encode the causal or statistical interaction between the components.

The simplest case is when G is an undirected graph. This happens when information about the directionality of the interaction is unknown or irrelevant. For example, protein–protein interaction (PPI) graphs, encoding physical interactions among proteins, are often undirected in part due to the limitations of the current experimental technologies [29]. On the other hand, other types of molecular biological graphs such as transcriptional regulatory networks, metabolic networks, and signaling networks are represented by directed graphs in which each arc represents positive (also called excitory) or negative (also called inhibitory) regulation of one node by another node. This nature of regulation is formally incorporated in the graph-theoretic framework by allowing each arc $e \in E$ to have a label ℓ_e which is either 1 and -1, where a label of 1 (respectively, -1) represents a positive (respectively, negative) influence. In such a topological representation, a path P from node u to node v is often referred to as a *pathway*, and the excitory or inhibitory nature of P is specified by the *product* of labels $\Pi_{e \in P} \ell_e$ of edges in the path. Figure 5.1 shows an illustration of such a labeled directed graph. The graph-theoretic nature of such a representation allows for efficient *computational complexity* analysis and algorithmic design by using techniques

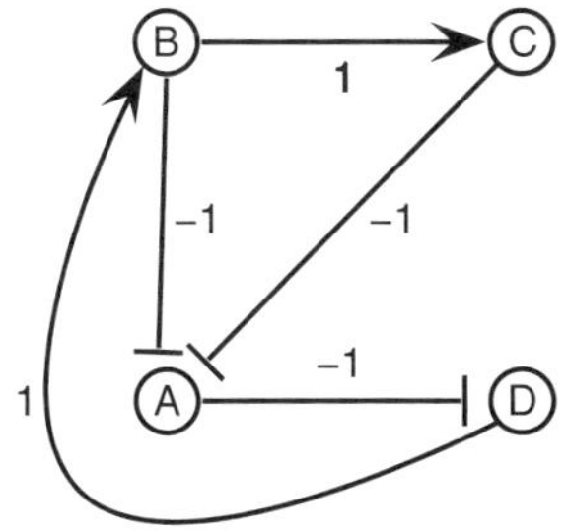

FIGURE 5.1 Illustration of the labeled directed graph representation for a molecular interaction network. The arc B ⊣ A indicates a negative influence of B on A; that is, an increase in the amount of protein B causes a decrease in the amount of protein A. The pathway B → C → A ⊣ D induces a positive influence of B on D since the product of labels of its arcs is $1 \times (-1) \times (-1) = 1$.

from theoretical computer science. A major disadvantage of such a representation is that it considers the interactions to be static and time-invariant; that is, all arcs in the graph are assumed to be present *simultaneously at any time*.

Another relevant static model is the Boolean circuits model. Now, the state of each node is binary (0 or 1), and each node computes a Boolean function of the states of the nodes to which it is connected. Figure 5.2 illustrates a Boolean model of three nodes involving three proteins and two genes. The Boolean model is restricted in the sense that the state of any node is restricted to be binary and each node can compute only a Boolean function. Boolean models are discussed further in Section 5.2.

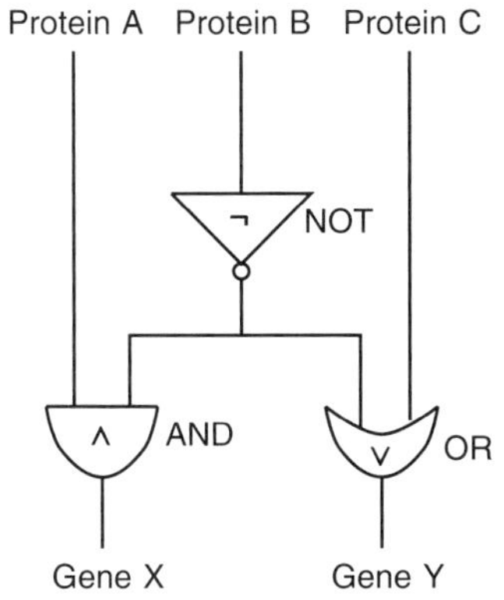

FIGURE 5.2 A Boolean circuit composed of logical AND, OR, and NOT gates that encode relationships between three proteins and two genes. For example, either Protein B must be absent or Protein C must be present (or both) to activate Gene Y.

5.1.2 Dynamical Representation

This type of representation, unlike the ones in the previous section, incorporates the (discrete or continuous) time-varying behavior of different molecular components in the network and thus provides a *more accurate* representation of the underlying biological mechanism. Dynamical representations are very suitable for simulating the biological system under study via different choices of values for parameters corresponding to unknown system characteristics or environmental conditions and then comparing the simulated dynamics with experimental measurements to refine the parameters of the model further.

A widely used continuous-time representation of this type is obtained via systems of ordinary differential equations (ODE) of the following nature in the formalism of control theory [76,77]:

$$\begin{aligned}
\dot{x}_1 &= \frac{\mathrm{d}x_1(t)}{\mathrm{d}t} = f_1\big(x_1(t), x_2(t), \ldots, x_n(t), u_1(t), u_2(t), \ldots, u_m(t)\big) \\
\dot{x}_2 &= \frac{\mathrm{d}x_2(t)}{\mathrm{d}t} = f_2\big(x_1(t), x_2(t), \ldots, x_n(t), u_1(t), u_2(t), \ldots, u_m(t)\big) \\
&\vdots \\
\dot{x}_n &= \frac{\mathrm{d}x_n(t)}{\mathrm{d}t} = f_n\big(\underbrace{x_1(t), x_2(t), \ldots, x_n(t)}_{\text{states}}, \underbrace{u_1(t), u_2(t), \ldots, u_m(t)}_{\text{inputs}}\big)
\end{aligned} \tag{5.1}$$

or, in more concise vector form

$$\dot{\mathbf{x}} = \mathbf{f}(\mathbf{x}, \mathbf{u}).$$

In this formalism, $\mathbf{x}(t) = \big(x_1(t), x_2(t), \ldots, x_n(t)\big)$ indicates the concentration of n molecular components at time t, each of $f_1, f_2, \ldots, f_n$ are functions of n variables, and $\mathbf{u}(t) = \big(u_1(t), u_2(t), \ldots, u_m(t)\big)$ are m inputs corresponding to external stimuli to the cellular system. In the concise vector form of representation, $\mathbf{f}$ is the vector of functions and the *dot* over the vector of variables $\dot{\mathbf{x}}$ indicates the vector of time derivatives of each component of the vector $\mathbf{x}(t)$. Many applications of this formalism require a few *mild* technical assumptions, such as $\big(x_1(t), x_2(t), \ldots, x_n(t)\big)$ must evolve in an open subset of $\mathbb{R}^n$, the f_i's must be differentiable, and/or solutions of the above system of ODE must be defined for *all* $t \geq 0$. In addition, for the purpose of measurement, one usually designates a subset of the variables $x_1(t), x_2(t), \ldots, x_n(t)$ as *outputs* whose values can be recorded by reporting devices such as florescent proteins. Variations or specific cases of the above general formalisms, based on different natures of the system dynamics, are also possible. Some examples are:

- The time variable could be *continuous* (e.g., given via ODE as above, or via delay equations) or *discrete* (e.g., given via difference equations or discretization of continuous variables). For discrete time systems, the left-hand side of Eq. (5.1), namely the $\frac{\mathrm{d}x_i(t)}{\mathrm{d}t}$ term, is replaced by $x_i(t+1)$ where $t+1$ is the next time step after t. Also, for discrete systems, two choices are possible for updating the values of the $x_i(t+1)$'s from the corresponding set of values of $x_i(t)$'s at every discrete time instance: a *synchronous* update (in which all $x_i(t+1)$'s update *simultaneously*) or an *asynchronous* update (in which a *selected* $x_i(t+1)$ updates and the rest remains the same).

- The state variables could be *continuous*, *discrete*, or *hybrid* (i.e., some discrete and some continuous).
- The model could be *deterministic* or *probabilistic* (e.g., the functions f_i in the formulation are probabilistic functions).

In addition, one can also devise hybrid models—for example, by combining continuous and discrete time-scales, or by combining continuous and discrete time variables.

5.1.3 Topological Representation of Dynamical Models

Often in the study of dynamical representation of a biological system, it is possible to relate its dynamical properties under investigation by associating the dynamics with a corresponding topological representation. For example, one such version that will be very useful later in Section 6.5 in studying the ''monotonicity'' of the dynamics of a biological system is obtained by a *signed graph* representation in the following manner [9,22,75]. Consider the time-varying system defined by Eq. (5.1) without the external inputs $u_1, u_2, \ldots, u_n$. For notational convenience, let $\mathbf{x}(t) = \big(x_1(t), x_2(t), \ldots, x_n(t)\big)$. Suppose that, for each i and j, either $\frac{\partial f_j}{\partial x_i} \geq 0$ for all $\mathbf{x}(t)$ or $\frac{\partial f_j}{\partial x_i} \leq 0$ for all $\mathbf{x}(t)$. Then, the system modeled by (5.1) can be associated with a directed interaction graph G (referred to as the ''associated signed graph'') in the following manner:

- The set of nodes is $\{x_1, \ldots, x_n\}$.
- If $\frac{\partial f_j}{\partial x_i} \geq 0$ for all $\mathbf{x}(t)$ and $\frac{\partial f_j}{\partial x_i} > 0$ for some $\mathbf{x}(t)$, then there is an arc $e_{i,j}$ in G directed from x_i to x_j with $\ell_{e_{i,j}} = 1$.
- If $\frac{\partial f_j}{\partial x_i} \leq 0$ for all $\mathbf{x}(t)$ and $\frac{\partial f_j}{\partial x_i} < 0$ for some $\mathbf{x}(t)$, then there is an arc $e_{i,j}$ in G directed from x_i to x_j with $\ell_{e_{i,j}} = -1$.

5.2 BOOLEAN INTERACTION NETWORKS

The well-studied Boolean interaction network model is generally used in analyzing the dynamics of gene regulatory networks in which the gene expression levels are *binarized*; that is, the expression levels are either 0 (indicating not expressed) or 1 (indicating expressed). Such a model was first proposed by Kauffman in 1969 as random models of genetic regulatory networks [46] and can be formally defined as follows. A Boolean variable is a variable that is either 0 or 1, and a function $f\colon \{0,1\}^n \mapsto \{0,1\}$ over n Boolean variables is called a Boolean function.

Definition 5.1 *A Boolean network $\langle \vec{s}, \vec{f} \rangle$ consists of the following components:*

(a) *A Boolean state vector $\vec{s} = (s_1, s_2, \dots, s_n) \in \{0,1\}^n$.*

(b) *A global activation function vector $\vec{f} = (f_1, f_2, \dots, f_n)$, where each $f_i\colon \{0,1\}^n \in \{0,1\}$ is a Boolean function.*

(c) *An update rule that specifies the dynamics of the network. Let $s_i(t)$ and $\vec{s}(t)$ denote the values of s_i and $\vec{s}$ at time t, respectively. An update rule specifies how the value of each $s_i(t+1)$ is computed from $\vec{s}(t)$. Two popular update rules are as follows.*

Synchronous update: $\vec{s}(t+1) = \big(f_1(\vec{s}(t)), f_2(\vec{s}(t)), \dots, f_n(\vec{s}(t)) \big)$; *that is, all nodes update their states simultaneously at time $t+1$ based on their states at time t.*

Asynchronous update: *A specific variable, say s_j, is selected such that $s_j(t) \neq f_j(\vec{s}(t))$, and then the new value of this variable is updated as $s_j(t+1) = f_j(\vec{s}(t))$ (if no such node exists, the update procedure terminates). In other words, $\vec{s}(t+1) = \big(s_1(t), \dots, s_{j-1}(t), f_j(\vec{s}(t)), s_{j+1}(t), \dots, s_n(t) \big)$. The variable s_j may be selected randomly among all possible candidate variables, or it may be selected based on some predefined rules such as the lexicographically first variable among all candidate variables.*

In the same spirit as in Section 5.1.3, one may also associate a directed graph $G_{\vec{s},\vec{f}} = \big(V_{\vec{s},\vec{f}}, E_{\vec{s},\vec{f}}\big)$ with a Boolean network $\langle \vec{s}, \vec{f} \rangle$ in the following manner:

- $V_{\vec{s},\vec{f}} = \{v_1, v_2, \dots, v_n\}$, where v_i is associated with the state variable s_i.
- $E_{\vec{s},\vec{f}}$ contains an arc (v_i, v_j) for each pair of indices i and j such that the function f_j depends on the state variable s_i, that is,

$$f_j(s_1, \dots, s_{i-1}, 0, s_{i+1}, s_{i+2}, \dots, s_n) \neq f_j(s_1, \dots, s_{i-1}, 1, s_{i+1}, s_{i+2}, \dots, s_n)$$

for some $s_1, \dots, s_{i-1}, s_{i+1}, s_{i+2}, \dots, s_n \in \{0,1\}$.

A Boolean network $\langle \vec{s}, \vec{f} \rangle$ in which $G_{\vec{s},\vec{f}}$ has no cycles is known as a *feed-forward* Boolean network; see Fig. 5.2 for an example.

In the next definition, the notation $\vec{s}(t+1) = \vec{f}\,(\vec{s}(t))$ is used to indicate that the state vector $\vec{s}(t+1)$ is obtained from $\vec{s}(t)$ by following an appropriate update rule.

Definition 5.2 (limit cycle) *A limit cycle of length k is an ordered sequence of state vectors $(\vec{s}_1, \vec{s}_2, \dots, \vec{s}_k)$ such that $\vec{s}_i(t+1) = \vec{f}\,(\vec{s}_{i-1}(t))$ for $i = 2, 3, \dots, k$, $\vec{s}_1 =$*

$$f_1 = s_1 \wedge \overline{s_3}$$

$$f_2 = s_2 \vee (\overline{s_3} \wedge s_1)$$

$$f_3 = s_1 \wedge s_2$$

(a)

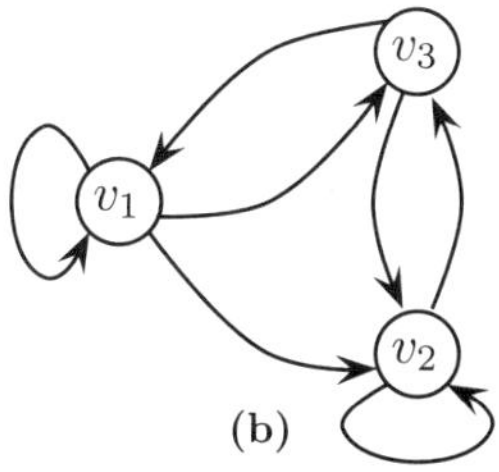

(b)

FIGURE 5.3 **(a)** A Boolean network with three binary states s_1, s_2, s_3. **(b)** The associated directed graph. A fixed point of the network is given by $\vec{s} = (s_1, s_2, s_3) = (0, 1, 0)$.

$\vec{f}\left(\vec{s}_k\right)$, *and* $\vec{s}_i \neq \vec{s}_j$ *for all* $i \neq j$. *A fixed-point or equilibrium point is a limit cycle of length* 1.

Limit cycles are also called ''attractors'' of a Boolean network. Figure 5.3 shows a Boolean network with its associated directed graph and an attractor.

Attractors are of considerable interest in Boolean genetic regulatory network models by associating them with different types of cells identified with specific *patterns of gene activities*. For example, reference 10 describes how to associate limit cycles with cellular cycles, and reference 37 associates fixed points with cell proliferation and apoptosis.

Boolean networks are not the main focus of this chapter, so we refer the reader to a textbook such as reference 19 for further discussions on this model.

5.3 SIGNAL TRANSDUCTION NETWORKS

Cells acquire many biological characteristics via complex interactions between its numerous constituents [1] and use signaling pathways and regulatory mechanisms to coordinate multiple functions, allowing them to respond to and acclimate to an ever-changing environment. Genes and gene products in cells interact on several levels. For example, at a *genomic level*, transcription factors can activate or inhibit the transcription of genes to give mRNA. Since these transcription factors are themselves products of genes, the ultimate effect is that genes regulate each other's expressions as part of a complex network. Similarly, proteins can participate in diverse post-translational interactions that lead to modified protein functions or to formation of protein complexes that have new roles. In many cases, different levels of interactions are integrated; for example, the presence of an external signal may trigger a cascade of interactions of different types. Recent advances in experimental methodologies in bioinformatics have led to development of genome-wide experimental methods resulting in identification of interactions among *thousands* of proteins, genes, and other components. Signal transduction network models discussed in this chapter are topological models that provide a concise summary of these interactions via *labeled* directed graphs. A major advantage of this model is that one can use powerful graph-theoretic techniques to analyze these networks. On the other hand, signal transduction networks only represent a network of possibilities, and not all edges are present and active *in vivo* in a given condition or in a given cellular location; therefore, an integration of time-dependent interaction may be necessary to more accurately predict the dynamical properties of these interactions.

Formally, a signal transduction network is defined by a edge-labeled directed graph $G = (V, E, \mathcal{L})$ where each node $v \in V$ represents an individual component of the cellular interaction, and each arc (directed edge) $(u, v) \in E$ indicates that node u has an influence on node v. The edge-labeling function $\mathcal{L}\colon E \mapsto \{-1, 1\}$ indicates the ''nature'' of the causal relationship for each arc, with $\mathcal{L}(u, v) = 1$ indicating that u has an *excitory* (positive) influence on v (e.g., increasing or decreasing the concentration of u increases or decreases the concentration of v, respectively), whereas $\mathcal{L}(u, v) = -1$, indicating that u has an *inhibitory* (negative) influence on v (e.g., increasing or decreasing the concentration of u decreases or increases the concentration of v, respectively). We will use the following notations and terminologies in the sequel.

- An excitory arc (u, v) will also be denoted by $u \overset{1}{\rightarrow} v$ or simply by $u \rightarrow v$ when the excitory nature is clear from the context. An inhibitory arc (u, v) will also be denoted by $u \overset{-1}{\rightarrow} v$ or $u \dashv v$.
- A pathway $\mathcal{P}$ refers to a directed path $u_1 \overset{y_1}{\rightarrow} u_2 \overset{y_2}{\rightarrow} \ldots u_{k-1} \overset{y_{k-1}}{\rightarrow} u_k$ in G of length $k - 1 \geq 1$ $(y_1, y_2, \ldots, y_{k-1} \in \{-1, 1\})$. The *parity* of $\mathcal{P}$ is $\mathcal{L}(\mathcal{P}) = \prod_{i=1}^{k-1} y_i \in \{-1, 1\}$. A path of parity 1 (respectively, -1) is called a path of *even* (respectively, *odd*) parity.
- The notation $u \overset{y}{\Rightarrow} v$ denotes a path from u to v of parity $y \in \{-1, 1\}$. If we do not care about the parity, we simply denote the path as $u \Rightarrow v$. Similarly, an arc will be denoted by $u \overset{y}{\rightarrow} v$ or $u \rightarrow v$.
- For a subset of arcs $E' \subseteq E$, reachable(E') is the set of all ordered triples (u, v, y) such that $u \overset{y}{\Rightarrow} v$ is a path of the arc-induced subgraph (V, E'). We will sometimes simply say $u \overset{y}{\Rightarrow} v$ is contained in E' to mean $u \overset{y}{\Rightarrow} v$ is a path of the subgraph (V, E').

5.3.1 Synthesizing Signal Transduction Networks

Large-scale repositories such as Many Microbe Microarrays, NASCArrays, or Gene Expression Omnibus contain expression information for thousands of genes under tens to hundreds of experimental conditions. Following the approach in references 4, 5, and 43, interaction information between components in these type of databases can be partitioned into three main categories.

(a) *Biochemical evidence that provides information on enzymatic activity or protein–protein interactions*—for example, binding of two proteins or a transcription factor activating the transcription of a gene or a chemical reaction with a single reactant and single product. These interactions are *direct* interactions.

(b) *Pharmacological evidence, in which a chemical is used either to mimic the elimination of a particular component or to exogenously provide a certain component*—for example, binding of a chemical to a receptor protein or observing gene transcription after exogenous application of a chemical. This type of experimental observation leads to observed relationships that are *not* direct interactions but indirect causal effects most probably resulting from a *chain* of interactions and reactions.

(c) *Genetic evidence of differential responses to a stimulus in wild-type organisms versus a mutant organism.* In a minority of cases this type of experimental observation may correspond to a single reaction (namely, when the stimulus is the reactant of the reaction, the mutated gene encodes the enzyme catalyzing the reaction and the studied output is the product of the reaction), but more often it is a chain of reactions.

As mentioned above, the last two types of experimental evidences may not give direct interactions but indirect double-causal relationships that correspond to reachability relationships in an (yet) unknown interaction network. Direct and indirect (pathway-level) information can be synthesized into a consistent network that maintains all the reachability relationships by the algorithm shown in Fig. 5.4. In Step 1, we incorporate biochemical interactions or single causal evidences as labeled arcs, noting the "mandatory arcs" corresponding to *confirmed* direct interactions. In Step 2, we incorporate double-causal relationships of the generic form $A \stackrel{x}{\to} (B \stackrel{y}{\to} C)$ by **(i)** adding a new arc $A \stackrel{x}{\to} B$ if $B \stackrel{y}{\to} C$ is a mandatory arc, **(ii)** doing nothing if existing paths in the network already explain the relationship, or **(iii)** adding a new "pseudo-node" and three new arcs. To correctly incorporate the parity of the $A \stackrel{x\,y}{\longrightarrow} C$ relationship, excitory $B \stackrel{y}{\to} C$ paths with $y = 1$ will be broken into two excitory edges, while inhibitory $B \stackrel{y}{\to} C$ paths with $y = -1$ will be broken into an excitory edge ($a = 1$) and an inhibitory edge ($b = -1$), summarized in a concise way by the equation $b = a\,b = y$.

A computer implementation of the framework in Fig. 5.4 needs an efficient algorithm for the following problem:

> *Given two nodes u_i and u_j and $y \in \{-1, 1\}$, does there exist a path from u_i to u_j of parity y?*

A straightforward solution is to adopt the so-called Floyd–Warshall transitive closure algorithm for directed graphs [20]. Let the nodes of G be $u_1, u_2, \ldots, u_n$, and let i be

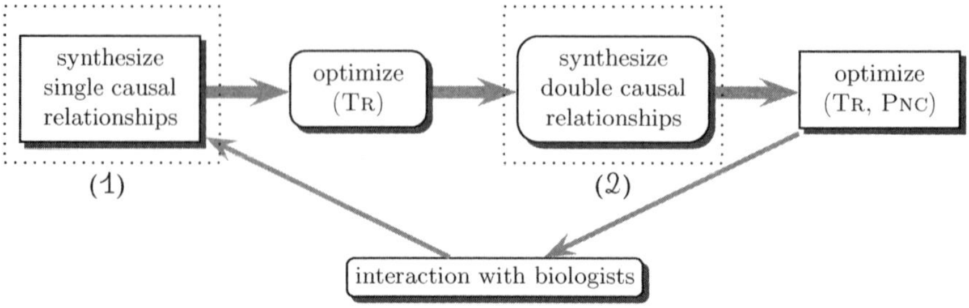

(1) [encoding single causal inferences (direct interactions)]

Biochemical and pharmacological evidences that define component-to-component relationships, namely relationships of the form "A promotes B" or "A inhibits B", are incorporated (in arbitrary order) as arcs $A \xrightarrow{1} B$ or $A \xrightarrow{-1} B$, respectively.

If the interaction is known to be a direct interaction with *concrete evidence*, then the arc is marked as a "mandatory arc". Let $\mathsf{E}_{\text{fixed}}$ denote the set of all mandatory arcs.

(2) [encoding double-causal evidences (indirect interactions)]

Consider each double-causal relationship $A \xrightarrow{x} (B \xrightarrow{y} C)$, where $x, y \in \{-1, 1\}$, in any arbitrary order. Add new nodes and/or arcs in the network based on the following cases:

- If $B \xrightarrow{y} C \in \mathsf{E}_{\text{fixed}}$, then add the arc $A \xrightarrow{x} B$.
- Otherwise, if there is *no* subgraph (in the network constructed so far) of the form $\boxed{\begin{array}{c} A \\ \Downarrow x \\ B \overset{a}{\Rightarrow} D \overset{b}{\Rightarrow} C \end{array}}$ for some node D where $b = a\,b = y$, then add to the network the subgraph $\boxed{\begin{array}{c} A \\ \downarrow x \\ B \xrightarrow{a} \mathrm{P} \xrightarrow{b} C \end{array}}$ where P is a new "pseudo-node", and $b = a\,b = y$.

FIGURE 5.4 An algorithmic framework for synthesizing signal transduction networks [5]. The optimization steps involving TR and PNC are explained in Section 5.3.3.

```
(* comment: initialization *)
for each x ∈ {−1,1} and for each i,j ∈ {1,2,...,n} do
   if u_i -x-> u_j ∈ E then N(i,j,0,x) = 1, P(i,j,k,x) = u_i -x-> u_j
                        else N(i,j,0,x) = 0, P(i,j,k,x) = ∅
   endif
endfor

(* comment: iterative calculation *)
for k = 1,2,...,n do
   for each i,j ∈ {1,2,...,n} and each x ∈ {−1,1} do
      if N(i,j,k − 1,x) = 1 then N(i,j,k,x) = 1 and P(i,j,k,x) = P(i,j,k − 1,x)
      else if ∃ y,z: N(i,k,k − 1,y) = N(k,j,k − 1,z) = 1 and yz = x
            then N(i,j,k,x) = 1
                 P(i,j,k,x) is P(i,k,k−1,y) followed by P(k,j,k−1,z)
            else N(i,j,k,x) = 0 and P(i,j,k,x) = ∅
         endif
      endif
   endfor
endfor
```

FIGURE 5.5 Dynamic programming algorithm to find all reachabilities.

called as the "index" of node u_i. Define the following quantities:

$$N(i,j,k,x) = \begin{cases} 1 & \text{if there is a path of parity } x \text{ from } u_i \text{ to } u_j \text{ using intermediate} \\ & \quad \text{nodes of indices no higher than } k, \\ 0 & \text{otherwise,} \end{cases}$$

$$P(i,j,k,x) = \begin{cases} \text{a path of parity } x \text{ from } u_i \text{ to } u_j & \text{if such a path exists,} \\ \emptyset & \text{otherwise.} \end{cases}$$

Then, the dynamic programming algorithm in Fig. 5.5 can be used to compute $N(i,j,k,x)$ and $P(i,j,k,x)$ for all i,j,k and x; the intuition behind the iterative calculations is shown pictorially in Fig. 5.6. The final answer is obtained by checking $N(i,i,n,x)$ and $P(i,i,n,x)$ for each $i \in \{1,2,\dots,n\}$ and each $x \in \{-1,1\}$. It is obvious that the algorithm takes $O\left(n^3\right)$ time and $O\left(n^3\right)$ space, both of which may be prohibitive for large networks. Exercise 5.11 explores some practical methods to reduce the running time and space requirements; further heuristic improvements are possible and were incorporated in the software reported in reference 43.

A software named NET-SYNTHESIS for synthesis of networks based on the framework in Fig. 5.4 was reported in reference 43; detailed discussion about usage of this software can be found in reference 6. The network synthesis framework in Fig. 5.4 was used successfully in [5,43] to synthesize a network of about 100 nodes for Abscisic Acid(ABA)-induced stomatal closure of the model plant *Arabidopsis*

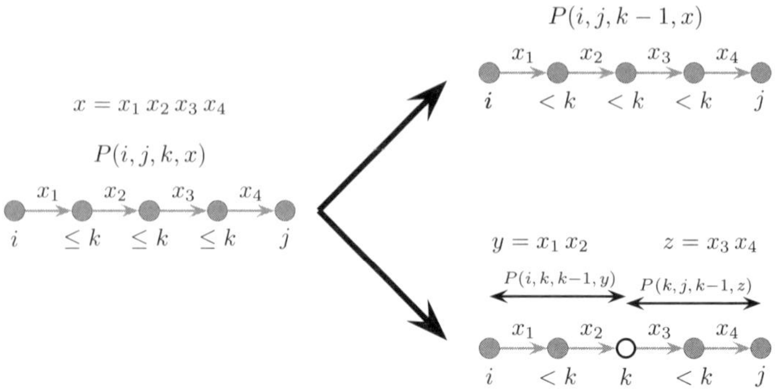

FIGURE 5.6 Pictorial illustration of the iterative calculations of the dynamic programming algorithm in Fig. 5.5.

Thaliana. The NET-SYNTHESIS software was further used in reference 87 to build a network model to study the signalling components that effect the survival of cytoxic T lymphocytes in LGL-leukemia.

5.3.2 Collecting Data for Network Synthesis

A starting point of gathering relevant data for synthesizing a signal transduction network is to read thoroughly relevant literatures concerning the signal transduction pathways of interest and then to assess if sufficient information is available such that network synthesis is indeed necessary. For example, if all that is known about a system is that component X activates component Y which in turn inhibits component Z, drawing a simple linear network and deducing that knockout of Y will eliminate signaling suffices, and a more formal analysis is hardly required. In assessing the literature, the focus should be specifically on experiments that provide information of the type relevant to network construction. Experiments that identify nodes belonging to a signaling pathway and their relationships include the following types of data.

(1) *In vivo* or *in vitro* experiments which show that the properties (e.g., activity or sub-cellular localization) of a protein change upon application of the input signal or upon modulation of components already *definitively* known to be associated with the input signal.

(2) Experiments that directly assay a small molecule or metabolite (e.g., imaging of cytosolic Ca^{2+} concentrations) and show that the concentration of that metabolite changes upon application of the input signal or modulation of its associated elements.

(3) Experiments that demonstrate physical interaction between two nodes, such as protein–protein interaction observed from yeast two-hybrid assays or *in vitro* or *in vivo* co-immunoprecipitation.

(4) Pharmacological experiments which demonstrate that the output of the pathway of interest is altered in the presence of an inhibitory agent that blocks signaling from the candidate intermediary node (e.g., a pharmacological inhibitor of an enzyme or strong buffering of an ionic species).

(5) Experiments which show that artificial addition of the candidate intermediary node (e.g., exogenous provision of a metabolite) alters the output of the signaling pathway.

(6) Experiments in which genetic knockout or over-expression of a candidate node is shown to affect the output of the signaling pathway.

Usually, **(1)–(3)** correspond to single causal (direct) inferences; **(3)** may also correspond to mandatory arcs. On the other hand, **(4)–(6)** usually correspond to double-causal (indirect) inferences.

Some choices may have to be made in distilling the relationships, especially in the case where there are *conflicting reports* in the literature. For example, suppose that in one report it is stated that proteins X and Y do *not* physically interact based on yeast two-hybrid analysis, while in another report it is stated that proteins X and Y *do* interact based on co-immunoprecipitation from the native tissue. One will then need to decide which information is more reliable and then proceed accordingly. Such aspects dictate that human intervention will inevitably be an important component of the literature curation process, as indicated by the ''interactions with biologists'' in Fig. 5.4, even as automated text search engines such as GENIES [31,40,59] grow in sophistication.

5.3.3 Transitive Reduction and Pseudo-node Collapse

The network obtained by steps (1) and (2) of the algorithm in Fig. 5.4 is ''not reduced'', that is, it may contains arcs and pseudo-nodes which can be systematically removed without changing the reachability relations between nodes (note, however, that no mandatory arc can be removed). In this section we describe two methods to find a *minimal* network, in terms of the number of pseudo-nodes and the number of nonmandatory arcs, that is consistent with all reachability relationships between non-pseudo-nodes. The algorithmic methodologies described are of two kinds:

- *transitive reduction* to reduce the number of nonmandatory arcs, and
- *pseudo-nodecollapse* to reduce the number of pseudo-nodes.

Applications of these methods do not necessarily imply that real signal transduction networks are the sparsest possible, but instead the goal is to minimize false positive (spurious) inferences even if risking false negatives; that is, the goal of these methods is to produce a network topology that is as close as possible to a *tree* topology while supporting all experimental observations. The implicit assumption of ''chain-like'' or ''tree-like'' topologies permeates the traditional molecular biology literature: Signal transduction and metabolic pathways are assumed to be close to linear chains, and

Instance: A directed graph $G = (V, E)$ with an arc labeling function $\mathcal{L}\colon E \mapsto \{-1, 1\}$, and a set of mandatory arcs $E_{\mathrm{m}} \subseteq E$.

Valid Solutions: A subgraph $G' = (V, E')$ where $E_{\mathrm{m}} \subseteq E' \subseteq E$, and reachable$(E')$ =reachable(E).

Objective: minimize $|E'|$.

FIGURE 5.7 The transitive reduction (TR) problem.

genes are assumed to be regulated by one or two transcription factors [1]. According to current observations, the reality is not far: *The average in/out degree of the transcriptional regulatory networks* [56,74] *and the mammalian signal transduction network* [58] *is close to* 1. Philosophically, the approach of obtaining a sparse network is similar to the *parsimony* approach used in the construction of phylogenies and elsewhere and can be linked to the ''Occam's razor'' principle.

***Transitive reduction* (TR).** The definition of the *transitive reduction* (TR) problem in the context of our signal transduction networks is shown in Fig. 5.7, and an illustration of a solution of this problem for an example network is shown in Fig. 5.8. Note that an exact or an approximate solution of the TR problem may *not* be unique; alternate solutions represent alternate interpretations of the same data. Transitive reduction are usually applied to the network synthesis procedure outlined in Fig. 5.4 after Step (1) and after Step (2).

The idea of a transitive reduction, in a more simplistic setting or integrated in an approach different from what has been described, also appeared in other papers than the ones already mentioned. For example, Wagner [85] proposed to find a network from the reachability information by constructing uniformly random graphs and scale-free networks in a range of connectivities and matching their reachability information to the range of gene reachability information found from yeast perturbation studies. In contrast, in the TR problem we consider the reachability information but with

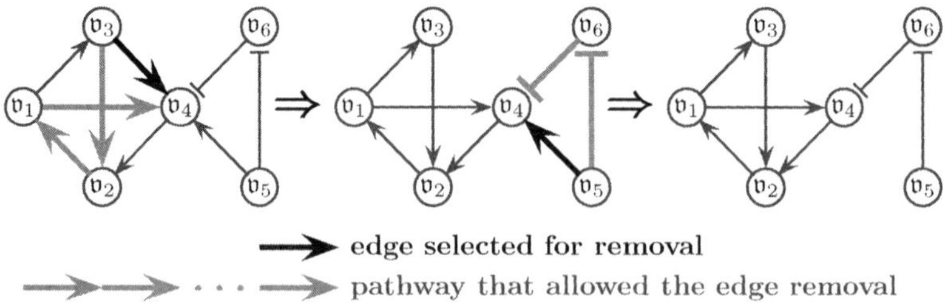

FIGURE 5.8 An example of obtaining a reduced network via transitive reduction. The obtained network is not minimal (see Exercise 5.4).

the *additional* information about the nature of interaction (excitory or inhibitory) for each path along with a subset of prespecified mandatory edges, and we wish to actually reconstruct the network and not only find a range of networks that have a given mean reachability. Another important difference between the two approaches is in the number of isolated nodes and weak subgraphs. Wagner's networks have a huge number of subnetworks, many of which are *isolated.* The network synthesis method in Fig. 5.4 does aim for sparseness, but does *not* allow isolated nodes if their reachabilities are nonzero. As another example, Chen et al. [18] used time-dependent gene expression information to determine candidate activators and inhibitors of each gene and then removed edges by assuming that no single gene functions both as activator and inhibitor.

Li et al. [57] used the greedy procedure in Fig. 5.9 for TR within the network synthesis procedure to *manually* create a network for ABA-induced stomatal closure.

This greedy procedure for TR is in fact optimal if the graph is a directed acyclic graph (DAG)—that is, if G has no cycles [4]. But even very special cases of TR—namely when all edges are excitory, $E_{\mathrm{m}} = \emptyset$, and G does not have a cycle of length more than 4—are known to be NP-hard [51], effectively ending the possibility of an efficient *exact* solution of TR for general graphs under the assumption of $\mathsf{P} \neq \mathsf{NP}$. Thus, the best that one could hope is that the greedy algorithm in Fig. 5.9 delivers a good *approximate* solution for TR on arbitrary graphs.

An approximation algorithm for a minimization problem has an approximation ratio of $\alpha \geq 1$ (or is an α-approximation) if it is a polynomial-time algorithm that *always* provides a solution with a value of the objective function that is *at most* α times the optimal value of the objective function (thus, for example, 1-approximation is an exact solution). The following result was shown in reference 5.

Theorem 5.1 [5] *The greedy algorithm in* Fig. 5.9 *is a* 3*-approximation.*

There are input instances of TR for which the greedy algorithm has an approximation ratio of *at least* 2; Fig. 5.10 shows such an example. The idea behind a proof of Theorem 5.1 is to first prove that the greedy procedure is a 3-approximation if the input graph G is strongly connected, and then to extend the result for the case when G may not be strongly connected. The idea behind a proof of 3-approximation of the greedy algorithm when G is strongly connected is as follows. Let $\mathsf{opt}(G)$ denote the number of edges in an optimal solution of TR for $G = (V, E)$. It is not difficult to observe that $\mathsf{opt}(G) \geq |V|$ (see Exercise 5.12). For a graph H, let H^{+1} be the

Definition an arc $u \overset{x}{\rightarrow} v$ is redundant if there is an alternate path $u \overset{x}{\Rightarrow} v$
Algorithm
    while there exists a redundant arc
        delete the redundant arc

FIGURE 5.9 A greedy algorithm to solve TR.

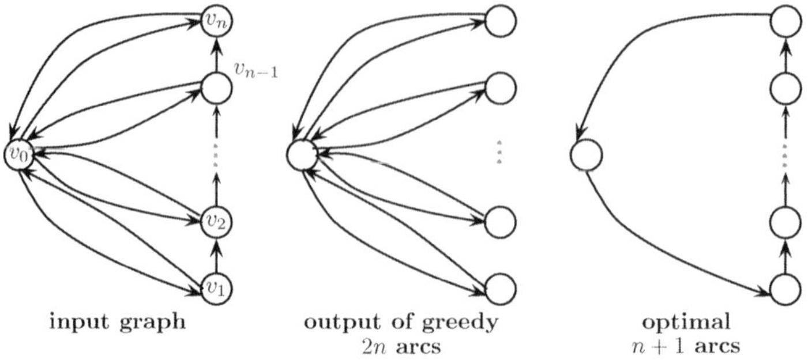

FIGURE 5.10 An example of a family of graphs for which the greedy algorithm has an approximation ratio of 2. The greedy algorithm may remove the arcs $v_i \rightarrow v_{i+1}$ for $i = 1, 2, \ldots, n-1$ providing a solution with $2n$ arcs, but an optimal solution with $n+1$ arcs is possible by selecting the arcs $v_0 \rightarrow v_1$, $v_i \rightarrow v_{i+1}$ for $i = 1, 2, \ldots, n-1$, and $v_n \rightarrow v_0$.

graph obtained from H by setting all arc labels to 1, and an arc e in H^{+1} is called *superfluous* if it would be removed by the greedy algorithm in H^{+1} but not in H. Let G_{greedy} be the graph obtained from G by the greedy algorithm. The proof uses the following sequence of steps.

- Show a 2-approximation for the case when $E_{\text{m}} = \emptyset$ in the following way:
  - **(i)** first show that G^{+1}_{greedy} contains at most 1 superfluous arc, and then
  - **(ii)** show that using **(i)** it follows that the number of arcs in G_{greedy} is at most $2|V| + 1$.
- Show that the constraint $E_{\text{m}} \neq \emptyset$ adds at most 1 to the approximation ratio.

Approximation algorithms with approximation ratios better than the greedy approach are possible and described in references 4, 13, and 43; currently the best possible approximation ratio achievable is $3/2$ as described in reference 13, but the algorithm is too complicated for efficient implementation. Exercise 5.13 explores a possible improvement over the greedy approach. A software for solving the TR problem was reported in references 6 and 43; extensive empirical evaluations in reference 43 showed that in practice the redundancy value calculated is almost always close to optimal.

Pseudo-node Collapse **(P_{NC})**. The definition of the *pseudo-node collapse* (PNC) problem in the context of our signal transduction network synthesis procedure is shown in Fig. 5.11. Intuitively, the PNC problem reduces the set of pseudo-nodes to a minimal set while maintaining that the graph is consistent with all experimental observations. As in the case of the TR problem, our goal is to minimize false-positive inferences of additional components in the network.

Instance: A directed graph $G = (V, E)$ with an arc labeling function $\mathcal{L}\colon E \mapsto \{-1, 1\}$, and a subset $V_{\text{pseudo}} \subset V$ of nodes called pseudo-nodes. For convenience, the nodes in $V \setminus V_{\text{pseudo}}$ are called "real" nodes.

Definitions:

- For any node $v \in V$, let $\mathsf{in}(v) = \{(u, x) \mid u \overset{x}{\Rightarrow} v, x \in \{-1, 1\}\} \setminus \{(v, 0)\}$ and $\mathsf{out}(v) = \{(u, x) \mid v \overset{x}{\Rightarrow} u, x \in \{-1, 1\}\} \setminus \{(v, 0)\}$.
- Collapsing two nodes u and v is permissible, provided that both are not real nodes, $\mathsf{in}(u) = \mathsf{in}(v)$, and $\mathsf{out}(u) = \mathsf{out}(v)$.
- If permissible, the collapse of two nodes u and v creates a new node w, makes every incoming (respectively, outgoing) arc to (respectively, from) either u or v an incoming (respectively, outgoing) arc from w, removes all parallel arcs that may result from the collapse operation, and also removes both the nodes u and v.

Valid Solutions: A graph $G' = (V', E')$ obtained from G by a sequence of permissible pseudo-node collapse operations.

Objective: minimize $|V'|$.

FIGURE 5.11 The pseudo-node collapse (PNC) problem [5].

Unlike the TR problem, the PNC problem can be easily solved in polynomial time in the following manner as outlined in reference 5. It is not difficult to see that the "permissibility of collapse" relation is in fact an *equivalence relation* on the set of nodes. Thus, we can partition the nodes into equivalence classes such that two nodes u and v are in the same partition provided $\text{in}(u) = \text{in}(v)$ and $\text{out}(u) = \text{out}(v)$. It can be easily seen that if two such nodes u and v in the same partition are collapsed into a new node w, then the resulting equivalence partition is the same as before except that the two nodes u and v are replaced by a new node w in the same equivalence partition. Thus, an optimal solution would consist of collapsing all pseudo-nodes with one arbitrary real-node (if it exists) in each equivalence partition.

Other Network Reduction Rules. In addition to TR and PNC, depending on the particular application in mind, it is possible to have additional rules to minimize the synthesized network. For example, [5] formalized the following additional rule in relation to enzyme-catalyzed reactions specific to the context of synthesizing a consistent guard cell signal transduction network for ABA-induced stomatal closure of Li et al. [57]. Li et al. [57] represent each of these reactions by two mandatory arcs, one from the reaction substrate to the product and one from the enzyme to the product. As the reactants (substrates) of the reactions in reference 57 are abundant, the only way to regulate the product is by regulating the enzyme. The enzyme, being a catalyst, is always promoting the product's synthesis; thus positive double-causal regulation of a product was interpreted as positive regulation of the enzyme, and negative indirect regulation of the product was interpreted as negative regulation of the enzyme. In our graph-theoretic terms, this leads to the following rule. We have a subset $E_{\text{enz}} \subseteq E_{\text{m}}$ of arcs that are all labeled 1. Suppose that we have a path $A \stackrel{a}{\rightarrow} X \stackrel{b}{\rightarrow} B$ and an arc $C \stackrel{1}{\rightarrow} B \in E_{\text{enz}}$. Then, one can identify the node C with node X by collapsing them together to a node X_C, and set the labels of the arcs $A \rightarrow X_C$ and $X_C \rightarrow B$ based on the following cases:

- If $a\,b = 1$, then $\mathcal{L}(A, X_C) = \mathcal{L}(X_C, B) = 1$.
- If $a\,b = -1$, then $\mathcal{L}(A, X_C) = 1$ and $\mathcal{L}(X_C, B) = -1$.

Other Applications of TR ***and*** PNC In addition to the network synthesis procedure in Fig. 5.4, the TR problem can be very useful in many other types of analysis of biological networks. For example, Wagner [85] applied a very special case of TR that included calculations of reachabilities only (without the activation/inhibition information) to determine network structure from gene perturbation data, and Chen et al. [18] used a so-called ''excess edge deletion'' problem to identify structure of gene regulatory networks. In Section 5.3.4, we will explain how TR can be used to provide a measure of redundancy for biological networks.

Although the original motivation in Section 5.3.1 for introducing pseudo-nodes was to represent the intersection of the two paths corresponding to 3-node inferences, PNC can also be used in a broader context of network simplification. In many large-scale regulatory networks, only a subset of the nodes are of inherent interest (e.g., because they are differentially expressed in different exogenous conditions), and the rest serve as *backgrounds* or mediators. One can therefore designate nodes of less interest or confidence as pseudo-nodes and then collapse them, thereby making the network among high-interest/confidence nodes easier to interpret. Using this idea, PNC with TR can be used to focus on specific pathways in disease networks to better understand the molecular mechanism of the onset of the disease and therefore help in *drug target designs*. This approach was used by Kachalo et al. [43] to focus on pathways that involve the 33 known T-LGL deregulated proteins in a cell-survival/cell-death regulation-related signaling network synthesized from the TRANSPATH 6.0 database. LGLs are medium-to large-size cells with eccentric nuclei and abundant cytoplasm. LGL leukemia is a disordered clonal expansion of LGL, and their invasions in the marrow, spleen, and liver. Currently, there is no standard therapy for LGL leukemia, and thus understanding the mechanism of this disease is crucial for drug and therapy development. Ras is a small GTPase which is essential for controlling multiple essential signaling pathways, and its deregulation is frequently seen in human cancers. Activation of H-Ras requires its farnesylation, which can be blocked by Farnesyltransferase inhibitiors (FTIs). This envisions FTIs as future drug target for anti-cancer therapies, and several FTIs have entered early phase clinical trials. One of these FTI is tipifarnib, which shows apoptosis induction effect to leukemic LGL *in vitro*. This observation, together with the finding that Ras is constitutively activated in leukemic LGL cells, leads to the hypothesis that Ras plays an important role in LGL leukemia and may function through influencing Fas/FasL pathway. The approach used by Kachalo et al. [43] was to focus special interest on the effect of Ras on apoptosis response through Fas/FasL pathway by designating nodes that correspond to "proteins with no evidence of being changed during this effect" as pseudo-nodes and simplifying the network via iterations of PNC and TR. Although performing comprehensive PNC in this manner may lead to a drastic reduction of the network, a drawback of such a dramatic simplification is that pairs of *incoherent* arcs (i.e., two parallel arcs with opposite labels) can appear among pairs of nodes. While incoherent paths between pairs of nodes are often seen in biological regulatory networks, interpretation of incoherent arcs is difficult without knowledge of the mediators of the two opposite regulatory mechanisms. Thus, optimal simplification in this manner may require more careful selection of pseudo-nodes in PNC algorithms.

5.3.4 Redundancy and Degeneracy of Networks

The concepts of degeneracy and redundancy are well known in information theory. Loosely speaking, *degeneracy* refers to structurally different elements performing the same function, whereas *redundancy* refers to identical elements performing the same function. In electronic systems, such a measure is useful in analyzing properties such as fault tolerance. It is an accepted fact that biological networks do *not* necessarily have the lowest possible degeneracy or redundancy; for example, the connectivities of neurons in brains suggest a high degree of redundancy [49]. However, as has been observed by researchers such as Tononi, Sporns, and Edelman [82], specific notions of degeneracy and redundancy have yet to be firmly incorporated into biological thinking, largely because of the lack of a formal theoretical framework. A further reason for the lack of incorporation of these notions in biological thinking is the lack of *efficient* algorithmic procedures for computing these measures for large-scale networks even when formal definitions are available. Therefore, studies of degeneracy and redundancy for biological networks are often done in a somewhat ad-hoc fashion [69]. There do exist notions of "redundancy" for *undirected* graphs based on clustering coefficients [12] or betweenness centrality measures [24]. However, such notions may not be appropriate for the analysis of biological networks where one must distinguish positive from negative regulatory interactions and where the study of dynamics of the network is of interest.

Information-Theoretic Degeneracy and Redundancy Measures. Formal information-theoretic definitions of degeneracy and redundancy for biological systems were proposed in references 80–82 based on the so-called *mutual-information* (MI) *content*. These definitions assume access to suitable perturbation experiments and corresponding accurate measurements of the relevant parameters. Consider a system consisting of n elements that produces a set of outputs $\mathcal{O}$ via a fixed connectivity matrix from a subset of these elements (see Fig. 5.12 for an illustration). The elements are described by a jointly distributed random vector X that represents steady-state activities of the components of the system. The degeneracy $\mathcal{D}(X\,;\,\mathcal{O})$ of the system is then expressed as the average mutual information (MI) shared between $\mathcal{O}$ and the "perturbed" bipartitions of X summed over all bipartition

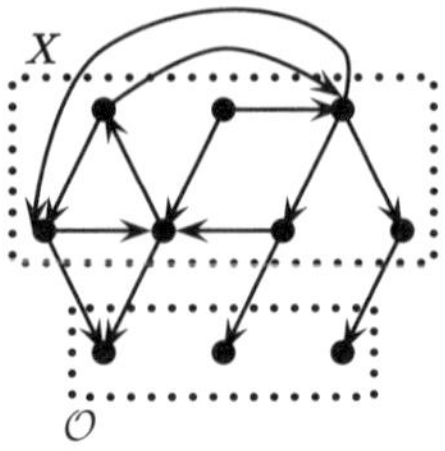

FIGURE 5.12 A system of seven elements.

sizes—that is, by the following equation:

$$\mathcal{D}(X\,;\,\mathcal{O}) \propto \sum_{k=1}^{n}\sum_{j}\left(\mathrm{MI}^P\left(X_j^k\,;\,\mathcal{O}\right) + \mathrm{MI}^P\left(X \setminus X_j^k\,;\,\mathcal{O}\right) - \mathrm{MI}^P\left(X\,;\,\mathcal{O}\right)\right).$$

In the above equation, X_j^k is a jth subset of X composed of k elements, and $\mathrm{MI}^P(\mathcal{A}\,;\,\mathcal{O})$ is the mutual information between a subset of elements $\mathcal{A}$ and an output set $\mathcal{O}$ when $\mathcal{A}$ is injected with a small fixed amount of uncorrelated noise. $\mathrm{MI}^P(\mathcal{A}\,;\,\mathcal{O})$ is given by the equation

$$\mathrm{MI}^P(\mathcal{A}\,;\,\mathcal{O}) = \mathcal{H}(\mathcal{A}) + \mathcal{H}(\mathcal{O}) - \mathcal{H}(\mathcal{A},\mathcal{O}),$$

where $\mathcal{H}(\mathcal{A})$ and $\mathcal{H}(\mathcal{O})$ are the entropies of $\mathcal{A}$ and $\mathcal{O}$ considered independently, and $\mathcal{H}(\mathcal{A},\mathcal{O})$ is the joint entropy of the subset of elements $\mathcal{A}$ and the output set $\mathcal{O}$. The above definition of the degeneracy measure is mathematically precise, but a significant computational difficulty in applying such a definition is that the number of possible bipartitions could be astronomically large even for a modest size network. For example, for a network with 50 nodes, the number of bipartitions is roughly $2^{50} > 10^{15}$. Measures avoiding averaging over all bipartitions were also proposed in reference 82, but the computational complexities and accuracies of these measures still remain to be thoroughly investigated and evaluated on larger networks.

Similarly, the redundancy $\mathcal{R}(X;\mathcal{O})$ of a system X can be defined as the difference between (a) the summed mutual information upon perturbation between all subsets of size up to 1 and $\mathcal{O}$ and (b) the mutual information between the entire system and $\mathcal{O}$, that is,

$$\mathcal{R}(X\,;\,\mathcal{O}) \propto \sum_{j=1}^{n}\mathrm{MI}^P\left(X_j^1\,;\,\mathcal{O}\right) - \mathrm{MI}^P\left(X\,;\,\mathcal{O}\right)$$

A shortcoming of this redundancy measure is that it only provides a number, but does not indicate which subset of elements are redundant. Identifying redundant elements is important for the interpretation of results and may also serve as an important step of the network construction and refinement process.

Topological Redundancy Measure Based on Transitive Reduction. Any good topological redundancy measure of degeneracy or redundancy should have a few desirable properties:

($\mathcal{P}$1) The measure must not just reflect simple connectivity properties such as degree-sequence or average degree, but also should incorporate *higher-order* connectivity properties.

($\mathcal{P}$2) The measure should not just provide a number, but also should indicate candidate subsets of components that are redundant or degenerate.

($\mathcal{P}$3) The measure must be efficiently computable so that it can be computed for large-scale networks.

Based on the TR problem, Albert et al. [3] proposed a new topological measure of redundancy that is amenable to efficient algorithmic analysis. Note that any solution of TR does not change pathway level information of the network since it removes only those arcs from one node to another when similar alternate pathways exist, thus truly removing redundant connections. Thus, $|E'|/|E|$ provides a measure of global compressibility of the network, and a topological redundancy measure $\mathcal{R}_{T_R}$ can be defined as

$$\mathcal{R}_{T_R} = 1 - \frac{|E'|}{|E|},$$

where the $|E|$ term in the denominator of the definition is just a "min–max normalization" of the measure [36] to ensure that $0 < \mathcal{R}_{T_R} < 1$. Note that the higher the value of $\mathcal{R}_{T_R}$, the more redundant the network. The TR problem used in computing $\mathcal{R}_{T_R}$ actually finds a subset of redundant arcs and, in the case of multiple minimal networks of similar quality, can also find multiple distinct subsets of redundant arcs by randomization of the greedy selection step in the algorithm in Fig. 5.9. $\mathcal{R}_{T_R}$ also satisfies Property ($\mathcal{P}$1) since paths of *arbitrary* length are considered for removal of an arc and thus, for example, $\mathcal{R}_{T_R}$ is *not* necessarily correlated to either the degree sequence (cf. Exercise 5.6a) or the average degree (cf. Exercise 5.6b) of the network.

Based on the evaluations of the above redundancy measure on seven biological networks of various types, Albert et al. [3] provided a few interesting biological and computational conclusions such as the following:

- $\mathcal{R}_{T_R}$ can be computed quickly for large networks and is statistically significant.
- Transcriptional networks are less redundant than signalling networks.
- Topological redundancy of the *C. elegans* metabolic network is largely due to its inclusion of *currency metabolites* such as ATP and ADP.

FIGURE 5.13 Equivalence of dynamical properties may depend on node functions.

- Calculations of $\mathcal{R}_{\mathrm{T_R}}$ and corresponding minimal networks provide insight into a predicted orientation of protein–protein interaction networks by determining whether the predicted oriented network has a level of redundancy similar to those in known related biological networks.

Correlation Between $\mathcal{R}_{\mathrm{T_R}}$ and Network Dynamics It is of interest to determine if a topologically minimal network has similar dynamical or functional properties as the original network such as stability and response to external inputs, when such properties are available for the original network. When the network has designated outputs or read-outs, such as gene expression rates in transcriptional networks, it may be of interest to characterize the behavior of these outputs as a function of the inputs. A topologically minimal network such as the one used in this section does have the same input–output connectivity as the original network, and thus the excitory or inhibitory influence between each input–output pair is preserved. Such a reduced network is minimal in an information-theoretic sense: *Any network with the same input–output behavior must be of at least this size.*

However, one may ask *if a topologically minimal network also has a ''similar'' output behavior as the original one for the same input*? For some dynamical properties, the question can be answered; for example, a correlation of $\mathcal{R}_{\mathrm{T_R}}$ with the monotonicity of dynamics is explored in Section 6.5. In general, however, such a question does not have a well-defined answer since the dynamics depend on what type of functions are used to combine incoming connections to nodes and the ''time delay'' in the signal propagation, both of which are omitted in the graph-theoretic representation of signal transduction networks. Therefore, deleting redundant arcs may result in functionalities that *may or may not be the same*. For example, consider the two networks shown in Fig. 5.13 in which network **(ii)** has a redundant connection a → c. The functions of these two circuits could be different, however, depending on the function used to combine the inputs a → c and b → c in network **(ii)** (cf. Exercise 5.7). However, despite the fact that a minimal network may not preserve *all* dynamic properties of the original one, a significant application of finding minimal networks lies precisely in allowing one to identify redundant connections (arcs). In this application, one may focus on investigating the functionalities of these redundant arcs—for example, identifying the manner in which their effect is cumulated with those of the other regulators of their target nodes could be a key step toward understanding the behavior of the entire network. Thus, the measure developed here is of general interest as it not only provides a quantified measure of overall redundancy of the network, but also allows identification of redundancies and hence helps direct future research toward the understanding of the functional significance of the redundant arcs.

5.3.5 Random Interaction Networks and Statistical Evaluations

A comprehensive statistical evaluation of any network measure such as the redundancy value $\mathfrak{R}_{T_R}$ discussed in the previous section requires the following ingredients:

(𝔞) generation of an ensemble of random networks (the "null hypothesis model") on which the measure can be computed,

(𝔟) if necessary, appropriate normalization of values of the measure on random networks to correct statistical bias, and

(𝔠) checking appropriate statistical correlation ("null hypothesis testing") between these random network measures with the measure on the given network.

Null Hypothesis Models. Ideally, if possible, it is preferable to use an *accurate generative* null model for highest accuracy, since such a model may be amenable to more rigorous mathematical analysis [68]. For signaling and transcriptional biological networks Albert et al. [5], based on extensive literature review of similar kind of biological networks in prior literature, arrived at the characteristics of a generative null model that is described below. One of the most frequently reported topological characteristics of biological networks is the distribution of in-degrees and out-degrees of nodes, which may be close to a power law or a mixture of a power law and an exponential distribution [2,33,55]. Specifically, in biological applications, metabolic and protein interaction networks are heterogeneous in terms of node degrees and exhibit a degree distribution that is a mixture of a power law and an exponential distribution [2,33,41,55,58], whereas transcriptional regulatory networks exhibit a power-law out-degree distribution and an exponential in-degree distribution [56,74]. Thus, usually a researcher is expected to use his or her judgement to select an appropriate degree distribution and other necessary parameters that is consistent with the findings in prior literature for an accurate generative null model. Based on the known topological characterizations, Albert et al. [5] arrived at the following degree distributions for generating random transcriptional and signaling networks:

- The distribution of in-degree is truncated *exponential*, namely, Pr[in-degree= x]= $c_1\, e^{-c_1 x}$ with $1/2 < c_1 < 1/3$ and $1 \leq x \leq 12$.
- The distribution of out-degree is governed by a truncated *power-law*, namely, for some constant $2 < c_2 < 3$:
  - Pr[out-degree= 0] $\geq c_2$, and
  - for $1 \leq x \leq 200$, Pr[out-degree=x] = $c_2\, x^{-c_2}$.
- Parameters in the distribution are adjusted to make the sum of in-degrees of all nodes equal to the sum of out-degrees of all nodes, and the expected number of arcs in the random network is the same as that in G.

- The percentages for activation, inhibition, and mandatory edges in the random network are the same as in G and are distributed over the arcs randomly.

Several methods are known in the literature to generate random directed graphs with specific degree distributions. For example, Newman, Strogatz, and Watts [68] suggest the following method. Suppose that the parameters of the degree distributions are appropriately adjusted such that the averages of the in-degree distribution and the out-degree distribution are almost the same. Let $v_1, v_2, \ldots, v_n$ be the set of n nodes with in-degrees $d_1^{\text{in}}, d_2^{\text{in}}, \ldots, d_n^{\text{in}}$ and out-degrees $d_1^{\text{out}}, d_2^{\text{out}}, \ldots, d_n^{\text{out}}$, respectively. Then, randomly pick two nodes $v_i \neq v_j$ such that $d_i^{\text{in}}, d_j^{\text{out}} > 0$, add the arc (v_j, v_i), and decrease both d_i^{in} and d_j^{out} by 1. Repeat the procedure until the in-degree and out-degree of every node are zero. See reference 66 for some other methods.

For the case when generative null models are not possible, alternate methods are available to generate random networks that preserve the first-order topological characteristics (such as the degree sequence) but randomize higher-order statistics (such as distribution of paths). We review two such alternate methods below.

The Markov-chain algorithm for generating random networks [44] starts with the given network $G = (V, E)$ and repeatedly swaps randomly chosen similar pairs of connections as illustrated in Fig. 5.14. This kind of algorithm was used in papers such as references 3 and 74. The percentage of arcs swapped depends on the application; for example, Shen-Orr et al. [74] considered swapping about 25% of the arcs.

Newman and others in several publications [34,54,64,65,67] have suggested using the following null model $\mathcal{G}$ generated from the degree-distribution of the given graph $G = (V, E)$. $\mathcal{G}$ is a random graph with the same set of nodes as the given graph G. Every possible directed edge (u, v) in $\mathcal{G}$ is selected with a probability of $p_{u,v} = \frac{d_u^{\text{out}} d_v^{\text{in}}}{|E|}$, where d_u^{out} and d_v^{in} are the out-degree of node u and the in-degree of node v, respectively, in G. The original definition also allows selection of self-loops,

repeat
  choose two arcs $a \xrightarrow{x} b, c \xrightarrow{y} d \in E$ randomly
    $(x, y \in \{-1, 1\})$
  if $x = y$ and $a \neq c$ and $b \neq d$
    and $a \xrightarrow{x} d \notin E$ and $c \xrightarrow{y} b \notin E$
   then
    add the arcs $a \xrightarrow{x} d$ and $c \xrightarrow{y} b$
    remove the arcs $a \xrightarrow{x} b$ and $c \xrightarrow{y} d$
  endif
until a specified percentage of arcs of G
  has been swapped

(a)

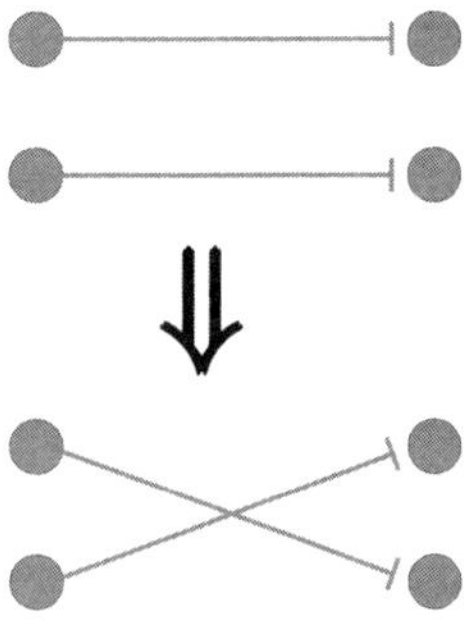

(b)

FIGURE 5.14 **(a)** The Markov-chain algorithm for generating random networks by arc swapping. **(b)** A pictorial illustration of arc swapping.

that is, edges (u, v) with $u = v$; in that case this null model preserves *in expectation* the distribution of the degrees of each node in the given graph G, that is,

$$\begin{aligned}\textstyle\sum_{v \in V} p_{u,v} &= d_u^{\text{out}}, \\ \textstyle\sum_{v \in V} p_{v,u} &= d_u^{\text{in}}.\end{aligned} \tag{5.2}$$

This null model is very popular in many biological applications [34,35,71].

Correction of Bias of Empirical Null Model Attributes. This is a very standard task in statistical data mining, and the reader is referred to suitable textbooks such as reference 53. We illustrate one such method for the correction of bias in computation of $\mathcal{R}_{\mathrm{T_R}}$ for random networks. Suppose that we generated p random networks with redundancy values $\mathcal{R}_{\mathrm{T_R}}^1, \mathcal{R}_{\mathrm{T_R}}^2, \ldots, \mathcal{R}_{\mathrm{T_R}}^p$ and let μ and σ be the mean and standard deviation of these p values. We can first compute the standardized value $\dfrac{\mathcal{R}_{\mathrm{T_R}}^j - \mu}{\sigma}$ for the observed value $\mathcal{R}_{\mathrm{T_R}}^j$. Then, we can calculate the standardized range (difference between maximum and minimum) of these standardized values and normalize the standardized values by dividing them by this standardized range.

Null Hypothesis Testing. Given the p random measures, say $r_1, r_2, \ldots, r_p$, and the value of measure for the given network, say r, this step essentially involves determining the probability that r can be generated by a distribution that fits the data points $r_1, r_2, \ldots, r_p$. There are a variety of standard statistical tests (e.g., one-sample student's t-test) that can be used for this purpose.

5.4 REVERSE ENGINEERING OF BIOLOGICAL NETWORKS

In previous sections we discussed how signal transduction networks can be synthesized based on available interaction data. However, in many situations, such interaction data are unavailable. Instead, for many biological systems, data are available about some *characteristics* of the system. Informally, the ''reverse engineering'' problem for biological networks is to unravel the interactions among the components of the network based on such observable data about the characteristics of the system, and it may be difficult to approach by means of standard statistical and machine learning approaches such as clustering into co-expression patterns. Information on direct functional interactions throws light upon the possible mechanisms and architecture underlying the observed behavior of complex molecular networks, but an intrinsic difficulty in capturing such interactions in cells by traditional genetic experiments, RNA interference, hormones, growth factors, or pharmacological interventions is that any perturbation to a particular gene or signaling component may rapidly propagate throughout the network, thus causing global changes which cannot be easily distinguished from direct (local) effects. Thus, a central goal in reverse engineering problems in biology is to use the observed global responses (such as steady-state changes in concentrations of activated activities of proteins, mRNA levels, or transcription rates) in order to infer the local interactions between individual nodes. This section focuses on several computational issues in reverse engineering of a biological system and, quoting reference 23, can be very broadly described as *the problem of analyzing a given system in order to identify, from biological data, the components of the system and their relationships*.

In general, application of a specific reverse engineering method depends upon several factors such as:

- (Q1) The type of interactions to be reconstructed—for example, statistical correlations, causal relationships, and so on.
- (Q2) The type of data or experiments that the modeler has access to—for example, input–output behavior on perturbations, gene expression measurements, and so on.
- (Q3) The quality of the expected output of the method—for example, the nature (excitory or inhibitory) of the interaction versus the actual strengths of the interactions.

For example, in references 11, 26, 32, 63, 70, 72, 86 and 88, interactions represent statistical correlation between variables whereas in references 23, 30, 38, 39, and 52, interactions represent causal relationships among nodes. Depending upon the type of network analyzed, quality and availability of data, network size, and so forth, different reverse engineering methods offer different advantages and disadvantages relative to each other.

In the next two sections we discuss two reverse engineering approaches. The first one is the so-called *modular response analysis* approach that relies on availability of perturbation experiments. The second approach is a collection of parsimonious *combinatorial* approaches that rely on the availability of time-varying measurements of relevant parameters of the system.

5.4.1 Modular Response Analysis Approach

The *modular response analysis* (MRA) approach for reverse engineering of networks was originally introduced in references 47 and 48 and further elaborated upon in references 8, 21, 78 and 79. In this approach, the architecture of the network is inferred on the basis of observed global responses (namely, the steady-state concentrations in changes in the phosphorylation states or activities of proteins, mRNA levels, or transcription rates) in response to experimental perturbations (representing the effect of hormones, growth factors, neurotransmitters, or pharmacological interventions). The MRA technique was employed in reference 73 in order to discover positive and negative feedback effects in the Raf/Mek/Erk MAPK network in rat adrenal pheochromocytoma (PC-12) cells to uncover connectivity differences depending on whether the cells are stimulated by epidermal growth factor (EGF) or by neuronal growth factor (NGF), with perturbations consisting of downregulating protein levels

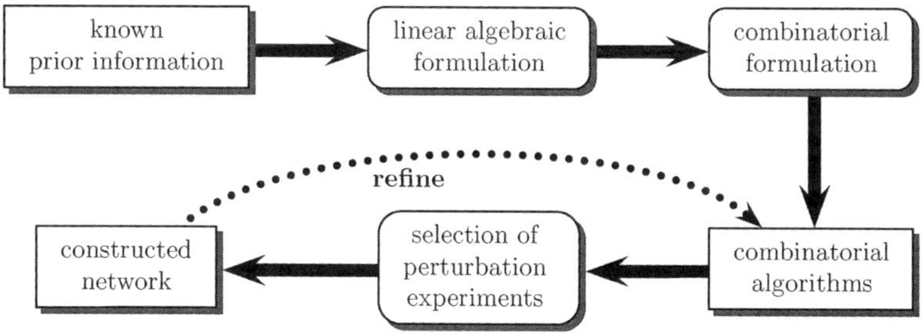

FIGURE 5.15 A schematic diagram for the overview of the MRA approach.

by means of RNAi. A schematic diagram for the MRA approach is shown in Fig. 5.15. Before describing the methodology, we first describe the ingredients of the method.

The Model (Q1). The model of the biological system considered in the MRA approach is the differential equation model described in Section 5.1.2 via Eq. (5.1). The n state variables $x_1(t), x_2(t), \ldots, x_n(t)$ of this dynamical system, collected into a time-dependent vector $\mathbf{x}(t) = \left(x_1(t), \ldots, x_n(t)\right)$, represent quantities that can be *measured*, such as the levels of activity of selected proteins or transcription rates of certain genes. The parameters (inputs) u_i's, collected into a vector $\mathbf{u} = \left(u_1, \ldots, u_m\right)$, represent quantities that can be perturbed, perhaps indirectly, such as total levels of proteins whose half-lives are long compared to the rate at which the variables evolve, but, once changed, they remain constant for the duration of the biological experiment. A basic assumption in this approach is that states converge to steady-state values, and these are the values used for network reconstruction (but see reference 78 for a time-dependent analysis). There is also a reference value $\bar{\mathbf{u}}$ of $\mathbf{u}$, which represents "wild-type" (i.e., normal) conditions, and a corresponding steady state $\bar{x}$; mathematically, $\mathbf{f}\left(\bar{\mathbf{x}}, \bar{\mathbf{u}}\right) = 0$. A mild technical assumption that is necessary is that, for each vector of parameters $\mathbf{u}$ in a neighborhood of $\bar{\mathbf{u}}$, there is a unique steady state $\xi(\mathbf{u})$ of the system where ξ is a differentiable function.

Required Biological Experiments (Q2). The required experimental protocol allows one to perturb *any one* of the parameters, say u_j, while leaving the remaining ones constant. For the perturbed vector $\mathbf{u} \approx \bar{\mathbf{u}}$, measurements are done of the perturbed steady-state vector $\mathbf{x} = \xi(\mathbf{u})$, which is an unique function of $\mathbf{u}$. This read-out might be done through methods such as Western blots or DNA microarrays. When the parameter u_j is perturbed, we compute the n "sensitivities"

$$b_{i,j} = \frac{\partial \xi_i}{\partial u_j}\left(\bar{\mathbf{u}}\right) \approx \frac{\left(\, \xi_i(\bar{\mathbf{u}} + u_j\, e_j) - \xi_i(\bar{\mathbf{u}}) \,\right)}{\bar{u}_j - u_j} \qquad \text{for } i = 1, 2, \ldots, n,$$

where $e_j \in \mathbb{R}^m$ is the jth canonical basis vector. We arrange these numbers into an $n \times m$ matrix $B = \left(b_{i,j}\right)$. This approach makes a *general position assumption* that all subsets of n columns of B are linearly independent; this entails *no* theoretical loss of generality since the entries of B correspond to experimental data; however, in actual implementations this may lead to numerical instabilities (see reference 78 for an analysis of numerical aspects as well as the effect of errors and noise).

Quality of Output of the Approach (Q3). This approach is expected to obtain information regarding the *signs* and *relative magnitudes* of the partial derivative $\frac{\partial f_i}{\partial x_j}(\bar{\mathbf{x}}, \bar{\mathbf{u}})$, which quantifies the direct effect of a variable x_j upon another variable x_i; for example, $\frac{\partial f_i}{\partial x_j} > 0$ means that x_j has a catalytic (excitory) effect upon the rate of formation of x_i, while $\frac{\partial f_i}{\partial x_j} < 0$ indicates inhibition.

Known Prior Information. The critical assumption is that, while one may not know the algebraic form of the vector field $\mathbf{f}$ in system (5.1), often it is known which parameters p_j directly affect which variables x_i. For example, x_i may be the level of activity of a particular protein and p_j might be the total amount (active plus inactive) of that protein in a cell. This prior information is summarized by a *binary* matrix $C^0 = \left(c^0_{ij}\right) \in \{0,1\}^{n \times m}$, where "$c^0_{i,j} = 0$" means that p_j does *not* appear in the equation for $\frac{dx_i(t)}{dt}$, that is, $\frac{\partial f_i}{\partial p_j} \equiv 0$.

Let $A = \frac{\partial \mathbf{f}}{\partial \mathbf{x}}$ be the Jacobian matrix with respect to state variables, and let C be the *negative* of $\frac{\partial \mathbf{f}}{\partial \mathbf{u}}$, the Jacobian matrix with respect to the parameters. Since $\mathbf{f}(\xi(\mathbf{u}), \mathbf{u})$ is identically zero, we may take derivatives with respect to $\mathbf{u}$ and use the chain rule of derivatives to obtain that $C = AB$.

The experimental design question we need to address is as follows. We wish to obtain *as much information as possible* about the matrix A. However, each parameter perturbation experiment involves a *cost* (resources, experimental difficulties, etc.), which we would like to *minimize*. We can think of these experiments as "queries" that return a column B_i of B if the ith parameter u_i is perturbed. Observe that the matrix C^0 tells us which rows of A have zero inner product with which B_i. This leads us to the linear algebraic question shown in Fig. 5.16.

The question is: *Exactly how much information about A we can obtain in the above formulation*? Notice that there are always *intrinsic* limits to what can be accomplished: If we multiply each row of A by some nonzero number, then the zero structure of C is unchanged. To help the reader appreciate this question, we discuss the following concrete example from reference 14. Consider the following instance of the linear algebraic formulation:

$$C^0 = \begin{pmatrix} 0 & 1 & 0 & 1 & 1 \\ 1 & 1 & 1 & 0 & 0 \\ 0 & 0 & 1 & 0 & 1 \end{pmatrix}, \quad A = \begin{pmatrix} -1 & 1 & 3 \\ 2 & -1 & 4 \\ 0 & 0 & -1 \end{pmatrix}, \quad B = \begin{pmatrix} 4 & 3 & 37 & 1 & 10 \\ 4 & 5 & 52 & 2 & 16 \\ 0 & 0 & -5 & 0 & -1 \end{pmatrix}$$

Instance: two matrices $A \in \mathbb{R}^{n\times n}$ and $B \in \mathbb{R}^{n\times m}$ such that

- A is unknown;
- B is *initially* unknown but each of its m columns, denoted as $B_1, B_2, \ldots, B_m$, can be retrieved with a *unit-cost query*,
- the columns of B are in general position, and
- the *zero structure* of the matrix $C = AB = (c_{i,j})$ is known, *i.e.*, a binary matrix $C^0 = (c^0_{i,j}) \in \{0,1\}^{n\times m}$ is given such that $c_{i,j} = 0$ for each i, j for which $c^0_{i,j} = 0$.

Goal: obtain as much information as possible about A while performing as few queries as possible.

FIGURE 5.16 Linear algebraic formulation of the experimental design question for the MRA approach.

and suppose that we perform four queries corresponding to the columns 1, 3, 4, and 5 of B to obtain the following data:

$$\begin{pmatrix} 4 & 37 & 1 & 10 \\ 4 & 52 & 2 & 16 \\ 0 & -5 & 0 & -1 \end{pmatrix}. \tag{5.3}$$

Let us first attempt to identify the first row A_1 of A. The first row of the matrix C_0 tells us that the vector A_1 is orthogonal to the first and second columns of (5.3) (which are the same as the first and third columns of B, respectively). This is the *only* information about A that we have available, and it is *not* enough information to *uniquely* determine A_1, because there is an entire line that is orthogonal to the plane spanned by these two columns. However, we can still find *some* nonzero vector in this line and conclude that A_1 is an unknown multiple of this vector. This nonzero vector may be obtained by simple linear algebra manipulations (cf. Exercise 5.14). Similarly, we may use the last two columns of (5.3) to estimate the second row of A, again only up to a multiplication by a constant, and the first and third columns of (5.3) to estimate the third row of A. Thus, as in the example, the best that we can hope for is to identify the rows of A up to scalings, or in other words the signs of the entries of A, which is precisely what the MRA approach promises.

At this point, two important questions remain to be answered:

(a) How much prior information via the matrix C^0 (i.e., how many zero entries in C^0) is needed to determine A, exactly or almost exactly, up to scalings?

($\mathscr{b}$) Assuming we have sufficient prior information as suggested by an answer to ($\mathscr{a}$), how do we find a minimal or near-minimal set of perturbation experiments to determine A?

To assist us in answering these questions, we reformulate the linear algebraic version of the problem in Fig. 5.16 in the following manner. Let A_i denote the ith row of A. Then the specification of C^0 amounts to the specification of *orthogonality relations* $A_i \cdot B_j = 0$ for each pair i and j for which $c^0_{i,j} = 0$. Thus, if we query the columns of B indexed by $\mathcal{J} = \{j_1, \ldots, j_\ell\} \subseteq \{1, 2, \ldots, m\}$, then letting $\mathcal{J}_i = \{j \mid j \in \mathcal{J} \text{ and } c^0_{i,j} = 0\}$, we get

$$A_i \in \mathcal{H}^{\perp}_{\mathcal{J},i},$$
$$\mathcal{H}_{\mathcal{J},i} = \text{span}\{B_j : j \in \mathcal{J}_i\},$$

where "$\perp$" indicates the *orthogonal complement* operation. Now if the set of indices $\mathcal{J}$ has the property

$$\forall i = 1, 2, \ldots, n\colon\ |\mathcal{J}_i| \geq n - k \tag{5.4}$$

for some given integer k, then, via the general position assumption, the space $\mathcal{H}_{\mathcal{J},i}$ has dimension of *at least* $n - k$, and hence the space $\mathcal{H}^{\perp}_{\mathcal{J},i}$ has a dimension of *at most* k. We now discuss the answers to questions ($\mathscr{a}$) and ($\mathscr{b}$) for various values of k.

The Case of $k = 1$. This is the most desirable special case since the dimension of $\mathcal{H}^{\perp}_{\mathcal{J},i}$ being at most 1, implies that each A_i is *uniquely* determined up to a scalar multiple, providing the best possible answer to question ($\mathscr{a}$). Assuming that the degenerate case $\mathcal{H}^{\perp}_{\mathcal{J},i} = \{0\}$ does not hold (which would determine $A_i = 0$), once an arbitrary nonzero element v in the line $\mathcal{H}^{\perp}_{\mathcal{J},i}$ has been picked, there are only two sign patterns possible for A_i (the pattern of v and that of $-v$). If, in addition, one knows at least one nonzero sign in A_i, then the sign structure of the whole row has been *uniquely* determined. For the MRA setup, typically one such sign is indeed known; for example, the ith element of each A_i is known to be negative as it represents a degradation rate. Thus, to settle question ($\mathscr{b}$), we need to solve the following combinatorial problem:

$$\begin{aligned} &\text{find } \mathcal{J} \subseteq \{1, 2, \ldots, m\} \text{ such that} \\ &|\mathcal{J}| \text{ is minimum} \\ &\forall i = 1, 2, \ldots, n\colon\ |\mathcal{J}_i| \geq n - 1 \end{aligned} \tag{5.5}$$

If queries have *variable* costs (i.e., different experiments have a different associated cost), this problem must be modified to that of minimizing a suitable linear combination of costs, instead of the number of queries.

***The More General Case of* $k > 1$.** More generally, suppose that the queries that we performed satisfy the constraint $|\mathcal{J}_i| \geq n - k$ with $k > 1$ but still small k. It is *no*

Problem name:	$\mathbf{SC}_\gamma$
Instance $<n, m, \gamma>$:	universe $U = \{u_1, u_2, \ldots, u_n\}$
	sets $S_1, S_2, \ldots, S_m \subseteq U$ with $\cup_{j=1}^{m} S_j = U$
	positive integer ("coverage factor") γ
Valid Solution:	a subset of indices $I \subseteq \{1, 2, \ldots, m\}$ such that
	$\forall u_j \in U: \left\| i \in I \; : \; u_j \in S_i \right\| \geq \gamma$
Objective:	minimize $\|I\|$

FIGURE 5.17 A combinatorially equivalent reformulation of (5.6).

longer true that there are only two possible sign patterns for any given A_i. However, Berman et al. [14–16] show that the number of possibilities is still small. We refer the reader to reference 14–16 for precise bounds that answers question ($\mathfrak{a}$). To answer question ($\mathfrak{b}$), we need to solve a more general version of Problem (5.5), namely the following:

$$\begin{aligned} &\text{find } \mathcal{J} \subseteq \{1, 2, \ldots, m\} \text{ such that} \\ &\quad |\mathcal{J}| \text{ is minimum} \\ &\quad \forall\, i = 1, 2, \ldots, n \colon\; |\mathcal{J}_i| \geq n - k. \end{aligned} \tag{5.6}$$

Algorithmic aspects of the Mra ***approach*** It is possible to show by a simple transformation that (5.6) can be written down in the form of the combinatorial problem in Fig. 5.17 (cf. Exercise 5.15). Problem $\mathbf{SC}_\gamma$ is in fact the (unweighted) *set multicover* problem well known in the combinatorial algorithms community [84]; $\mathbf{SC}_1$ is simply called the *set cover* problem. Let $\alpha = \max_{i \in \{1,2,\ldots,m\}} \{|S_i|\}$ denote the maximum number of elements in any set. Two well-known algorithms for solving $\mathbf{SC}_1$ are as follows:

- The greedy approach shown in Fig. 5.18a that repeatedly selects a new set that covers a maximum number of ''not yet covered'' elements. This algorithm is known to have an approximation ratio of $(1 + \ln \alpha)$ [42,84].
- The randomized algorithm shown in Fig. 5.18b is a Δ-approximation with high probability where $\mathbb{E}[\Delta] = O(\log n)$. The first step in the algorithm is to solve a linear program (LP) that is a relaxation of an integer linear program (ILP) for $\mathbf{SC}_1$. The best way to understand the LP formulation is to imagine as if each variable x_j is binary (i.e., takes a value of 0 or 1) and interpret the 1 and the 0 value as the set S_j being selected or not. There are efficient solutions for any LP problem [45,60]. However, a solution of the LP may provide non-binary real fractional values for some variables, and the remaining steps are ''randomized rounding'' steps that transform these real fractional values to binary values (0 or 1).

It is also known that the above approximation bounds for $\mathbf{SC}_1$ are in fact the *best possible* theoretical guarantees for $\mathbf{SC}_1$ [28,83].

```
(* comment: greedy approach *)

I = ∅, uncovered = U
while uncovered ≠ ∅ do
    select an index j ∈ {1,2,...,m} \ I
      that maximizes |uncovered ∩ S_j|
    uncovered = uncovered \ S_j
    I = I ∪ {j}
endwhile
```

(a)

```
(* comment: randomized approach *)

solve the following linear program (LP)
    minimize Σ_{j=1}^{m} x_j
    subject to ∀ u_i ∈ U: Σ_{j: u_i ∈ S_j} x_j ≥ 1
                        ∀ j ∈ {1,2,...,m}: x_j ≥ 0
let the solution vector be (x_1^*, x_2^*, ..., x_n^*)
I = ∅
repeat  2 ln n  times
    for i = 1,2,...,n do
        if i ∉ I then
            put i in I with probability x_i^*
    endfor
endrepeat
```

(b)

FIGURE 5.18 Two well-known algorithms to solve $\mathbf{SC}_1$ [84].

The greedy approach shown in Fig. 5.18a can be easily generalized for $\gamma > 1$ by selecting at each iteration a new set that covers the maximum number of those elements that has not been covered at least γ times yet; the resulting algorithm still has an approximation ratio of $(1 + \ln \alpha)$. An improved randomized approximation algorithm for $\mathbf{SC}_\gamma$ when $\gamma > 1$ was provided by Berman et al. [14] and is shown in Fig. 5.19. This algorithm has an approximation ratio of Δ where

$$\mathbb{E}[\Delta] \leq \begin{cases} 1 + \ln \alpha & \text{if } \gamma = 1 \\ \left(1 + e^{-\frac{\gamma-1}{5}}\right) \ln\left(\frac{\alpha}{\gamma-1}\right) & \text{if } \frac{\alpha}{\gamma-1} \geq e^2 \text{ and } \gamma > 1 \\ \min\left\{2 + 2\,e^{-\frac{\gamma-1)}{5}},\, 2 + \left(e^{-2} + e^{-9/8}\right)\frac{\alpha}{\gamma}\right\} & \text{if } \frac{1}{4} < \frac{\alpha}{\gamma-1} < e^2 \text{ and } \gamma > 1 \\ 1 + 2\sqrt{\alpha/\gamma} & \text{if } \frac{\alpha}{\gamma-1} \leq 1/4 \text{ and } \gamma > 1. \end{cases}$$

Note that if $k = 1$, then $\gamma = n - 1 \gg \alpha$ and consequently $\Delta \approx 1$; thus, in this case the algorithm in Fig. 5.19 returns an *almost* optimal solution *in expectation*.

5.4.2 Parsimonious Combinatorial Approaches

In this section, we discuss a few parsimonious combinatorial approaches for reverse engineering causal relationships between components of a biological system. These methods make use of time-varying measurements of relevant variables (e.g., gene expression levels) of the system. The causal relationships reconstructed can be synthesized to a Boolean circuit, if needed.

select a positive constant $c \geq 1$ as follows:

$$c = \begin{cases} \ln \alpha, & \text{if } \gamma = 1 \\ \ln\left(\frac{\alpha}{\gamma-1}\right), & \text{if } \frac{\alpha}{\gamma-1} \geq e^2 \text{ and } \gamma > 1 \\ 2, & \text{if } \frac{1}{4} < \frac{\alpha}{\gamma-1} < e^2 \text{ and } \gamma > 1 \\ 1 + \sqrt{\alpha/\gamma}, & \text{otherwise} \end{cases}$$

solve the following LP

$$\begin{aligned} &\text{minimize } \textstyle\sum_{j=1}^{m} x_j \\ &\text{subject to } \forall u_i \in U: \textstyle\sum_{j:\, u_i \in S_j} x_j \geq \gamma \\ &\qquad \forall j \in \{1, 2, \ldots, m\}: x_j \geq 0 \end{aligned}$$

let the solution vector be $(x_1^*, x_2^*, \ldots, x_n^*)$

form the indices of a family of sets $I_0 = \{j : cx_j^* \geq 1\}$

form the indices of a family of sets $I_1 \subseteq \{1, 2, \ldots, m\} \setminus I_0$ by selecting a set S_j for $j \in \{1, 2, \ldots, m\} \setminus I_0$ with probability cx_j^*

form the indices of a family of sets I_2 by greedy choices:
if an $u_i \in U$ belongs δ sets in $\cup_{j \in I_0 \cup I_1} S_j$ and $\delta < \gamma$
then choose any $\gamma - \delta$ sets containing u_i from the remaining sets whose index is not in $I_0 \cup I_1$
endif

$I = I_0 \cup I_1 \cup I_2$

FIGURE 5.19 Improved randomized approximation algorithm for $\mathbf{SC}_\gamma$ [14].

Before going into the details of the method and its ingredients, we explain a key idea behind the method using an example. Suppose that we are measuring the expression levels of 5 genes $\mathcal{G}_1$, $\mathcal{G}_2$, $\mathcal{G}_3$, $\mathcal{G}_4$ and $\mathcal{G}_5$ to determine their causal relationships. Consider two adjacent times of measurements, say $t = \tau$ and $t = \tau + 1$, and suppose that the measurements are as in Fig. 5.20**a**.

Suppose that we wish to reconstruct a causal relationship of every other gene to gene $\mathcal{G}_5$ that is consistent with this given data. Note that the variable x_5 changes its value between the two successive time intervals, and this change *must* be caused by other variables. The variable x_2 retains the same value 1 during the two successive time intervals, so the change of value of x_5 must be caused by at least one of the three other variables x_1, x_3, and x_4. Intuitively, this explanation of observed data makes the assumption that the regulatory network for $x_1, \ldots, x_n$ can be viewed as a *dynamical system* that is described by a function $\mathbf{f}: \mathbf{X}^n \mapsto \mathbf{X}^n$ that transforms an "input" state $\mathbf{x} = (x_1, x_2, \ldots, x_n) \in \mathbf{X}^n \subset \mathbb{N}^n$ of the network at a time step to an "output" state $\mathbf{x}' = (x_1', \ldots, x_n')$ of the network at the *next* time step, and a directed edge $x_i \to x_j$ in the graph for the network topology of this dynamical system indicates that the value of x_j under the application of $\mathbf{f}$ depends on the value of x_i. Thus, our reconstructed

		measurements	
		$t = \tau$	$t = \tau + 1$
variables	x_1	3	2
	x_2	1	1
	x_3	1	0
	x_4	0	2
	x_5	0	1

(a)

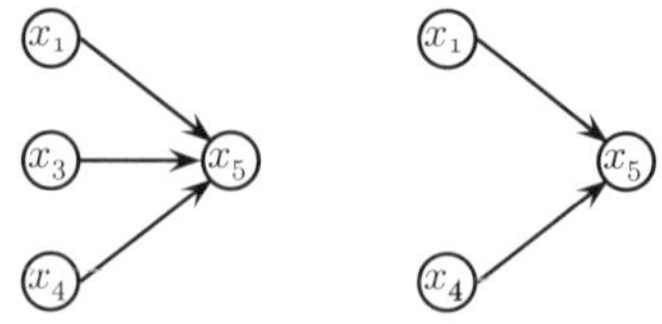

$x_5 = x_1 \vee \overline{x_3} \vee x_4$ (b)

$x_5 = x_1 \vee x_4$ (c)

FIGURE 5.20 **(a)** Measurements of expression levels of 5 genes $\mathcal{G}_1$, $\mathcal{G}_2$, $\mathcal{G}_3$, $\mathcal{G}_4$, and $\mathcal{G}_5$ at two successive time steps; variable x_i corresponds to gene $\mathcal{G}_i$. **(b)** A causal relationship and Boolean formula that explains the causal relationship of other variables to x_5 based only on the data shown in **(a)**. **(c)** Another causal relationship and Boolean formula for $\mathcal{G}_5$ that is consistent with the data in **(a)**.

causal relationships can be summarized by the arcs shown in Fig. 5.20b, where the presence of an arc, say $x_1 \rightarrow x_5$, indicates that x_1 may have an influence on x_5. A Boolean formula that explains this can also be obtained.

Notice that a reconstructed causal relationship may *not* be unique. For example, Fig. 5.20c shows that omitting the dependency of x_5 on x_3 still explains the observed data in Fig. 5.20a correctly. In fact, the most that such a reconstruction method can claim is that x_5 *should depend on at least one of the three variables* x_1, x_3, *or* x_4. The combinatorial approaches discussed in this section are of *parsimonious* nature in the sense that they attempt to choose a network with a *minimal* number of causal relationships that explains the given data; for the example in Fig. 5.20a such a method will select only one of the three possible causal relationships. A minimal network can be recovered by using the so-called *hitting set* problem. However, as we have already observed in Section 5.3.4, biological networks do not necessarily have the lowest possible redundancy, and thus we will need suitable modifications to the approach to introduce redundancy.

We now discuss more details of the two combinatorial approaches based on the above ideas which have been used by other researchers.

Approach by Ideker et al. [38]. In this approach, a set of sparse networks, each from a different perturbation to the genetic network under investigation, is generated, the *binarized* steady-state of gene expression profiles of these networks are observed, and then, using the idea explained in the previous section, a *set* of Boolean networks consistent with an observed steady state of the expression profiles is estimated. Next, an "optimization step" involving the use of an *entropy-based* approach is used to select an *additional* perturbation experiment in order to perform a selection from the set of predicted Boolean networks. Computation of the sparse networks relies upon the hitting set problem. To account for the nonminimal nature of real biological measurements over time t

$$X = \begin{array}{c} \begin{matrix} 0 & 1 & 2 & 3 \end{matrix} \\ \begin{pmatrix} 0 & 1 & 2 & 0 \\ 0 & 0 & 3 & 1 \\ 1 & 0 & 0 & 0 \\ 0 & 1 & 2 & 1 \\ 0 & 1 & 1 & 1 \end{pmatrix} \end{array}$$

(a)

$$\mathcal{G} = \left\{ \mathcal{G}_1, \mathcal{G}_2, \mathcal{G}_3, \mathcal{G}_4, \mathcal{G}_5 \right\}$$

$$T_1^2 = \emptyset,\ T_2^2 = \left\{ \mathcal{G}_1, \mathcal{G}_4, \mathcal{G}_5 \right\},\ T_3^2 = \left\{ \mathcal{G}_1, \mathcal{G}_4 \right\}$$

(b)

FIGURE 5.21 **(a)** Data matrix $X = \left(x_{i,j}\right)$ (quantized to four values) for measurement of expression levels of $m = 5$ genes at $n + 1 = 4$ time points. **(b)** The universe and sets corresponding to gene $\mathcal{G}_2$ in the hitting set formulation of Fig. 5.22a.

networks, one can modify the hitting set algorithm to add redundancies *systematically* by allowing additional parameters to control the extra connections.

Approach by Jarrah et al. [39]. This approach uses one or more time courses of observed data on gene expression levels. Such data can be represented by an $m \times (n+1)$ matrix $X = \left(x_{i,j}\right)$, where m is number of variables (molecular species) and $n + 1$ is the number of points of times at which observations were made. For example, $m = 5$ species data (quantized to four values) at $n + 1 = 4$ time instances are shown in Fig. 5.21a. Suppose that we wish to construct a causal relation for gene $\mathcal{G}_i$ corresponding to the ith row of X. Consider two successive times $t = j - 1$ and $t = j$, and let

$$T_j^i = \begin{cases} \left\{ \mathcal{G}_\ell \,\middle|\, x_{\ell,j-1} \neq x_{\ell,j} \text{ and } \ell \neq i \right\}, & \text{if } x_{i,j-1} \neq x_{i,j}, \\ \emptyset & \text{otherwise.} \end{cases}$$

Then, at least one gene in T_j^i must have a causal effect on $\mathcal{G}_i$. Repeating the argument for $j = 1, 3, \ldots, n$, we arrive at the so-called hitting set problem shown in Fig. 5.22a.

HS is the *combinatorial dual* of the $\mathbf{SC}_1$ problem introduced in Fig. 5.17. The duality transformation between **HS** and $\mathbf{SC}_1$ is given by (5.7) (cf. Exercise 5.10):

$\mathbf{SC}_1$	**HS**	**Condition for equivalence**
$U = \left\{u_1, u_2, \ldots, u_n\right\}$ $S_1, S_2, \ldots, S_m \subseteq U$	$\mathcal{G} = \left\{\mathcal{G}_1, \ldots, \mathcal{G}_m\right\}$ $T_1^i, T_2^i, \ldots, T_n^i \subseteq \mathcal{G}$	$\forall \ell, j : u_\ell \in S_j \equiv \mathcal{G}_j \in T_\ell^i$

(5.7)

Problem name: Hitting set (**HS**)

Instance $<m,n>$: universe $\mathcal{G} = \{\mathcal{G}_1, \ldots, \mathcal{G}_m\}$
sets $T_1^i, T_2^i, \ldots, T_n^i$
with $\cup_{j=1}^n T_j^i \subseteq \mathcal{G}$

Valid Solution: $\mathcal{F} \subseteq \mathcal{G}$ such that
$\forall 1 \leq j \leq n: |\mathcal{F} \bigcap T_j^i| \geq 1$

Objective: minimize $|\mathcal{F}|$

(a)

$\mathcal{F} = \emptyset$, un-hit $= \{1, 2, \ldots, n\}$
while un-hit $\neq \emptyset$ **do**
select $\mathcal{G}_\ell \in \mathcal{G} \setminus \mathcal{F}$ that
maximizes $\sum_{q \in \text{unhit}} \left|\{\mathcal{G}_\ell\} \cap T_q\right|$
$\mathcal{F} = \mathcal{F} \cup \{\mathcal{G}_\ell\}$
un-hit $=$ un-hit $\setminus \bigcup_{|\{\mathcal{G}_\ell \cap T_q\}|>0} \{q\}$
endwhile

(b)

FIGURE 5.22 **(a)** A hitting set formulation of the combinatorial approach for gene $\mathcal{G}_i$. **(b)** A greedy algorithm for **HS** that iteratively selects a new element of the universe that hits a maximum number of sets not hit yet.

Thus, to solve **HS**, we can simply transform it to an equivalent instance of $\mathbf{SC}_1$ and use the methods described before. A more direct greedy approach to solve **HS** is shown in Fig. 5.22b: We repeatedly pick a new element that is contained in a maximum number of sets none of whose elements have been selected yet. This greedy algorithm can be shown to have an approximation ratio of μ, where $\mu = \max_{1 \leq \ell \leq m} \left| j \mid \mathcal{G}_\ell \in T_j^i \right|$ is the maximum number of sets in which an element belongs [84].

Finally, to allow for nonminimal nature of real biological networks, redundancy can be easily introduced by allowing additional edges in the network in a ''controlled'' manner that is consistent with the given data. For this purpose, we need to generalize the **HS** problem to ''multi-hitting'' version when more than one element from each may need to be picked. Formally, let $\vec{\gamma} = (\gamma_1, \gamma_2, \ldots, \gamma_n) \in \mathbb{N}^n$ be a vector whose components are positive integers; the positive integer γ_j indicates that at least γ_j elements need to be selected from the set T_j^i. Then, **HS** can be generalized to the problem $\mathbf{HS}_{\vec{\gamma}}$ by replacing the each constrain $|\mathcal{F} \bigcap T_j^i| \geq 1$ in Fig. 5.22a by $|\mathcal{F} \bigcap T_j^i| \geq \gamma_j$. It is easy to modify the greedy algorithm for **HS** in Fig. 5.22b to find a valid solution of $\mathbf{HS}_{\vec{\gamma}}$.

Comparative Analysis of the Two Approaches. DasGupta et al. [23] compared the two combinatorial reverse engineering approaches discussed in Section 5.4.2 and Section 5.4.2 on the following two biological networks:

(1) An artificial gene regulatory network with external perturbations generated using the software package in reference 61. The interactions between genes in this regulatory network are phenomenological and represent the net effect of transcription, translation, and post-translation modifications on the regulation of the genes in the network.

(2) Time courses from a Boolean network of segment polarity genes responsible for pattern formation in the embryo of *Drosophila melanogaster* (fruit fly). This Boolean model, based on the binary ON/OFF representation of mRNA and protein levels of five segment polarity genes, was validated and analyzed by Albert and Othmer [7].

For the approach by Ideker et al., the redundancy vector $\vec{\gamma}$ used was of the form $(r, r, \ldots, r)$ for $r = 1, 2, 3$. For network **(1)**, Jarrah et al.'s method obtained better results than Ideker et al.'s method, although both fare very poorly. In contrast, for network **(2)**, Jarrah et al.'s method could not obtain any results after running their method for over 12 hours, but Ideker et al.'s method was able to compute results for such network in less than 1 minute, and the results of Ideker et al.'s method improved slightly for larger values of r. The reader is referred to reference 23 for further details about the comparison.

5.4.3 Evaluation of Quality of the Reconstructed Network

By the very nature, the reverse-engineering problems are highly "ill-posed" in the sense that solutions are *far from unique*. This lack of uniqueness stems from the many sources of uncertainty such as

- measurement error,
- *stochasticity* of molecular processes,
- hidden variables—that is, lack of knowledge of *all* the molecular species that are involved in the behavior being analyzed.

Thus, reverse-engineering methods can at best provide *approximate* solutions for the network that is to be reconstructed, making it difficult to evaluate their performance through a theoretical study. Instead, their performance is usually assessed *empirically* in the following two ways.

Experimental Testing of Predictions. After a network has been reconstructed, the newly found interactions or predictions can be tested experimentally for network topology and network dynamics inference, respectively.

Benchmark Testing. This type of performance evaluation consists on measuring how *close* the reverse engineering method under investigation is from recovering a *known* network, usually referred to as the "gold standard" network. In the case of dynamical models, one evaluates the ability of the method of interest to reproduce observations that were not taken into account in the *training* phase involved in the construction of the model. For methods that only reconstruct the network topology, a variety of standard metrics, such as the ones described below, may be applied.

Metrics for Network Topology Benchmarking. Let $\mathcal{G}$ be the graph representing the network topology of a chosen gold standard network, and Let $\widetilde{\mathcal{G}}$ be the graph representing the network topology inferred by the reverse engineering method. Each interaction η in $\widetilde{\mathcal{G}}$ can be classified into one of the following four classes when comparing to the gold standard:

True positive (TP)**:** η exists both in $\mathcal{G}$ and in $\widetilde{\mathcal{G}}$.
False positive (FP)**:** η exists in $\widetilde{\mathcal{G}}$ but not in $\mathcal{G}$.
True negative (TN)**:** η does not exist both in $\mathcal{G}$ and in $\widetilde{\mathcal{G}}$.
False negative (FN)**:** η does not exist in $\widetilde{\mathcal{G}}$ but exists in $\mathcal{G}$.

Let n_{TP}, n_{FP}, n_{TN}, n_{FN} and n_{total} be the number of true positives, false positives, true negatives, false negatives, and total number of possible interactions in the network, respectively. Four standard metrics for benchmarking are as follows:

$$\text{recall rate or true positive rate TPR} = \frac{n_{\text{TP}}}{n_{\text{TP}}+n_{\text{FN}}},$$

$$\text{false positive rate FPR} = \frac{n_{\text{FP}}}{n_{\text{FP}}+n_{\text{TN}}},$$

$$\text{accuracy rate ACC} = \frac{n_{\text{TP}}+n_{\text{TN}}}{n_{\text{total}}},$$

$$\text{precision or positive predictive value PPV} = \frac{n_{\text{TP}}}{n_{\text{TP}}+n_{\text{FP}}}.$$

Since reverse-engineering problems are under-constrained, usually the network reconstruction algorithm will have one or more free parameters that helps to select a best possible prediction. For example, components of redundancy vector $\vec{\gamma}$ in the hitting set approach could be a set of such type of parameters. In this case, a more objective evaluation of performance needs to involve a *range* of parameter values. A standard approach to evaluate performance across the range of parameters is the *receiver operating characteristic* (Roc) method based on the plot of FPR versus TPR values. The resulting Roc *plot* depicts relative trade-offs between true positive predictions and false positive prediction across different parameter values; see Fig. 5.23 for an illustration. An alternative plot is the *recall precision plot* obtained by plotting TPR versus PPV values.

Examples of Gold Standard Networks. We give two examples of gold standard networks that can be used for benchmark testing of reverse engineering methods. Further discussion on generation of gold standard networks can be found later in Section 6.3.3.1.

(i) Gene regulatory networks with external perturbations can be generated from the differential equation models using the software package in reference 61.

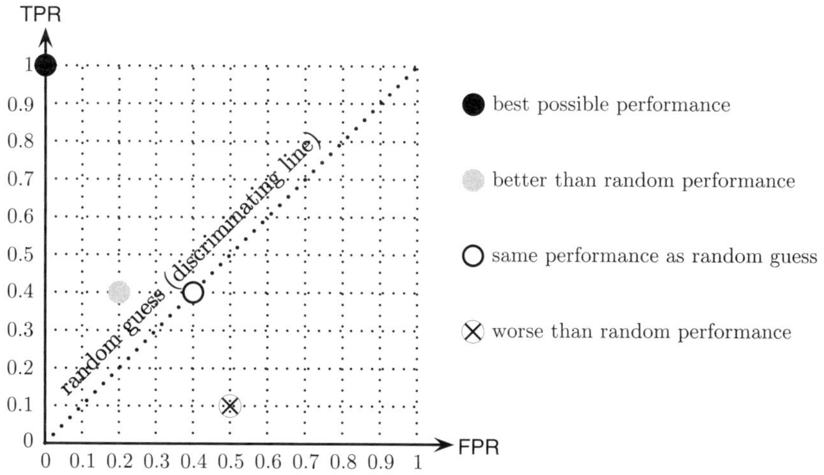

FIGURE 5.23 Two-dimensional ROC space obtained by plotting FPR versus TPR values.

(ii) Time courses can be generated from the Boolean model of network of segment polarity genes involved in pattern formation in the *Drosophila melanogaster* embryo. This model was proposed by Albert and Othmer [7]; the network for each cell has 15 nodes.

REFERENCES

1. B. Alberts. *Molecular Biology of the Cell.* Garland Publishers, New York, 1994.
2. R. Albert and A.-L. Barabási. Statistical mechanics of complex networks. Rev. Modern Phys., **74**(1):47–97, 2002.
3. R. Albert, B. DasGupta, A. Gitter, G. Gürsoy, R. Hegde, P. Pal, G. S. Sivanathan and E. Sontag. A new computationally efficient measure of topological redundancy of biological and social networks. *Phys. Rev. E*, **84**(3):036117, 2011.
4. R. Albert, B. DasGupta, R. Dondi and E. Sontag. Inferring (biological) signal transduction networks via transitive reductions of directed graphs. *Algorithmica*, **51**(2):129–159, 2008.
5. R. Albert, B. DasGupta, R. Dondi, S. Kachalo, E. Sontag, A. Zelikovsky, and K. Westbrooks. A novel method for signal transduction network inference from indirect experimental evidence. *J. Comput. Biol.*, **14**(7):927–949, 2007.
6. R. Albert, B. DasGupta, and E. Sontag. Inference of signal transduction networks from double causal evidence. In *Methods in Molecular Biology: Topics in Computational Biology*, 673, D. Fenyo, editor, Chapter 16, Springer, New York, 2010.
7. R. Albert and H. Othmer. The topology of the regulatory interactions predicts the expression pattern of the segment polarity genes in *Drosophila melanogaster. J. Theoret. Biol.*, **223**:1–18, 2003.

8. M. Andrec, B. N. Kholodenko, R. M. Levy, and E. D. Sontag. Inference of signaling and gene Regulatory networks by steady-state perturbation experiments: Structure and accuracy. *J. Theoret. Biol.*, **232**:427–441, 2005.
9. D. Angeli and E.D. Sontag. Monotone control systems. *IEEE Trans. Automatic Control*, **48**:1684–1698, 2003.
10. J. Aracena, M. González, A. Zuñiga, M. Méndez, and V. Cambiazo. Regulatory network for cell shape changes during *Drosophila* ventral furrow formation. *J. Theoret. Biol.*, **239**(1):49–62, 2007.
11. M. J. Beal and F. Falciani. A Bayesian approach to reconstructing genetic regulatory networks with hidden factors. *Bioinformatics*, **21**(3):349–356, 2005.
12. N. Beckage, L. Smith, and T. Hills. Semantic network connectivity is related to vocabulary growth rate in children. In *Annual Meeting of the Cognitive Science Society*, pp. 2769–2774, 2010.
13. P. Berman, B. DasGupta, and M. Karpinski. Approximating transitive reduction problems for directed networks, Algorithms and Data Structures Symposium, F. Dehne, M. Gavrilova, J.-R. Sack and C. D. Tóth, eds., *LNCS* **5664**:74–85, 2009.
14. P. Berman, B. DasGupta, and E. Sontag. Randomized approximation algorithms for set multicover problems with applications to reverse engineering of protein and gene networks. *Discrete Appl. Math.*, **155**(6–7):733–749, 2007.
15. P. Berman, B. DasGupta and E. Sontag. Computational complexities of combinatorial problems with applications to reverse engineering of biological networks. In F.-Y. Wang and D. Liu, editors, *Advances in Computational Intelligence: Theory and Applications,* Series in Intelligent Control and Intelligent Automation, Vol. 5, World Scientific publishers, Singapur, 2007.
16. P. Berman, B. DasGupta, and E. Sontag. Algorithmic issues in reverse engineering of protein and gene networks via the modular response analysis method. *Ann. NY Acad. Sci.*, **1115**:132–141, 2007.
17. J. A. Bondy and U. S. R. Murty. *Graph Theory*, Springer, New York, 2008.
18. T. Chen, V. Filkov, and S. Skiena. Identifying gene regulatory networks from experimental data. *Third Annual International Conference on Computational Moledular Biology*, 94–103, 1999.
19. D.-Z. Cheng, H. Qi, and Z. Li. *Analysis and Control of Boolean Networks: A Semitensor Product Approach.*, Springer, London, 2011.
20. T. H. Cormen, C. E. Leiserson, R. L. Rivest, and C. Stein. *Introduction to Algorithms.* The MIT Press, Cambridge, MA, 2001.
21. E. J. Crampin, S. Schnell, and P. E. McSharry. Mathematical and computational techniques to deduce complex biochemical reaction mechanisms. *Prog. Biophysic. Mol. Biol.*, **86**:77–112, 2004.
22. B. DasGupta, G. A. Enciso, E. Sontag, and Y. Zhang. Algorithmic and complexity results for decompositions of biological networks into monotone subsystems. *Biosystems*, **90**(1):161–178, 2007.
23. B. DasGupta, P. Vera-Licona, and E. Sontag. Reverse engineering of molecular networks from a common combinatorial approach. In *Algorithms in Computational Molecular Biology: Techniques, Approaches and Applications*, Chapter 40, pp. 941–955, John Wiley & Sons, Hoboken, NJ, 2011.

24. L. Dall'Asta, I. Alvarez-Hamelina, A. Barrata, A. Vázquezb, and A. Vespignania. Exploring networks with traceroute-like probes: Theory and simulations. Theoret. Comput. Sci., **355**:6–24, 2006.

25. E. H. Davidson. *The Regulatory Genome*, Academic Press, New York, 2006.

26. N. Dojer, A. Gambin, A. Mizera, B. Wilczynski, and J. Tiuryn. Applying dynamic Bayesian networks to perturbed gene expression data. *BMC Bioinform.*, **7**(1):249, 2006.

27. G. Enciso and E. Sontag. On the stability of a model of testosterone dynamics, *J. Math. Biol.*, **49**:627–634, 2004.

28. U. Feige. A threshold for approximating set cover. *J. ACM*, **45**:634–652, 1998.

29. S. Fields. High-throughput two-hybrid analysis: the promise and the peril. *FEBS J.*, **272**(21):5391–5399, 2005.

30. N. Friedman, M. Linial, I. Nachman and D. Pe'er. Using Bayesian networks to analyze expression data. *J. Comput. Biol.*, **7**(3–4):600–620, 2000.

31. C. Friedman, P. Kra, H. Yu, M. Krauthammer, and A. Rzhetsky. GENIES: A natural-language processing system for the extraction of molecular pathways from journal articles. *Bioinformatics*, **17**(Suppl 1):S74–82, 2001.

32. A. de la Fuente, N. Bing, I. Hoeschele and P. Mendes. Discovery of meaningful associations in genomic data using partial correlation coefficients, Bioinformatics, **20:** 3565–3574, 2004.

33. L. Giot, J. S. Bader, C. Brouwer, A. Chaudhuri, B. Kuang, Y. Li, Y. L. Hao, C. E. Ooi, B. Godwin, E. Vitols, G. Vijayadamodar, P. Pochart, H. Machineni, M. Welsh, Y. Kong, B. Zerhusen, R. Malcolm, Z. Varrone, A. Collis, M. Minto, S. Burgess, L. McDaniel, E. Stimpson, F. Spriggs, J. Williams, K. Neurath, N. Ioime, M. Agee, E. Voss, K. Furtak, R. Renzulli, N. Aanensen, S. Carrolla, E. Bickelhaupt, Y. Lazovatsky, A. DaSilva, J. Zhong, C. A. Stanyon, R. L. Finley, K. P. White, M. Braverman, T. Jarvie, S. Gold, M. Leach, J. Knight, R. A. Shimkets, M. P. McKenna, J. Chant, and J. M. Rothberg. A protein interaction map of *Drosophila melanogaster. Science*, **302**(5651):1727–1736, 2003.

34. M. Girvan and M. E. J. Newman. Community structure in social and biological networks. *Proc. Nat. Acad. Sci. USA*, **99**:7821–7826, 2002.

35. R. Guimerà, M. Sales-Pardo, and L. A. N. Amaral. Classes of complex networks defined by role-to-role connectivity profiles, *Nature Physics*, **3**:63–69, 2007.

36. J. Hann and M. Kamber. *Data Mining: Concepts and Techniques*. Morgan Kaufman Publishers, Burlington, MA, 2000.

37. S. Huang. Gene expression profiling, genetic networks and cellular states: an integrating concept for tumorigenesis and drug discovery. *J. Mol. Med.*, **77**(6):469–480, 1999.

38. T. E. Ideker, V. Thorsson, and R. M. Karp. Discovery of regulatory interactions through perturbation: Inference and experimental design. *Pacific Symp. Biocomput.*, **5**:305–316, 2000.

39. A. S. Jarrah, R. Laubenbacher, B. Stigler and M. Stillman. Reverse-engineering polynomial dynamical systems, *Adv. Appl. Math.*, **39**(4):477–489, 2007.

40. L. J. Jensen, J. Saric and P. Bork. Literature mining for the biologist: from information retrieval to biological discovery, *Nature Rev. Genet.*, **7**(2):119–129, 2006.

41. H. Jeong, B. Tombor, R. Albert, Z. N. Oltvai, and A.-L. Barabási. The large-scale organization of metabolic networks. *Nature*, **407**:651–654, 2000.

42. D. S. Johnson. Approximation algorithms for combinatorial problems. *J. Comput. Syst. Sci.*, **9**:256–278, 1974.

43. S. Kachalo, R. Zhang, E. Sontag, R. Albert, and B. DasGupta. NET-SYNTHESIS: A software for synthesis, inference and simplification of signal transduction networks. *Bioinformatics*, **24**(2):293–295, 2008.

44. R. Kannan, P. Tetali, and S. Vempala. Markov-chain algorithms for generating bipartite graphs and tournaments. *Random Structures and Algorithms*, **14**:293–308, 1999.

45. N. Karmarkar. A new polynomial-time algorithm for linear programming. *Combinatorica*, **4**:373–395, 1984.

46. S. A. Kauffman. Metabolic stability and epigenesis in randomly constructed genetic nets. *J. Theore. Biol.*, **22**:437–467, 1969.

47. B. N. Kholodenko, A. Kiyatkin, F. Bruggeman, E. D. Sontag, H. Westerhoff, and J. Hoek. Untangling the wires: A novel strategy to trace functional interactions in signaling and gene networks. *Proc. Nat. Acad. Sci. USA*, **99**:12841–12846, 2002.

48. B. N. Kholodenko and E. D. Sontag. Determination of Functional Network Structure from Local Parameter Dependence Data, arXiv physics/0205003, May 2002.

49. B. Kolb and I. Q. Whishaw. *Fundamentals of Human Neuropsychology*. Freeman, New York, 1996.

50. S. Khuller, B. Raghavachari and N. Young. On strongly connected digraphs with bounded cycle length. *Discrete Appl. Math.*, **69**(3):281–289, 1996.

51. S. Khuller, B. Raghavachari and N. Young. Approximating the minimum equivalent digraph, *SIAM J. Comput.*, **24**(4):859–872, 1995.

52. B. Krupa. On the number of experiments required to find the causal structure of complex systems. *J. Theore. Biol.*, **219**(2):257–267, 2002.

53. R. J. Larsen and M. L. Marx. *An Introduction to Mathematical Statistics and Its Applications*, 3rd edition, Prentice Hall, Englewood Cliffs, NJ, USA, 2000.

54. E. A. Leicht and M. E. J. Newman. Community structure in directed networks. *Phy. Rev. Lett.*, **100**:118703, 2008.

55. S. Li, C. M. Armstrong, N. Bertin, H. Ge, S. Milstein, M. Boxem, P.-O. Vidalain, J.-D. J. Han, A. Chesneau, T. Hao, D. S. Goldberg, N. Li, M. Martinez, J.-F. Rual, P. Lamesch, L. Xu, M. Tewari, S. L. Wong, L. V. Zhang, G. F. Berriz, L. Jacotot, P. Vaglio, J. Reboul, T. Hirozane-Kishikawa, Q. Li, H. W. Gabel, A. Elewa, B. Baumgartner, D. J. Rose, H. Yu, S. Bosak, R. Sequerra, A. Fraser, S. E. Mango, W. M. Saxton, S. Strome, S. van den Heuvel, F. Piano, J. Vandenhaute, C. Sardet, M. Gerstein, L. Doucette-Stamm, K. C. Gunsalus, J. W. Harper, M. E. Cusick, F. P. Roth, D. E. Hill, and M. Vidal. A map of the interactome network of the metazoan *C. elegans*. *Science*, **303**:540–543, 2004.

56. T. I. Lee, N. J. Rinaldi, F. Robert, D. T. Odom, Z. Bar-Joseph, G. K. Gerber, N. M. Hannett, C. T. Harbison, C. M. Thompson, I. Simon, J. Zeitlinger, E. G. Jennings, H. L. Murray, D. B. Gordon, B. Ren, J. J. Wyrick, J.-B. Tagne, T. L. Volkert, E. Fraenkel, D. K. Gifford, and R. A. Young. Transcriptional regulatory networks in *Saccharomyces cerevisiae*. *Science*, **298**(5594):799–804, 2002.

57. S. Li, S. M. Assmann, and R. Albert. Predicting essential components of signal transduction networks: A dynamic model of guard cell abscisic acid signaling. *PLoS Biol.*, **4**(10):e312, 2006.

58. A. Ma'ayan, S. L. Jenkins, S. Neves, A. Hasseldine, E. Grace, B. Dubin-Thaler, N. J. Eungdamrong, G. Weng, P. T. Ram, J. J. Rice, A. Kershenbaum, G. A. Stolovitzky,

R. D. Blitzer and R. Iyengar. Formation of regulatory patterns during signal propagation in a mammalian cellular network. *Science*, **309**(5737):1078–1083, 2005.

59. E. M. Marcotte, I. Xenarios, and D. Eisenberg. Mining literature for protein–protein interactions. *Bioinformatics*, **17**(4):359–363, 2001.
60. N. Megiddo. Linear programming in linear time when the dimension is fixed, *J. ACM*, **31**:114–127, 1984.
61. P. Mendes. Biochemistry by numbers: Simulation of biochemical pathways with Gepasi 3. *Trends Biochem. Sci.*, **22**:361–363, 1997.
62. J. D. Murray. *Mathematical Biology, I: An Introduction*. Springer, New York, 2002.
63. N. Nariai, Y. Tamada, S. Imoto, and S. Miyano. Estimating gene regulatory networks and protein–protein interactions of *Saccharomyces cerevisiae* from multiple genome-wide data. *Bioinformatics*, **21**(suppl 2):ii206–ii212, 2005.
64. M. E. J. Newman. The structure and function of complex networks. *SIAM Rev.*, **45**: 167–256, 2003.
65. M. E. J. Newman. Detecting community structure in networks. *Eur. Phys. J. B*, **38**:321–330, 2004.
66. M. E. J. Newman and G. T. Barkema. *Monte Carlo Methods in Statistical Physics*, Oxford University Press, New York, 1999.
67. M. E. J. Newman and M. Girvan. Finding and evaluating community structure in networks. *Phys. Rev. E*, **69**:026113, 2004.
68. M. E. J. Newman, S. H. Strogatz and D. J. Watts. Random graphs with arbitrary degree distributions and their applications. *Phys. Rev. E*, **64**(2):026118–026134, 2001.
69. J. A. Papin and B. O. Palsson. Topological analysis of mass-balanced signaling networks: a framework to obtain network properties including crosstalk. *J. Theoret. Biol.*, **227**(2):283–297, 2004.
70. I. Pournara and L. Wernisch. Reconstruction of gene networks using Bayesian learning and manipulation experiments. *Bioinformatics*, **20**(17):2934–2942, 2004.
71. E. Ravasz, A. L. Somera, D. A. Mongru, Z. N. Oltvai and A.-L. Barabási. Hierarchical organization of modularity in metabolic networks. *Science*, **297**(5586):1551–1555, 2002.
72. J. J. Rice, Y. Tu, and G. Stolovitzky. Reconstructing biological networks using conditional correlation analysis, *Bioinformatics*, **21**(6):765–773, 2005.
73. S. D. M. Santos, P. J. Verveer, and P. I. H. Bastiaens. Growth factor-induced MAPK network topology shapes Erk response determining PC-12 cell fate. *Nature Cell Biol*, **9**:324–330, 2007.
74. S. S. Shen-Orr, R. Milo, S. Mangan, and U. Alon. Network motifs in the transcriptional regulation network of *Escherichia coli*. *Nature Genet.* **31**:64–68, 2002.
75. H. L. Smith. Monotone Dynamical Systems, Providence, R.I., AMS 1995.
76. E. D. Sontag. *Mathematical Control Theory: Deterministic Finite Dimensional Systems*. Springer, New York, 1998.
77. E. D. Sontag. Molecular systems biology and control. *Europ. J. Control*, **11**(4–5):396–435, 2005.
78. E. D. Sontag, A. Kiyatkin and B. N. Kholodenko. Inferring dynamic architecture of cellular networks using time series of gene expression, protein and metabolite data. Bioinformatics, **20**:1877–1886, 2004.

79. J. Stark, R. Callard and M. Hubank. From the top down: Towards a predictive biology of signalling networks. *Trends in Biotechnol.*, **21**:290–293, 2003.
80. G. Tononi, O. Sporns, and G. M. Edelman. A measure for brain complexity: Relating functional segregation and integration in the nervous system. *Proc. Nat. Acad. Sci. USA*, **91**(11):5033–5037, 1994.
81. G. Tononi, O. Sporns, and G. M. Edelman. A complexity measure for selective matching of signals by the brain. *Proc. Nat. Acad. Sci. USA*, **93**:3422–3427, 1996.
82. G. Tononi, O. Sporns, and G. M. Edelman. Measures of degeneracy and redundancy in biological networks. *Proc. Nat. Acad. Sci. USA*, **96**:3257–3262, 1999.
83. L. Trevisan. Non-approximability results for optimization problems on bounded-degree instance. Thirty-third ACM Symposium on Theory of Computing, pp. 453–461, 2001.
84. V. Vazirani. *Approximation Algorithms*, Springer-Verlag, Berlin, 2001.
85. A. Wagner. Estimating coarse gene network structure from large-scale gene perturbation data. *Genome Res.*, **12**(2):309–315, 2002.
86. J. Yu, V. Smith, P. Wang, A. Hartemink, and E Jarvis. Advances to Bayesian network inference for generating causal networks from observational biological data. *Bioinformatics*, **20**:3594–3603, 2004.
87. R. Zhang, M. V. Shah, J. Yang, S. B. Nyland, X. Liu, J. K. Yun, R. Albert, and T. P. Loughran, Jr. Network model of survival signaling in LGL leukemia. *Proc. Nat. Acad. Sci. USA*, **105**(42):16308–16313, 2008.
88. M. Zou and S. D. Conzen. A new dynamic Bayesian network (DBN) approach for identifying gene regulatory networks from time course microarray data. *Bioinformatics*, **21**(1):71–79, 2005.

EXERCISES

5.1 Suppose that your data suggest that Protein A inhibits the transcription of the RNA that codes Protein B, whereas Protein B in turn enhances the production of Protein A. Give a two node regulatory network that is consistent with this information.

5.2 Consider the following biological model of testosterone dynamics [27,62]:

$$\frac{\mathrm{d}x_1}{\mathrm{d}t}(t) = \frac{A}{K + x_3(t)} - b_1 x_1(t),$$

$$\frac{\mathrm{d}x_2}{\mathrm{d}t}(t) = c_1 x_1(t) - b_2 x_2(t),$$

$$\frac{\mathrm{d}x_3}{\mathrm{d}t}(t) = c_2 x_2(t) - b_3 x_3(t),$$

where A, K, b_1, b_2, b_3, c_1, and c_2 are positive constants. Draw the associated signed graph of this model.

5.3 Consider five genes $\mathcal{G}_1$, $\mathcal{G}_2$, $\mathcal{G}_3$, $\mathcal{G}_4$ and $\mathcal{G}_5$, and suppose that each gene switches its state if the remaining four genes are not in identical states.

(**a**) Write down the Boolean network and its associated directed network for the interaction of the five genes described above.

(**b**) Does the Boolean network in 5.3 has a fixed point? If so, show one fixed point.

5.4 Show that the reduced network in Fig. 5.8 is not minimal by giving a minimal network in which *more* than two arcs are removed.

5.5 Consider the following subset of the evidence gathered for the signal transduction network responsible of abscicic acid induced closure of plant stomata [57]:

ABA ⊣ NO
ABA → PLD
ABA → GPA1
ABA → PLC
GPA1 → PLD
PLD → PA
NO ⊣ KOUT
KOUT → Closure
PA → Closure
PLC → (ABA → KOUT)
PLD → (ABA → Closure)

and suppose that $E_{\rm m} = \{\text{GPA1} \to \text{PLD}, \text{KOUT} \to \text{Closure}\}$. Follow the network synthesis approach in Fig. 5.4 to synthesize a minimal network.

5.6 [3] Consider the three signal transduction networks shown in Fig. 5.24.

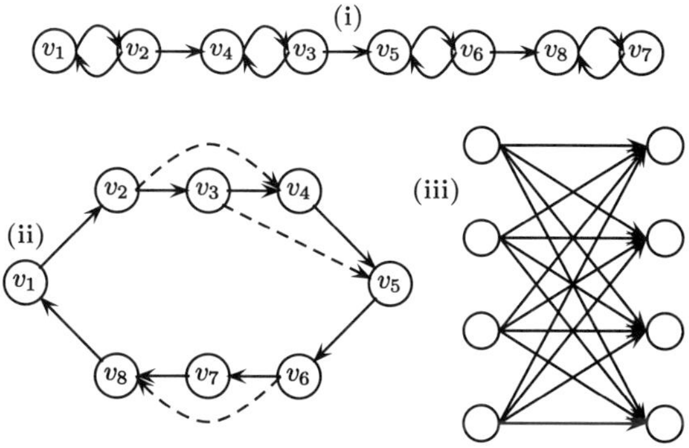

FIGURE 5.24 Three n-node graphs (shown for $n = 8$) discussed in Exercise 5.6.

(**a**) Verify that the two networks shown in (**i**) and (**ii**) has the same in-degree and out-degree sequence, but $\mathfrak{R}_{\text{T}_\text{R}} = 0$ for the network in (**i**) and $\mathfrak{R}_{\text{T}_\text{R}} = 3/11$ for the network in (**ii**). Can you give an exact expression of the value of $\mathfrak{R}_{\text{T}_\text{R}}$ for the network in (**ii**) when generalized to have n nodes?

(**b**) Verify that higher average degree does not necessarily imply higher values of $\mathfrak{R}_{\text{T}_\text{R}}$ by showing that the network in (**ii**), generalized on n nodes, has an average degree below 2, but the graph in (**iii**), generalized on n nodes, has an average degree of $n/2$ but a redundancy value of 0.

5.7 Argue that the dynamics of the two networks in Fig. 5.13 compute different functions if an OR gate is used to combine the inputs to node c, and all arcs have the same time delay.

5.8 Prove Eq. (5.2).

5.9 Convince yourself that Problem (5.6) is indeed a correct formulation for the case of arbitrary k.

5.10 Show the validity of the equivalence shown in (5.7) between $\mathbf{SC}_1$ and **HS**. In other words, show that, assuming the equivalence conditions hold, $S_{i_1}, S_{i_2}, \ldots, S_{i_t}$ is a valid solution of $\mathbf{SC}_1$ *if and only if* $\mathcal{G}_{i_1}, \mathcal{G}_{i_2}, \ldots, \mathcal{G}_{i_t}$ is a valid solution of **HS**.

5.11 This problems explores some possibilities to improve the time and space requirements of the dynamic programming algorithm in Fig. 5.5.

(**a**) Suppose that we remove the third index (corresponding to the variable k in the loop) from the variable N and P; that is, instead of $N(i, j, k, x)$ we simply use $N(i, j, x)$ and so on. Note that the space requirement is now $O\left(n^2\right)$ instead of $O\left(n^3\right)$. Will the algorithm still compute all reachabilities correctly?

(**b**) In practice, the variable k in the **for** loop may not need to go until it reaches the value of n. Suggest how we can abort the loop for $k = \ell$ based on the calculations performed during the executing of the loop with $k = \ell - 1$.

5.12 In Section 5.3.3 we claimed that $\mathsf{opt}(G) \geq |V|$. Prove this.

5.13 Consider the special case of the TR problem when *every arc is excitory* (i.e., when $\mathcal{L}(u, v) = 1$ for every arc $(u, v) \in E$). The goal of this problem is to design an algorithm for this special case that improves upon the approximation ratio of 3 of the greedy approach as stated in Theorem 5.1.

(**a**) [5] Show that the problem can be solved in polynomial time if the given graph G has no cycles of with more than 3 arcs.

(**b**) By contraction of an arc $(u, v) \in E$ we mean merging nodes u and v to a single node, and deleting any resulting self-loops or parallel arcs. By contraction of a cycle, we mean contraction of every arc of the cycle; see Fig. 5.25 for an illustration. Consider the following algorithm:

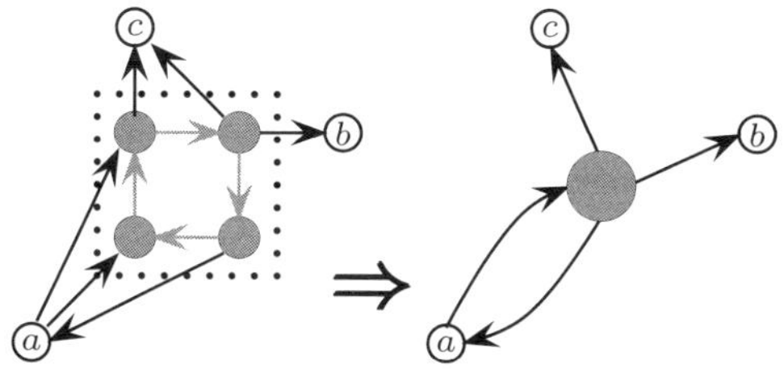

FIGURE 5.25 Contraction of a cycle of length 4.

```
select an integer constant k > 3
for i = k, k − 1, ..., 4
      while G contains a cycle C of at least i edges
            contract C and select the edges of C in our solution
      endwhile
endfor
(* comment: now G contains no cycle of more than 3 edges *)
use the algorithm in part a) to solve the TR problem on G exactly
    and select the edges in this exact solution
output all selected edges as the solution
```

(i) Show that the above algorithm returns a valid solution of TR.

(ii) [51] Show that if $E_m = \emptyset$ and G is strongly connected then the above algorithm has an approximation ratio of $\beta = \frac{\pi^2}{6} - \frac{1}{36} + \frac{1}{k(k-1)} \approx 1.617 + \frac{1}{k(k-1)}$.

(c) Use part **b) (ii)** to provide an algorithm with an approximation ratio of $1 + \beta$ if $E_m \neq \emptyset$.

5.14 [14] In the example explaining the linear algebraic formulation of MRA on page 163, we claimed that we can find a nonzero vector by simple linear algebra manipulations such that A_1 is an unknown multiple of this vector. Show how to find such a nonzero vector.

5.15 [14] Show that Problem (5.6) as mentioned on page 165 can be written down as a problem of the form shown in Fig. 5.17.

6

DYNAMICAL SYSTEMS AND INTERACTION NETWORKS

In this chapter, we view biological models in the framework of dynamical systems. In this framework, the biological system has a vector $\mathbf{x}(t) = \left(x_1(t), x_2(t), \dots, x_n(t)\right)$ of n time-dependent state variables (e.g., indicating the concentration of the n proteins in the model at time t), a vector $\mathbf{u}(t) = \left(u_1(t), u_2(t), \dots, u_m(t)\right)$ of m time-dependent input variables for the external stimuli to the system, and n functions $f_1, f_2, \dots, f_n$, where f_i governs the evolution of the state variable x_i. Different types of system dynamics can be obtained by varying the nature of these variables and functions.

Discrete Versus Continuous-State Variables $x_1, x_2, \dots, x_n$. The state variables could be *continuous* (i.e., real-valued) or *discrete* (e.g., binary). When state variables are continuous, the evolution of cellular components are assumed to be continuous functions of time. In contrast, when the state variables are discrete, each component is assumed to have a small number of qualitative states, and the regulatory interactions are typically described by logical functions such as Boolean circuits; examples of such models are discussed in references 2, 19, 21, 34, 42, 60, 72, 84; and 102.

Discrete Versus Continuous-Time Variable t. The time variable t can be *continuous* or *discrete*. Continuous-time models can describe dynamics such as mass-action kinetics (e.g., see [50,81,100]) via differential equations (such as via system (5.1) discussed in Chapter 5) or by delay equations if delays are important. Discrete-time models involve difference equations, or they may arise as quantized descriptions of continuous-variable systems (e.g., see reference 37). For example, the discrete time

Models and Algorithms for Biomolecules and Molecular Networks, First Edition. Bhaskar DasGupta and Jie Liang.

version of system (5.1) of Chapter 5 using a simple ‘‘Jacobi-type’’ iteration with synchronous updates can be written down as

$$\begin{aligned}
x_1(t+1) &= f_1\big(x_1(t), x_2(t), \ldots, x_n(t), u_1(t), u_2(t), \ldots, u_m(t)\big),\\
x_2(t+1) &= f_2\big(x_1(t), x_2(t), \ldots, x_n(t), u_1(t), u_2(t), \ldots, u_m(t)\big),\\
&\ \ \vdots\\
x_n(t+1) &= f_n\big(x_1(t), x_2(t), \ldots, x_n(t), u_1(t), u_2(t), \ldots, u_m(t)\big)
\end{aligned} \tag{6.1}$$

or, in more concise vector form,

$$\mathbf{x}(t+1) = \mathbf{f}(\mathbf{x}(t), \mathbf{u}(t)) \quad \text{or} \quad \mathbf{x}^{+} = \mathbf{f}(\mathbf{x}, \mathbf{u}).$$

Communication Delays. Delays may result in cellular systems due to differences in times for transcription, translation, degradation and other biochemical processes. Delays can be implemented, for example, by appropriate modifications of Eqs. (5.1) or (6.1). For example, in the context of Eq. (6.1) suppose that the output of variable x_i is delayed by δ time units to reach variable x_j. Then, the modified equation for change of x_j can be written as

$$x_j(t+1) = f_j\big(\ \ldots\ldots, x_i(t+1-\delta), \ldots\ldots\ \big).$$

It is not difficult to see that delays may affect the dynamic behavior in a nontrivial manner (cf. Exercise 6.1).

Deterministic Versus Stochastic Dynamics. In a deterministic model, the rules of the evolution of the system is deterministic in nature. In contrast, stochastic models may address the deviations from population homogeneity by transforming reaction rates into probabilities and concentrations into numbers of molecules (e.g., see reference 79).

It is also possible to consider ‘‘hybrid’’ models that may combine continuous and discrete time-scales or continuous and discrete variables (e.g., see references 4, 13, 20, 22, and 40). The choice of a model for a specific application depends on several factors such as simplicity of the model, accuracy of prediction, and computational aspects for simulating the model. Sometimes more than one model may be used to model the *same* biological process at different levels. For example, the segment polarity gene network has been investigated using a continuous-state model with 13 equations and 48 unknown kinetic parameters in reference 100 and also using a synchronous Boolean model in reference 2.

6.1 SOME BASIC CONTROL-THEORETIC CONCEPTS

We illustrate some basic control-theoretic concepts and definitions used in the study of dynamical systems. Consider the discrete-time model, a special case of the system (6.1) with the explicit variable vector $z \in \mathbb{R}^p$ for the p output measurements included, as shown below:

$$\begin{pmatrix} x_1(t+1) \\ x_2(t+1) \\ \vdots \\ x_n(t+1) \end{pmatrix} = A \begin{pmatrix} x_1(t) \\ x_2(t) \\ \vdots \\ x_n(t) \end{pmatrix} + B \begin{pmatrix} u_1(t) \\ u_2(t) \\ \vdots \\ u_m(t) \end{pmatrix},$$

$$\begin{pmatrix} z_1(t) \\ z_2(t) \\ \vdots \\ z_p(t) \end{pmatrix} = C \begin{pmatrix} x_1(t) \\ x_2(t) \\ \vdots \\ x_n(t) \end{pmatrix} \tag{6.2}$$

or, in simple vector notation,

$$\mathbf{x}(t+1) = A\mathbf{x}(t) + B\mathbf{u}(t), \qquad \mathbf{z}(t) = C\,\mathbf{x}(t),$$

where $A \in \mathbb{R}^{n \times n}$, $B \in \mathbb{R}^{n \times m}$ and $C \in \mathbb{R}^{p \times n}$.

Recall that a directed graph of n nodes is strongly connected if and only if, for any pair of nodes u and v, there exists a path from u to v using at most $n-1$ edges. An analog of similar kind of behavior is captured by the definition of controllability of a dynamical system.

Definition 6.1 (Controllability) *The system* (6.2) *is said to be controllable if, for every initial condition* $\mathbf{x}(0)$ *and every vector* $\mathbf{y} \in \mathbb{R}^n$*, there exist a finite time* t_0 *and input (control) vectors* $\mathbf{u}(0), \mathbf{u}(1), \ldots, \mathbf{u}(t_0) \in \mathbb{R}^m$ *such that* $\mathbf{y} = \mathbf{x}(t_0)$ *when the system is started at* $\mathbf{x}(0)$.

Intuitively, controllability means the ability to reach any final state $\mathbf{y}$ from any initial state $\mathbf{x}(0)$ by time t_0 without posing any conditions of the trajectory of the dynamics or any constraints on the input vectors $\mathbf{u}(t)$. For system (6.2), a necessary and sufficient condition for controllability can be easily deduced in the following manner. We first unwind the evolution of the system by using the recurrence equations:

$$\begin{aligned}
\mathbf{x}(1) &= A\mathbf{x}(0) + B\mathbf{u}(0), \\
\mathbf{x}(2) &= A\mathbf{x}(1) + B\mathbf{u}(1) = A^2\mathbf{x}(0) + AB\mathbf{u}(0) + B\mathbf{u}(1), \\
\mathbf{x}(3) &= A\mathbf{x}(2) + B\mathbf{u}(2) = A^3\mathbf{x}(0) + A^2B\mathbf{u}(0) + AB\mathbf{u}(1) + B\mathbf{u}(2), \\
&\vdots \\
\mathbf{y} = \mathbf{x}(n) &= A^n\mathbf{x}(0) + A^{n-1}B\mathbf{u}(0) + A^{n-2}B\mathbf{u}(1) + A^{n-3}B\mathbf{u}(2) + \\
&\qquad \ldots + AB\mathbf{u}(n-2) + B\mathbf{u}(n-1).
\end{aligned}$$

Thus, for the system to be controllable, there must be a solution to the following set of linear equations where we treat $\mathbf{u}(0), \mathbf{u}(1), \ldots, \mathbf{u}(n-1)$ as unknowns:

$$\left(B \;\; AB \;\; A^2B \ldots \ldots A^{n-1}B\right) \begin{pmatrix} \mathbf{u}(n-1) \\ \mathbf{u}(n-2) \\ \vdots \\ \mathbf{u}(1) \\ \mathbf{u}(0) \end{pmatrix} = \mathbf{y} - A^n\,\mathbf{x}(0).$$

This is clearly possible only when the $n \times nm$ controllability matrix (also called the Kalman matrix) $\mathcal{Z} = \left(B \;\; AB \;\; A^2B \ldots \ldots A^{n-1}B\right)$ has a full rank of n.

Another useful control-theoretic aspect of dynamical systems is the *observability* property. Intuitively, observability is a quantification of how well the internal states of a system can be distinguished or identified by observing its external outputs. There are several possibilities for such a quantitative definition; below we state one such definition. Let the notation $\mathbf{z}(t)\big|_{\mathbf{x},\mathbf{u}_0,\ldots,\mathbf{u}_{t-1}}$ denote the value of the output vector $\mathbf{z}(t)$ with the initial state vector $\mathbf{x}(0) = \mathbf{x}$ and the input vectors $\mathbf{u}(j) = \mathbf{u}_j$ for $j = 0, 1, 2, \ldots, t-1$.

Definition 6.2 (Observability) *A state vector* $\mathbf{x} \in \mathbb{R}^n$ *of the system* (6.2) *is called unobservable over a time interval* $\left[0, \tau\right]$ *if*

$$\left(\forall\, \mathbf{u}_0, \mathbf{u}_1, \ldots, \mathbf{u}_{\tau-1} \in \mathbb{R}^m\right) : \mathbf{z}(t)\big|_{\mathbf{x},\mathbf{u}_0,\ldots,\mathbf{u}_{\tau-1}} = \mathbf{z}(t)\big|_{\mathbf{0},\mathbf{u}_0,\ldots,\mathbf{u}_{\tau-1}},$$

where $\mathbf{0} = 0^n$. *System* (6.2) *is said to be observable if it has no unobservable state.*

It is not very difficult to see that if the system (6.2) is observable as defined in Definition 2, then this also implies that the initial state $\mathbf{x}(0)$ can be uniquely determined. Exercise 6.8 asks the reader to verify that if a system is observable, then all states are observable within the time duration $\left[0, n\right]$. Using a similar type of recurrence unravelling as was done for the controllability case, it can be see that system (6.2) is observable if and only if the matrix $\left(\, C \;\; CA \;\; CA^2 \;\ldots\; CA^{n-1}\right)$ has a rank of n.

In this remainder of this chapter, we discuss several dynamical models that are typically used in modeling cellular processes and interactions.

6.2 DISCRETE-TIME BOOLEAN NETWORK MODELS

Discrete-time Boolean networks were introduced in Section 5.2. In our dynamical systems framework, a discrete-time Boolean network is described by setting time variables t to have discrete values such as $0, 1, 2, \ldots$ in a system of difference equations such as (6.1), where $\mathbf{x}(t) = \left(x_1(t), x_2(t), \ldots, x_n(t)\right) \in \{0, 1\}^n$ for all t, and $f_1, f_2, \ldots, f_n$ are Boolean functions. For example, the reverse engineering method

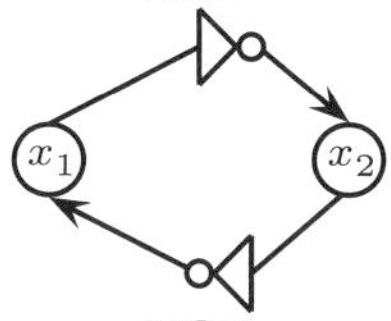

FIGURE 6.1 A Boolean network of two species interaction.
$(x_1, x_2) = (0, 1)$ and $(x_1, x_2) = (1, 0)$ are the fixed points of this network. If synchronous update rule is used, then the Boolean system is defined by

$$x_1(t+1) = \overline{x_2(t)}, \qquad x_2(t+1) = \overline{x_1(t)}$$

and thus the state vector $(1, 1)$ does not lead to a fixed point. On the other hand, suppose that the following asynchronous update rule is used:

$$x_1(t+1) = \begin{cases} \overline{x_2(t)}, & \text{if } t \text{ is even,} \\ x_1(t), & \text{otherwise} \end{cases} \qquad x_2(t+1) = \begin{cases} \overline{x_1(t)}, & \text{if } t \text{ is odd,} \\ x_2(t), & \text{otherwise,} \end{cases}$$

then the state vector $(1, 1)$ leads to the fixed point $(0, 1)$.

of Jarrah *et al.* in Section 5.4.2.2 reconstructed a Boolean network, and the model for the segment polarity gene network in *Drosophila melanogaster* given in reference 2 is a Boolean network model.

Note that the dynamics in (6.1) are described by *simultaneously* updating the state of all the variables in the network (the so-called ''synchronous update'' rule). In reality, however, the rates of transcription, translation, degradation and other biochemical processes can vary widely from gene to gene, thus giving rise to the so-called ''asynchronous update'' rule where not all variables update themselves at each time step. Asynchronous behavior can be formally incorporated in a Boolean model by following the formalism of Thomas [95]. In this formalism, each component of the cellular system is associated with two variables, an usual state variable, and an image variable corresponding to each logical function in the Boolean model (which takes appropriate state variables as its input); and depending on the asynchronous update rule, the value of the image variable is copied (or not copied) to the corresponding state variable. The fixed points (also called steady states) of a Boolean model remain *the same* regardless of the update method [16, page 431]. But, other aspects of the dynamic behavior of a Boolean network may drastically change by using asynchronous as opposed to synchronous updates, resulting in the same initial state leading to different steady states or limit cycles (see Fig. 6.1 for an illustration). On the other hand, asynchronous update rules provide more realistic modeling of biological processes by allowing individual variability of the rate of each process and thus allowing for decision reversals. For example, it becomes possible to model the phenomena of overturning of mRNA decay when its transcriptional activator is synthesized, a phenomena that synchronous updates *cannot* model. Several asynchronous versions of the synchronous Boolean model for the segment polarity network in reference 2 appear in references 21 and 22.

6.3 ARTIFICIAL NEURAL NETWORK MODELS

Artificial neural network (ANN) models were originally used to model the connectivities of neurons in brains; thus, in contrast to cellular signaling networks, ANN models generally have a *less direct* one-to-one correspondence to biological data. In the framework of dynamical systems, ANN models are discrete-time dynamical systems where each function f_i applies a suitable transformation to a linear combination of its inputs; that is, f_i is of the form

$$f_i\big(x_1(t), x_2(t), \ldots, x_n(t)\big) = g\left(\sum_{j=1}^{n} w_{i,j}x_j - \theta_i\right)$$

for some (common) global function $g\colon \mathbb{R} \mapsto \mathbb{R}$, where the $w_{i,j}$'s and θ_i's are real or integral-valued given parameters. The global function g is commonly known as the ''activation'' function or the ''gate function'' in the ANN literature. Some popular choices for the function g are the *Heaviside* or *threshold* function $\mathcal{H}$ and the *sigmoidal* function σ (illustrated in Fig. 6.2); other choices for the activation function include the *cosine squasher*, the *gaussian*, the *hyperbolic tangent*, generalized *radial basis functions*, *polynomials* and *trigonometric polynomials*, *splines*, and *rational functions*. Often an ANN is pictorially represented by a directed graph G in which there is a node v_i corresponding to each variable x_i, the node v_i is labeled with its ''threshold'' parameter θ_i, and a directed arc (v_i, v_j) is labeled by its ''weight'' $w_{i,j}$; see Fig. 6.3 for an illustration. The topology of G is usually referred to as the *architecture* of the ANN. An ANN is classified as a *feed-forward* or *recurrent* type depending on whether G is acyclic or not. For acyclic feed-forward ANNs, the *depth*

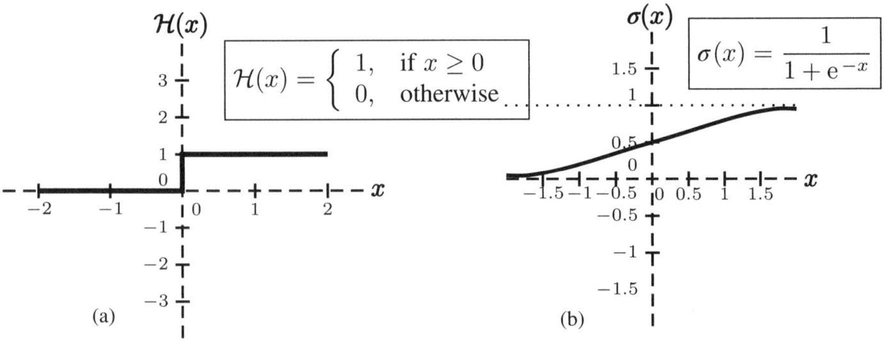

FIGURE 6.2 **(a)** The threshold gate function. **(b)** The sigmoidal gate function.

$$x_1(t+1) = 1/\left(1+e^{-(5x_2(t)-7x_3(t)+25)}\right)$$
$$x_2(t+1) = 1/\left(1+e^{12x_3(t)}\right)$$
$$x_3(t+1) = 1/\left(1+e^{-(2x_1(t)+2x_2(t)-19)}\right)$$

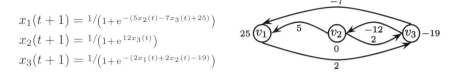

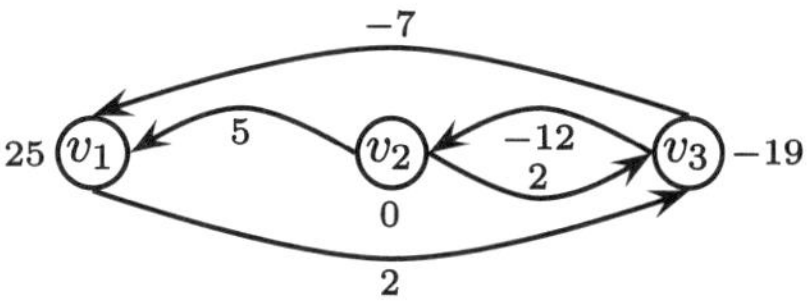

FIGURE 6.3 A discrete-time sigmoidal neural network and its graphical representation.

of the network is usually defined to be the length (number of nodes) in a longest directed path in the directed graph representation of the network.

Often ANN models are used for computing a function. For this purpose, one designates a specific variable $x_p \in \{x_1, x_2, \ldots, x_n\}$ as the output of the function; if computation of a Boolean function is desired and the variables are real-valued, then the output of the function is obtained by a suitable discretization of the output variable x_p—for example, designating $\mathcal{H}(x_p - \theta)$ as the output of the function for some suitable threshold value θ. In this setting, the ANN is said to compute a function $\mathfrak{f}(t)$ of the input $\mathbf{x}(0) = (x_1(0), x_2(0), \ldots, x_n(0))$ if $x_p(t) = \mathfrak{f}(x_1(0), x_2(0), \ldots, x_n(0))$.

6.3.1 Computational Powers of ANNs

It is not difficult to see that Boolean networks are special cases of the ANN models with threshold or sigmoidal functions (cf. Exercise 6.3). The simplest type of feed-forward ANN is the classical *perceptron* model [73] consisting of $n+1$ variables $x_1, x_2, \ldots, x_{n+1}$ with a corresponding update rule of $x_{n+1}(t+1) = \mathcal{H}(w_1x_1(t) + w_2x_2(t) + \cdots + w_nx_n(t) - \theta)$ for some real numbers $w_1, w_2, \ldots, w_n$ and θ. Such a model has a rather limited computational power [73]. However, for more complex feed-forward ANNs, computational powers for ANN models far exceed than that of the Boolean networks discussed in Sections 5.2 and 6.2. Computational powers of ANNs depend heavily on the nature of the activation function as well as whether the network is feed-forward or recurrent.

***Feed-Forward* ANNs.** Threshold networks—that is, feed-forward ANNs with threshold gate function, have been studied in considerable details by the theoretical computer science research community, and upper and lower bounds on depth and number of nodes (variables) required for these types of circuits to compute various Boolean functions (such as computing the parity function and computing the multiplication of binary numbers) have been obtained in computational complexity research areas; for example, see references 46, 52, 77, and 80.

However, it is more common in practice to use ANNs with continuous activation functions such as the sigmoidal function. It is known that feed-forward ANNs of *only* depth 2 but with arbitrarily large number of variables (nodes) can approximate *any* real-valued function up to *any* desired accuracy using a continuous gate function such as the sigmoidal function [26,61]; however, these proofs are mostly *nonconstructive*.

It is known that feed-forward ANNs with sigmoidal gate function are more powerful in computing functions than are feed-forward ANNs having the same number of variables with the Heaviside gate function [29,67]. A more detailed theoretical comparison of the computational powers of ANNs with various types of activation functions are provided in references 28 and 29; in particular, DasGupta and Schnitger [29] showed that any polynomial of degree n with polynomially bounded coefficients can be approximated with *exponential* accuracy by depth 2 feed-forward sigmoidal ANNs with a polynomial number of nodes.

***Recurrent* ANNs** Investigations of the computational powers of recurrent ANNs were initiated in references 86 and 87; reference 85 provides a thorough discussion of recurrent ANNs and analog computation in general. Recurrent ANNs gain considerably more computational power compared to the feed-forward ANNs with increasing computation time. In the following discussion, for the sake of concreteness, assume that the initial values of the variables at $t = 1$ are binary, the sigmoid function is used as the gate function, and the output of the ANN is binarized for computing a function on its inputs. If all the weights and thresholds are integers, then the recurrent ANNs turn out to be equivalent to finite automata and thus they recognize exactly the class of *regular language* over the binary alphabet $\{0, 1\}$. In contrast, recurrent ANNs whose weights and thresholds are rational numbers are, up to a polynomial time computation, equivalent to a Turing machine. Irrational weights provide an even *further boost* in computation power in the sense that, if the ANNs are allowed exponential computation time, then arbitrary Boolean functions (including noncomputable functions) are recognizable [86].

Finally, a precise theory of what can be computed by *noisy* ANNs has not very satisfactorily developed yet. Especially when dealing with continuous variables, one should also allow for noise due to continuous-valued disturbances, which may lead to stringent constraints on what can be effectively computed [68].

6.3.2 Reverse Engineering of ANNs

In the context of ANN models, reverse engineering is most commonly performed in a *learning theory* framework. In this framework, the architecture of the unknown ANN is given *a priori*, and the goal is to "learn" or reverse-engineer the weights and thresholds of all nodes in the architecture. Assume for concreteness that the ANN is used to compute an unknown Boolean function $\mathfrak{f}(\mathbf{x}) = \mathfrak{f}(x_1, x_2, \ldots, x_n) : \{0,1\}^n \mapsto \{0,1\}$, $x_{\text{out}} \in \{x_1, x_2, \ldots, x_n\}$ is the output variable that provides the output at a specific time $t_{\text{out}} > 0$, and the set of all possible inputs to the ANN come from some probability distribution $\mathcal{D}$. Just as our reverse-engineering problems for Boolean or signal transduction networks in Section 5.4 used biological experiments such as perturbation or knock-out experiments to generate the necessary data, it is assumed that we have available a set of s input-output pairs (called *samples*) $(\mathbf{x}_1, \mathfrak{f}(\mathbf{x}_1)), (\mathbf{x}_2, \mathfrak{f}(\mathbf{x}_2)), \ldots, (\mathbf{x}_s, \mathfrak{f}(\mathbf{x}_s))$, where the inputs $\mathbf{x}_1, \mathbf{x}_2, \ldots, \mathbf{x}_s$ are drawn from the set of all possible inputs based on the distribution $\mathcal{D}$. The reverse engineering process involves determining the values of weights and thresholds of the ANN based on these s samples. The final outcome of such a process provides a function $\widehat{\mathfrak{f}}(x_1, x_2, \ldots, x_n)$ computed by the ANN based on the determined valued of the weights and the thresholds. Let the *error probability* be defined as

$$\text{error-prob}_{\mathcal{D}} \stackrel{\text{def}}{\equiv} \Pr_{\mathcal{D}}\left[\widehat{\mathfrak{f}}(\mathbf{x}) \neq \mathfrak{f}(\mathbf{x})\right],$$

where the notation $\Pr_{\mathcal{D}}$ is used the clarify that probabilities are calculated with respect to the distribution $\mathcal{D}$. Typically, the reverse engineering process is considered to be possible (or, in the terminologies of learning theory, the function $\mathfrak{f}$ is *probably–approximately–correctly learnable*) if there exists a finite s such that for *every* $0 < \varepsilon, \delta < 1$ the following holds:

$$\Pr_{\mathcal{D}}\left[\text{errror-prob}_{\mathcal{D}} > \varepsilon\right] < \delta. \tag{6.3}$$

In the case when reverse engineering is possible, the function $s(\varepsilon, \delta)$ which provides, for any given ε and δ, the *smallest* possible s such that Eq. (6.3) holds, is called the *sample complexity* of the class of functions in which $\mathfrak{f}$ belongs. It can be proved that learnability is equivalent to finiteness of a combinatorial quantity called the *Vapnik–Chervonenkis* (Vc) *dimension* $\nu_{\mathcal{F}}$ of the class of functions $\mathcal{F}$ in which $\mathfrak{f}$ belongs. In fact, $s(\varepsilon, \delta)$ is bounded by a polynomial in $1/\varepsilon$ and $1/\delta$ and is proportional to $\nu_{\mathcal{F}}$ in the following sense [18,98]:

$$s(\varepsilon, \delta) = \mathrm{O}\left(\left(\nu_{\mathcal{F}}/\varepsilon\right)\log(1/\varepsilon) + (1/\varepsilon)\log(1/\delta)\right).$$

A specific set of weights and threshold values, denoted by $w^*_{i,j}$'s and θ^*_i's, can be computed by solving the so-called "loading problem" [30] that fits these samples by solving the following set of equations:

$$\widehat{\mathfrak{f}}(\mathbf{x}_i) = \mathfrak{f}(\mathbf{x}_i), \qquad i = 1, 2, \ldots, s. \tag{6.4}$$

Generalizations to the reverse engineering of *real-valued* (as opposed to Boolean) functions, by evaluation of the so-called "pseudo-dimension" are also possible (e.g., see references 53 and 65). The Vc dimension appears to grow only moderately with the complexity of the topology and the gate function for many types of ANNs. For example:

- The Vc dimension of the class of functions computed by any feed-forward ANN with α programmable parameters and the threshold activation function is $\Theta(\alpha \log \alpha)$ [12].
- Maass [66] and Goldberg and Jerrum [45] showed polynomial upper bounds of the Vc dimension of the class of functions computed by feed-forward ANNs with piecewise-polynomial activation functions.
- Karpinski and Macintyre [59] provided a $O(w^2 n^2)$ upper bound on the Vc dimension for the class of functions computed by any feed-forward ANN with the sigmoidal activation function, where w is the number of programmable parameters and n is the number of variables.

Unfortunately, solving Eq. (6.4) exactly to find an appropriate set of weights and thresholds may turn out to be NP-hard even for a simple 3-variable ANN with threshold or piecewise-linear gate function [17,31]. Thus, in practice, various types of optimization heuristics such as *back-propagation* [83] are used to obtain an approximate solution.

6.3.3 Applications of ANN Models in Studying Biological Networks

Although ANN models were originally inspired by the interconnectivity patterns of neurons in brains, they have also been successfully applied to model genetic circuits. We review two such applications below.

Application in Reverse Engineering of Gene Regulatory Networks [62]. In this application, an ANN model is to used to reverse engineer a genetic regulatory network from steady-state wild-type versus mutant gene expression data. We described a framework for reverse engineering in Section 5.4.2 to construct the causal relationships between genes from such data. As we commented in Section 5.4.3.2, testing such methods requires the generation of the so-called gold standard networks. Kyoda *et al.* [62] used an extended ANN model with a continuous gate function for obtaining the differential equation models of such gold-standard networks. An illustration of this application is shown in Fig. 6.4.

Perturbation experimental data via gene knock-outs can be generated from the model in the following manner. We may start with an equilibrium point, say $\bar{\mathbf{x}} = (\bar{x}_1, \bar{x}_2, \bar{x}_3, \bar{x}_4)$, of the model as the wild type. To knock out the gene v_i, we can remove the node v_i and its associated arcs from the network, start the network at $\bar{\mathbf{x}}$, and run the network dynamics until it reaches a new equilibrium point $\widetilde{\mathbf{x}} = (\widetilde{x}_1, \widetilde{x}_2, \widetilde{x}_3, \widetilde{x}_4)$. To generate time series data, we can simply start the network at a prescribed initial state and run it for a specific number of time steps.

Application in Crop Simulation Modeling. Accurate modeling of the genetic regulatory mechanism leading to flowering time is critical in crop modeling for finding time intervals during which growth and yield-generating processes operate. Welch *et al.* [101] provided a simple 8-node ANN with the sigmoidal gate function to model the control of inflorescence transition in the plant *Arabidopsis thaliana*.

6.4 PIECEWISE LINEAR MODELS

An affine map h over a vector space $\mathbf{z} = (z_1, z_2, \ldots, z_n)$ is given by $h(\mathbf{z}) = P\,\mathbf{z} + \mathbf{b}$, where P is a $n \times n$ matrix (called the *affine transformation matrix*) and $\mathbf{b} = (b_1, b_2, \ldots, b_n)$ is a vector. Piecewise linear (PL) systems, in the sense defined in reference 91, are discrete-time systems described by equations $\mathbf{x}(t+1) = P(\mathbf{x}(t), \mathbf{u}(t))$ (or, in more concise notation, "$\mathbf{x}^+ = P(\mathbf{x}, \mathbf{u})$") for which the transition mapping P is a PL map, that is, there is a decomposition of the state space $\mathbf{x}$ and the input vector

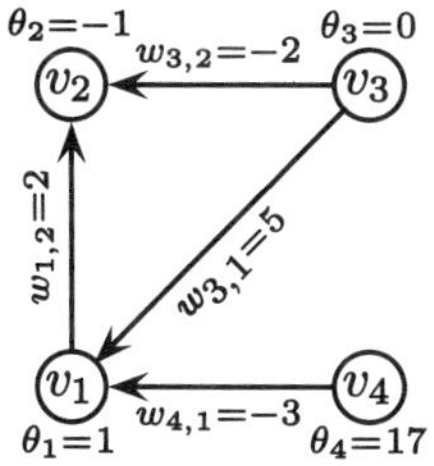

(a)

$$x_1(t+1) = R_1\, g\Big(5x_3(t) - 3x_4(t) + 1\Big) - \lambda_1\, x_1(t)$$
$$x_2(t+1) = R_2\, g\Big(2x_1(t) - 2x_3(t) + 1\Big) - \lambda_2\, x_2(t)$$
$$x_3(t+1) = R_3\, g\big(0\big) - \lambda_3\, x_3(t)$$
$$x_4(t+1) = R_4\, g\big(0\big) - \lambda_4\, x_4(t)$$

(b)

$$g(x) = \frac{1}{2}\left(1 + \frac{x}{\sqrt{x^2+1}}\right)$$
$$\lim_{x\to-\infty} g(x) = 0$$
$$\lim_{x\to\infty} g(x) = 1$$

(c)

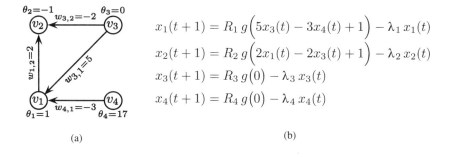

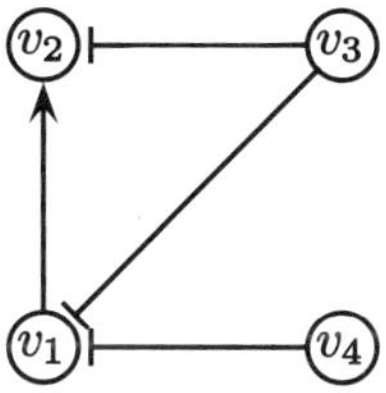

(d)

FIGURE 6.4 **(a)** A continuous-state discrete-time ANN with a continuous gate function g. **(b)** The difference equation model corresponding to the ANN in (a). R_i is the maximum rate of synthesis of gene v_i, λ_i is the degradation rate of the product from gene v_i, and the threshold θ_i summarizes the effect of general transcription factors on gene v_i. **(c)** The specific activation function g used by Kyoda et al. [62]. **(d)** The topological version of the model in **(b)** indicating excitory and inhibitory effects.

space **u** into *finitely* many pieces such that, in each of these pieces, the mapping P is given by an affine map. The decomposition is required to be *polyhedral*, meaning that each piece is described by a set of linear equalities and inequalities. Explicit constraints on controls and states are included as part of the specification of a PL system. Thus, the state space and input value sets are subsets of $\mathbb{R}^n$ and $\mathbb{R}^m$, respectively, indicating *a priori* restrictions on the allowed ranges of variables; and to make the dynamics remain piecewise linear, these sets are required to be definable in terms of a finite number of linear equalities and inequalities.

Linear systems arise in the special case in which there is just *one* region. But the PL system paradigm includes many more situations of interest, such as:

- linear systems such as $\mathbf{x}^+ = A\,\mathbf{x} + B\,\text{sat}\,(\mathbf{u})$ whose actuators are subject to saturation, where $\text{sat}\,(\mathbf{u}) = \text{sat}\,\big(u_1, u_2, \ldots, u_m\big) = (\tilde{u}_1, \tilde{u}_2, \ldots, \tilde{u}_m) = \tilde{\mathbf{u}}$ in which
$$\tilde{u}_i = \begin{cases} 1 & \text{if } u_i > 1, \\ -1 & \text{if } u_i < 1, \\ u_i & \text{otherwise,} \end{cases}$$
- switched systems $\mathbf{x}^+ = A_i\,\mathbf{x} + B_i\,\mathbf{u}$, where the choice of matrix pair (A_i, B_i) depends on a set of linear constraints on current inputs and states, and

- systems $\mathbf{x}^+ = \text{sat}\,(A\mathbf{x} + B\mathbf{u})$ for which underflows and overflows in state variables must be taken into account.

PL systems are the *smallest* class of systems which is closed under interconnections and which contains both linear systems and finite automata.

6.4.1 Dynamics of PL Models

It was shown in reference 14 that the general class of hybrid ''Mixed Logical Dynamical'' systems introduced in reference 15 is in a precise sense equivalent to the PL systems. Based on this equivalence, and using tools from piecewise affine systems, the Bemporad et al. [14] studied basic system-theoretic properties and suggested numerical tests based on mixed-integer linear programming for checking controllability and observability.

Another basic question that is often asked about the PL class of systems is the one regarding *equivalence*, that is, *given two systems, do they represent the same dynamics under a change of variables*? Indeed, as a preliminary step in answering such a question, one must determine if the *state spaces of both systems are isomorphic* in an ''appropriate sense.'' That is, one needs to know if an invertible change of variables is at all possible, and only then can one ask if the equations are the same. For classical finite-dimensional linear systems, this question is trivial since only the dimensions must match. For finite automata, similarly, the question is also trivial, because the cardinality of the state set is the only property that determines the existence of a relabeling of variables. For other classes of systems, however, the question is not as trivial, and single numbers such as dimensions or cardinalities may not suffice to settle this isomorphism (i.e., state-equivalence) problem. For example, if one is dealing with continuous-time systems defined by smooth vector fields on manifolds, the natural changes of variables are smooth transformations, and thus a system whose state space is a unit circle *cannot* be equivalent to a system whose state space is the real line, even though both systems have a dimension of 1. As another example, for systems whose variables are required to remain bounded—for instance, because of saturation effects—a state-space like the unit interval $[-1, 1]$ looks very different from the unbounded state-space $\mathbb{R}$, even though both have dimension 1.

To provide a precise definition of equivalence between two PL systems, we start with an equivalent alternate definition of PL sets and maps. The *PL subsets* of $\mathbb{R}^n$ are those belonging to the *smallest* Boolean algebra that contains all the open half-spaces of $\mathbb{R}^n$. A map $h\colon X \to Y$ between two PL subsets $X \subseteq \mathbb{R}^a$ and $Y \subseteq \mathbb{R}^b$ is a *PL map* if its graph is a PL subset of $\mathbb{R}^a \times \mathbb{R}^b$. By a *PL set*, one means a PL subset of some $\mathbb{R}^n$. Then, a PL system is a discrete-time system $\mathbf{x}^+ = P(\mathbf{x}, \mathbf{u})$ with PL state and input value sets and PL transition matrix P. Viewed in this manner, two PL sets X and Y are PL-isomorphic if there are PL maps $h_1 : X \to Y$ and $h_2 : Y \to X$ such that $h_1 \circ h_2$ and $h_2 \circ h_1$ are both equal to the identity mapping, that is, $y = h_1(x)$ is a bijective piecewise linear map.

An elegant geometric interpretation of PL-isomorphism was provided in reference 32 and is summarized below. Geometrically, a PL-isomorphism is a sequence of operations of the following type:

- Make a finite number of cuts along a set of lines (or segments).
- Apply an affine transformation to each piece (without dropping any lower-dimensional pieces).
- Paste it all together.

For example, consider the interior $\triangle$ of the triangle in $\mathbb{R}^2$ obtained as the interior of the convex hull of the points $(0,0)$, $(1,1)$ and $(2,0)$ and the interior $\square$ of the open square with points $(0,0)$, $(1,1)$, $(0,1)$, and $(1,0)$. Then, $\triangle$ is PL-isomorphic to $\square$ as shown below:

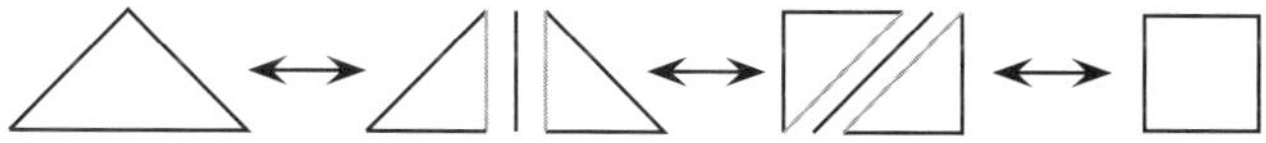

Sontag [92] introduced and developed a logic formalism of the PL systems (''PL algebra'') to help in classifying PL sets that are equivalent under isomorphism. The critical step in this classification was to associate to each PL set X a ''label'' with the property that two PL sets X and Y are isomorphic if and only if their labels are related in a certain manner. Subsequent algorithmic developments on this type of equivalence checking procedures are described in reference 32.

6.4.2 Biological Application of PL Models

PL systems are quite powerful in modeling biological systems since they may be used as identification models (by means of piecewise linear approximations), or as controllers for more general systems. Arbitrary interconnections of linear systems and finite automata can be modeled by PL systems, and vice versa. More precisely, given any finite automaton with state space Q, input alphabet Σ, and state transition function $\delta\colon Q \times \Sigma \to Q$, we allow the state $q \in Q$ of the finite automaton to control switching among $|Q|$ possible linear dynamics in the following manner:

$$\begin{aligned} \mathbf{x}(t+1) &= A_q\,\mathbf{x}(t) + B_q\,\mathbf{u}(t) + c_q, \\ q(t+1) &= \delta\big(\,q(t),\ h\big(\,\mathbf{x}(t), \mathbf{u}(t)\,\big)\,\big), \end{aligned} \tag{6.5}$$

where $A_1, A_2, \ldots, A_{|Q|} \in \mathbb{R}^{n\times n}$, $B_1, B_2, \ldots, B_{|Q|} \in \mathbb{R}^{n\times m}$, $c_1, c_2, \ldots, c_{|Q|} \in \mathbb{R}^n$, and $h\colon \mathbb{R}^n \times \mathbb{R}^m \to \Sigma$ is a PL map representing quantized observations of the linear systems.

We now discuss two specific applications of PL systems in modeling biological processes.

PL Models for Genetic Regulatory Networks. PL models for regulatory networks can be built starting with the formulations proposed originally by Glass and Kauffman [43]. Such an approach was illustrated and investigated by Casey, Jong, and Gouzé [20]. We illustrate the approach by starting with the special form of system (6.1):

$$\begin{aligned} x_1(t+1) &= f_1(\mathbf{x}(t)) + (1-\lambda_1)x_1, \\ x_2(t+1) &= f_2(\mathbf{x}(t)) + (1-\lambda_2)x_2, \\ &\vdots \\ x_n(t+1) &= f_n(\mathbf{x}(t)) + (1-\lambda_n)x_n, \end{aligned} \tag{6.6}$$

where we omitted the input vector under the assumption that the state vector $\mathbf{x}$ can itself be influenced externally. The parameter λ_i represents the degradation rate of the ith molecular component (e.g., protein). The function f_i representing the synthesis rate of the ith molecular component is approximated by the piecewise-linear function by taking a linear combination of the form

$$f_i(\mathbf{x}(t)) = \kappa_{i,1}\mathfrak{n}_{i,1}(\mathbf{x}(t)) + \kappa_{i,2}\mathfrak{n}_{i,2}(\mathbf{x}(t)) + \ldots + \kappa_{i,\ell}\mathfrak{n}_{i,\ell}(\mathbf{x}(t)),$$

where the $\kappa_{i,j} > 0$ parameters control the rates of regulation. The Boolean regulatory functions $\mathfrak{n}_{i,j}$ model the conditions of regulation of x_i by the remaining variables and is written as an appropriate combinations of the Heaviside (threshold) functions $\mathcal{H}(x_\ell - \theta_\ell)$'s and $1 - \mathcal{H}(x_\ell - \theta_\ell)$'s, for various $\ell \in \{1, 2, \ldots, n\}$, where θ_ℓ's are appropriate threshold concentration parameters.

As an illustration, suppose that we want to model the following regulatory mechanism: Gene x_1 is expressed at the rate of 0.23 if the concentration of protein x_2 is above 0.3 and the concentration of protein x_3 is below 0.9; otherwise gene x_1 is not expressed. This is obtained by the equation

$$f_i((x_1(t), x_2(t), x_3(t))) = 0.23\,\mathcal{H}(x_2 - 0.3)\left[1 - \mathcal{H}(x_3 - 0.9)\right].$$

It is easy to verify that the above kind of system partitions the entire n-dimensional state space into a finite number of n-dimensional hyper-rectangular regions ("pieces") and within each such region the behavior of the dynamics is linear. For example, consider the following system:

$$\begin{aligned} x_1^+ &= 0.1\,\mathcal{H}(0.3\,x_1)\left(1 - \mathcal{H}(0.2\,x_2)\right) + 0.7\,x_1, \\ x_2^+ &= 0.2\,\mathcal{H}(0.3\,x_1)\,\mathcal{H}(0.2\,x_2) + 0.9\,x_2 \end{aligned} \tag{6.7}$$

and suppose that $0 \leq x_1(t), x_2(t) \leq 1$ for all $t \geq 0$. Then, the two-dimensional state space is partitioned as shown in Fig. 6.5.

Hybrid Automata for Delta-Notch Protein Signaling. Several cellular processes, such as pattern formation due to lateral inhibition [41,64], involve the so-called *Delta-Notch protein signaling mechanism.* A model for the regulation of intracellular

if	$x_1(t) > 0.3$ and $x_2(t) < 0.2$
then	$x_1(t+1) = 0.1 + 0.7x_1(t)$
	$x_2(t+1) = 0.9\,x_2(t)$
if	$x_1(t) > 0.3$ and $x_2(t) > 0.2$
then	$x_1(t+1) = 0.7x_1(t)$
	$x_2(t+1) = 0.2 + 0.9\,x_2(t)$
if	$x_1(t) < 0.3$ and $x_2(t) > 0.2$
then	$x_1(t+1) = 0.7x_1(t)$
	$x_2(t+1) = 0.9\,x_2(t)$
if	$x_1(t) < 0.3$ and $x_2(t) < 0.2$
then	$x_1(t+1) = 0.7x_1(t)$
	$x_2(t+1) = 0.9\,x_2(t)$

Graphical illustration of the hyperplane (colored black) that shows $x_1(t+1)$ as a function of $x_1(t)$ and $x_2(t)$ when $0.3 < x_1(t) \leq 1$ and $0.2 < x_2(t) \leq 1$

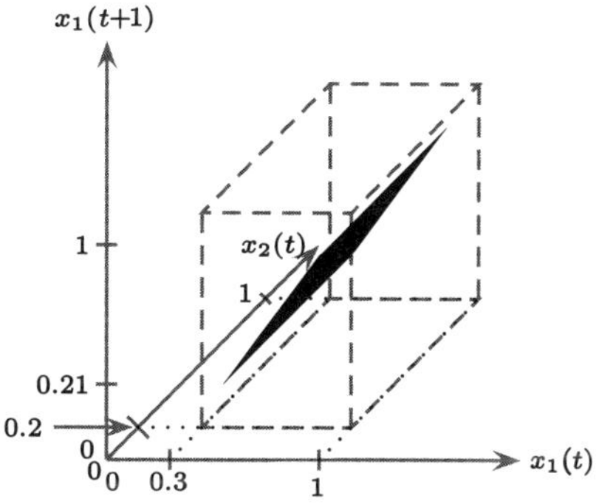

FIGURE 6.5 Rectangular partition of state space induced by system (6.7).

protein concentrations in this signaling mechanism through the feedback system was described in reference 40 using the experimentally observed rules for the associated biological phenomenon. In this model, each cell was implemented as a piecewise linear hybrid automaton with 4 states. In this section, we review this application after defining a piecewise linear hybrid automata.

Recall that a standard nondeterministic finite automata (NFA) is defined [56] as a 5-tuple $(Q, \Sigma, \delta, q_0, F)$, where Q is a finite set of states, Σ is a finite input alphabet, $\delta\colon Q \times \Sigma \mapsto 2^Q$ is the state transition map where 2^Q denotes the power set of Q, $q_0 \in Q$ is the initial state, and $\emptyset \subset F \subseteq Q$ is the set of final (accepting) states. See Fig. 6.6**a** for an illustration. This NFA model can be combined with a piecewise-linear dynamics to obtain a piecewise-linear hybrid automata (PLHA) [5,40] as discussed below.

In PLHA, the set of states Q are of two types: a set of m *discrete* states $Q^d = \{q_1, q_2, \ldots, q_m\}$ and a set $Q^c = \{x_1, x_2, \ldots, x_n\}$ of n *continuous* state variables such that every state variable $q \in Q^c$ assumes a real value. We also have a set of m *discrete* inputs $\Sigma = \{\sigma_1, \sigma_2, \ldots, \sigma_m\}$ and a set of t linear functions $\mathfrak{g}_1, \mathfrak{g}_2, \ldots, \mathfrak{g}_t\colon \mathbb{R}^{n+m} \mapsto \mathbb{R}$ where

$$\mathfrak{g}_j(x_1, x_2, \ldots, x_n, \sigma_1, \sigma_2, \ldots, \sigma_m) = \left(\sum_{k=1}^{n} p_{j,k} x_k\right) - p_{j,n+1} + \sum_{k=1}^{m} p'_{j,k} \sigma_k$$

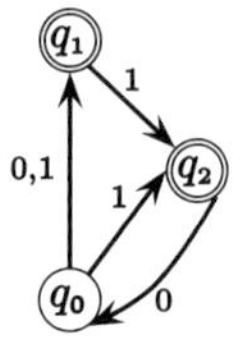

$Q = \{q_0, q_1, q_2\}$
$\Sigma = \{0, 1\}$
$F = \{q_1, q_2\}$

δ	0	1
q_0	$\{q_0\}$	$\{q_1, q_2\}$
q_1	—	$\{q_2\}$
q_2	$\{q_0\}$	—

(a)

$$Q^d = \{q_1, q_2\},\;\; Q^c = \{x_1, x_2\}$$
$$\mathfrak{g}_1(x_1, x_2) = x_1,\;\; \mathfrak{g}_2(x_1, x_2) = x_2$$
$$\mathfrak{g}_3(x_1, x_2) = x_1 + x_2 - 1,\;\; \mathfrak{g}_4(x_1, x_2) = x_2 - x_1$$
$$\mathbf{s}(q_1) = (>, >, \leq, >),\;\; \mathbf{s}(q_2) = (>, >, \leq, \leq)$$
$$Q_0^c = (0, 0.4),\;\; Q_0^d = \{q_1\}$$
$$a_{q_1,1} = 1/100,\, a_{q_1,2} = -1/100$$
$$a_{q_2,1} = 2/50,\, a_{q_2,2} = -1/100$$
$$\forall x_1 + x_2 = 1\colon \delta(q_1, x_1, x_2) = (q_2, x_1, x_2)$$

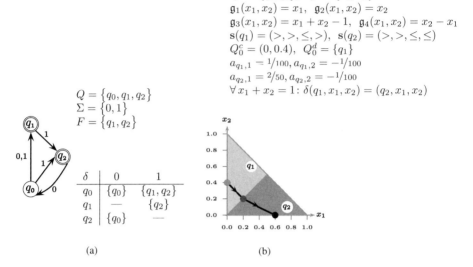

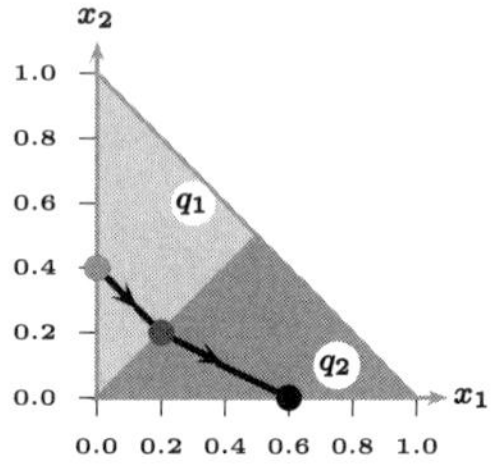

(b)

FIGURE 6.6 **(a)** An example of an NFA. The input $0101 \in \Sigma^*$ is accepted by the NFA since there is a directed path from the initial state q_0 to a final state q_2 labeled 0101. **(b)** An example of a piecewise-linear hybrid automata (PLHA) with two continuous-state variables and no inputs. A hypothetical trajectory of the dynamics is shown by thick black lines with arrows.

for some numbers $p_{j,1}, \ldots, p_{j,n+1}, p'_{j,1}, \ldots, p'_{j,m} \in \mathbb{R}$. These functions are used to define linear constraints that decide the boundaries of the piecewise-linear system and associate discrete states with these regions in the following manner. Let a "sign pattern" $\mathbf{s} = (s_1, s_2, \ldots, s_t) \in \{<, \leq, =, \geq, >, \sqcup\}^t$ denote the intersection of t half-spaces where the jth half-space is

$$\begin{aligned}
&\mathfrak{g}_j(x_1, x_2, \ldots, x_n, \sigma_1, \sigma_2, \ldots, \sigma_m) > 0 && \text{if } s_j \text{ is } > \\
&\mathfrak{g}_j(x_1, x_2, \ldots, x_n, \sigma_1, \sigma_2, \ldots, \sigma_m) \geq 0 && \text{if } s_j \text{ is } \geq \\
&\mathfrak{g}_j(x_1, x_2, \ldots, x_n, \sigma_1, \sigma_2, \ldots, \sigma_m) = 0 && \text{if } s_j \text{ is } = \\
&\mathfrak{g}_j(x_1, x_2, \ldots, x_n, \sigma_1, \sigma_2, \ldots, \sigma_m) \leq 0 && \text{if } s_j \text{ is } \leq \\
&\mathfrak{g}_j(x_1, x_2, \ldots, x_n, \sigma_1, \sigma_2, \ldots, \sigma_m) < 0 && \text{if } s_j \text{ is } < \\
&\mathfrak{g}_j(x_1, x_2, \ldots, x_n, \sigma_1, \sigma_2, \ldots, \sigma_m) = \sqcup && \text{if there is no constraint involving } s_j.
\end{aligned}$$

Then, each discrete state $q \in Q^d$ is associated with a *distinct* sign-vector $\mathbf{s}(q)$ such that the state q is active if all the corresponding half-space constraints are satisfied. We can then define adjacency of two states as being the adjacency of their corresponding geometric regions in the n-dimensional space, for example, two states may be adjacent if their sign patterns differ in *exactly* one coordinate. It is possible in practice to design algorithms to check this kind of geometric adjacency in the n-dimensional space [40]. The state transition map $\delta\colon Q^d \times Q^c \times \Sigma \mapsto 2^{Q^d \times Q^c}$ now involves both discrete and

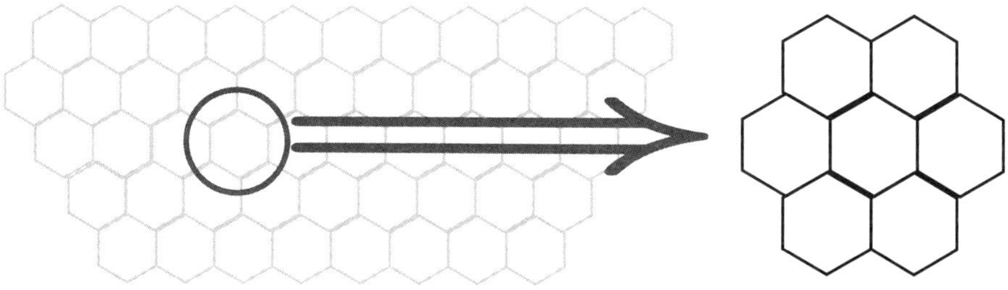

FIGURE 6.7 Hexagonal lattice of cells for Delta-Notch protein signaling. A cell with its six neighbors is shown amplified.

continuous states by allowing a switch from a discrete state to another adjacent discrete state at the boundary of the two regions corresponding to the two states. At the "inside" of the region for a discrete state q, the continuous state variables $\mathbf{x} = (x_1, x_2, \ldots, x_n)$ evolve in a linear manner given by

$$\frac{d\,\mathbf{x}(t)}{dt} = \begin{bmatrix} a_{q,1} \\ a_{q,2} \\ \cdots \\ a_{q,n} \end{bmatrix}$$

or

$$\begin{bmatrix} x_1(t+1) \\ x_2(t+1) \\ \cdots \\ x_n(t+1) \end{bmatrix} = \begin{bmatrix} x_1(t) \\ x_2(t) \\ \cdots \\ x_n(t) \end{bmatrix} + \begin{bmatrix} a_{q,1} \\ a_{q,2} \\ \cdots \\ a_{q,n} \end{bmatrix}$$

for some $a_{q,1}, a_{q,2}, \ldots, a_{q,n}$. The system is started by specifying an *initial condition* $Q_0^d \times Q_0^c$ of the discrete and continuous states. Figure 6.6**b** illustrates a PLHA.

More discussions related to various abstractions of hybrid automata and their decidability issues can be found in reference 5. In particular, the formalism described above allows *symbolic description* of many parameters without specific numerical instantiation; for example, the elements $a_{q,1}, \ldots, a_{q,n}$ can be represented symbolically using a few free parameters and constraints over these parameters. For example, an exponential decay with a rate of r for a continuous state variable $x_j \in Q^c$ in a discrete state $q \in Q^d$ can be obtained by specifying $a_{q,j} = r\,x_j + r'$ where r' determines the boundary condition.

Ghosh and Tomlin [40] used the PLHA model to study the Delta-Notch protein signaling mechanism that is responsible for many cellular processes. The cells are assumed to be packed in a hexagonal lattice with each cell (except those at the boundary of the lattice) having six neighboring cells (see Fig. 6.7). The parameters of the PLHA for each cell are as follows [40]:

$$Q^d = \{q_1, q_2, q_3, q_4\}, \qquad Q^c = (x_1, x_2) \in \mathbb{R}^2, \qquad \Sigma = \left\{ u_N = \sum_{i=1}^{6} x_{\text{Delta},i} \right\},$$

$$\mathfrak{g}_1(x_1, x_2, u_N) = -x_2 - h_D, \qquad \mathfrak{g}_2(x_1, x_2, u_N) = u_N - h_N,$$

$$\mathbf{s}(q_1) = (\,<,<\,), \qquad \mathbf{s}(q_2) = (\,\geq,<\,), \qquad \mathbf{s}(q_3) = (\,<,\geq\,), \qquad \mathbf{s}(q_4) = (\,\geq,\geq\,),$$

$$a_{q_1,1} = a_{q_3,1} = -\lambda_D\, x_1(t), \qquad a_{q_1,2} = a_{q_2,2} = -\lambda_N\, x_2(t), \qquad a_{q_2,1} = R_D - \lambda_D\, x_1(t),$$

$$a_{q_3,2} = a_{q_4,2} R_N - \lambda_N\, x_2(t), \qquad a_{q_4,1} = R_D - \lambda_D\, x_1(t),$$

$$Q_0^d = Q^d, \qquad Q_0^c = \{(x_1, x_2) \mid x_1, x_2 \in \mathbb{R}\}.$$

Biological interpretations of various parameters are as follows:

Q^d	:	Four discrete states to switch ON or OFF the production of the Delta protein or the Notch protein individually
x_1 and x_2	:	Concentrations of Delta and Notch proteins, respectively
u_D and u_N	:	Inputs to physically realize the biological constraint that production of Notch protein in a cell is turned ON by high levels of the Delta protein in the immediate neighbourhood of the cell, and the production of the Delta protein is switched ON by low levels of the Notch protein in the same cell
$x_{\text{Delta},i}$	:	Value of x_1 in the ith neighboring cell
λ_D and λ_N	:	Decay constants for Delta and Notch proteins, respectively
R_D and R_N	:	Constant production rates for Delta and Notch proteins, respectively
h_D and h_N	:	Switching thresholds for productions of Delta and Notch proteins, respectively

Note that cells influence and are influenced by their six neighboring cells via the input u_N and the variables $x_{\text{Delta},j}$ for $j = 1, 2, \ldots, 6$. Empirical estimates for the thresholds h_D and h_N are provided in reference 41. A major goal of Ghosh and Tomlin [40] in formulating the above PLHA model for Delta-Notch signalling was to determine initial conditions from which specific equilibrium points of the dynamics are reachable (the so-called "backward reachability" problem). To this effect, Ghosh and Tomlin [40] designed and implemented an efficient heuristic algorithm for the backward reachability problem for this model.

6.5 MONOTONE SYSTEMS

One approach to mathematical analysis of complex biological systems relies upon viewing them as made up of subsystems whose behavior is simpler and easier to understand. Coupled with appropriate interconnection rules, the hope is that emergent properties of the complete system can be deduced from the understanding of these subsystems. Diagrammatically, one can picture this as in Fig. 6.8, which shows a full system as composed of five interconnected subsystems.

An interesting class of biological systems with simpler behaved dynamics are systems with *monotone* dynamics (or, simply, the *monotone systems*) [54,55,89]. Monotone systems constitute a nicely behaved class of dynamical systems in several ways. For example, for these systems, pathological behaviors of dynamics (e.g., chaotic attractors) are ruled out. Even though they may have arbitrarily large dimensionality, monotone systems behave in many ways like one-dimensional systems—for example, bounded trajectories generically converge to steady states—and there are no stable oscillatory behaviors (limit cycles). Monotonicity with respect to orthant orders is equivalent to the nonexistence of negative loops in systems; analyzing the behaviors of such loops is a long-standing topic in biology in the context of regulation, metabolism, and development, starting from the work of Monod and Jacob in 1961 [74] and culminating in many subsequent works such as references 7, 9, 23, 63, 71, 78, 82, 90, and 96. An interconnection of monotone systems may or may not be monotone: "Positive feedback" preserves monotonicity, while "negative feedback" destroys it. Thus, oscillators such as circadian rhythm generators require negative feedback loops in order for periodic orbits to arise and hence are not themselves monotone systems; however they can be decomposed into monotone subsystems [10]. Theoretical characterizations of the behavior of non-monotone interconnections are available in references 6, 8, 35, 38, and 39.

A key point brought up in references 93 and 94 is that new techniques for monotone systems in many situations allow one to characterize the behavior of an entire system, based upon the "qualitative" knowledge represented by general network topology and the inhibitory or activating character of interconnections, combined with only a relatively *small amount of quantitative* data. The latter data may consist of steady-state responses of components (dose–response curves and so forth), and there is no need to know the precise form of dynamics or parameters such as kinetic constants in order to obtain global stability conclusions.

6.5.1 Definition of Monotonicity

Recall that a partial order $\preceq$ over a set U is a binary relation that is *reflexive* (i.e., $u \preceq u$ for every $u \in U$), *antisymmetric* (i.e., if $u \preceq u'$ and $u' \preceq u$ then $u = u'$), and transitive (i.e., if $u \preceq u'$ and $u' \preceq u''$ then $u \preceq u''$). For example, the $\div$ relation on the set of positive integers $\mathbb{N}$ defined by "$a \div b$ if and only if a is an integral multiple of b" is a partial order relation.

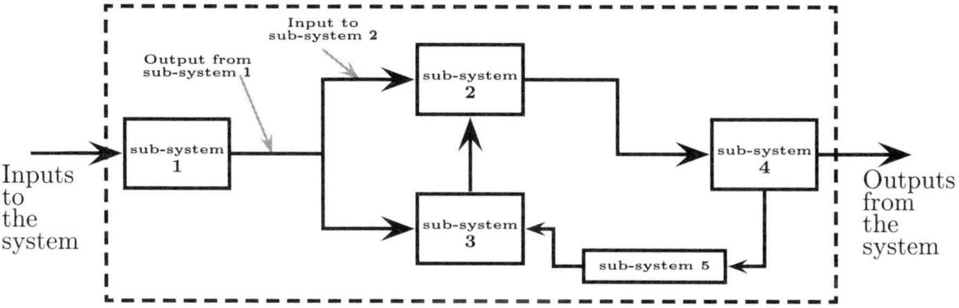

FIGURE 6.8 A system composed of five interconnected subsystems.

For easier understanding, we illustrate the definition of monotonicity of dynamics of a system when it is expressed via the differential equation model (5.1) with no inputs, that is,

$$\frac{\mathrm{d}x_i(t)}{\mathrm{d}t} = f_i\big(x_1(t), x_2(t), \ldots, x_n(t)\big), \qquad i = 1, 2, \ldots, n \\ \text{or, in concise vector notation, } \dot{\mathbf{x}} = f(\mathbf{x}). \tag{6.8}$$

However, much of the discussion also applies to more general types of dynamical systems such as delay-differential systems or certain systems of reaction-diffusion partial differential equations. We will discuss in Section 6.5.2 how to incorporate control inputs (and outputs) in our definitions of monotonicity.

An additional requirement that we need before giving the definition of a monotone system is that, for each i and j, either $\frac{\partial f_j}{\partial x_i} \geq 0$ for all $\mathbf{x}$ or $\frac{\partial f_j}{\partial x_i} \leq 0$ for all $\mathbf{x}$. In other words, in our model the direct effect that one given variable has over another variable is *unambiguous* in the sense that it is always inhibitory or always excitory. Thus, for example, if protein binds to the promoter region of another gene, we assume that it does so either to prevent the transcription of the gene or to facilitate it irrespective of the respective concentrations. As explained in detail in reference 22, this requirement is *not* a severe restriction. Firstly, note that this unambiguity assumption does *not* prevent a protein from having an *indirect* influence, through other molecules, that can ultimately lead to the opposite effect on a gene from that of a direct connection. Secondly, in biomolecular networks, ambiguous signs in Jacobians often represent heterogeneous mechanisms, and introducing a new species into the model (i.e., an additional variable for this intermediate form) reduces the original system to an equivalent new system in which the signs of the Jacobian entries are unambiguous. Finally, small-scale negative loops that are abundant in nature often represent fast dynamics which may be collapsed into self-loops via time-scale decomposition (singular perturbations or, specifically for enzymes, ‘‘quasi-steady state approximations’’) and hence may be viewed as diagonal terms, but the requirement of a fixed sign for Jacobian entries is *not* imposed on diagonal elements.

The dynamics of a monotone system preserves a *specific* partial order of its inputs over time. More precisely, monotonicity is defined as follows.

Definition 6.3 [55,89]. *Given a partial order $\preceq$ over $\mathbb{R}^n$, system* (6.8) *is said to be* monotone with respect to $\preceq$ *if*

$$\forall t \geq 0\colon\ \big(x_1(0), \ldots, x_n(0)\big) \preceq \big(y_1(0), \ldots, y_n(0)\big) \\ \Longrightarrow \big(x_1(t), \ldots, x_n(t)\big) \preceq \big(y_1(t), \ldots, y_n(t)\big),$$

where $\big(x_1(t), \ldots, x_n(t)\big)$ and $\big(y_1(t), \ldots, y_n(t)\big)$ are the solutions of (6.8) *with initial conditions $\big(x_1(0), \ldots, x_n(0)\big)$ and $\big(y_1(0), \ldots, y_n(0)\big)$, respectively.*

Of course, whether a system is monotone or not depends on the partial order being considered. We will consider the following partial order that has been investigated in previous research works such as references 9, 27, and 89.

Definition 6.4 (Orthant Order) *For a given* $s = (s_1, \ldots s_n) \in \{-1, 1\}^n$*, an orthant order* $\preceq_s$ *is a partial order over* $\mathbb{R}^n$ *defined by*

$$\forall \mathbf{x} = (x_1, x_2, \ldots, x_n) \in \mathbb{R}^n \, \forall \mathbf{y} = (y_1, y_2, \ldots, y_n) \in \mathbb{R}^n : \mathbf{x} \preceq_s \mathbf{y} \Longleftrightarrow \forall i : s_i \, x_i \le s_i \, y_i.$$

An example of orthant order is the "cooperative order," which is the partial order $\preceq_s$ for $s = (1, 1, \ldots, 1)$, that is, $\mathbf{x} \preceq_{1,1,\ldots,1} \mathbf{y} \Longleftrightarrow \forall i : x_i \le y_i$; in traditional computational geometry literature the cooperative order is also known as the *dominance* relationship between n-dimensional points [49].

In the rest of our discussions on monotone systems, we will use the term "monotone systems" with the assumption that the monotonicity is with respect to some fixed orthant order $\preceq_s$.

6.5.2 Combinatorial Characterizations and Measure of Monotonicity

The reader can easily verify the following characterization of monotone systems [89].

Kamke's Condition. Consider an orthant order $\preceq_s$ generated by $s = (s_1, \ldots, s_n)$. Then, the system (6.8) is monotone with respect to $\preceq_s$ if and only if

$$\forall 1 \le i \ne j \le n : \; s_i \, s_j \, \frac{\partial f_j}{\partial x_i} \ge 0.$$

Based on Kamke's condition, DasGupta *et al.* [27] discussed an elegant graph-theoretic characterizations of monotonicity of the system (6.8) that may be subjected to algorithmic analysis. For this characterization, we consider the system (6.8) and its associated signed (directed) graph $G = (V, E)$ as defined in Section 5.1.3. To provide an intuition behind the first characterization, consider the following biological system with three variables:

$$\frac{dx_1}{dt}(t) = \left(x_3(t)\right)^{-3} - x_1(t), \qquad \frac{dx_2}{dt}(t) = 2\,x_1(t) - 9\left(x_2(t)\right)^7, \qquad \frac{dx_3}{dt}(t) = x_2(t) - 3\,x_3(t). \tag{6.9}$$

The associated signed graph for this system is as shown below:

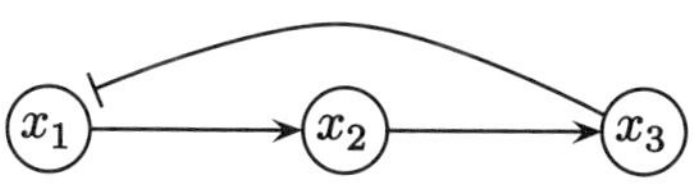

It is possible to show that the system (6.9) is *not* monotone with respect to $\preceq_s$ for any s (cf. Exercise 6.10). But, if we remove the term $\left(x_3(t)\right)^{-3}$ in the first equation, we obtain a system that *is* monotone with respect to $\preceq_{1,1,1}$. A cause of nonmonotonicity of the system is the existence of *sign-inconsistent* paths between two nodes in an *undirected* version of the signed graph — that is, the existence of both an activation and an inhibitory path between two nodes when *the directions of the arcs are ignored*. Define a closed *undirected chain* in G as a sequence of nodes $x_{i_1}, \ldots, x_{i_r}$ such that $x_{i_1} = x_{i_r}$ and such that for every $t = 1, \ldots, r-1$ either $(x_{i_t}, x_{i_{t+1}}) \in E$ or $(x_{i_{t+1}}, x_{i_t}) \in E$. The graph G is said to be *sign-consistent* if all paths between any two nodes have the same parity or, equivalently, all closed undirected chains in G have a parity of 1 (i.e., all closed undirected chains have an even number, possibly zero, of arcs labeled -1). The following characterization now holds [27,33,88].

(Combinatorial Characterization of Monotonicity)

System (6.8) is monotone with respect to some orthant order if and only if G is sign-consistent, that is, all closed undirected chains of its associated signed graph G have parity 1.

Figure 6.9 illustrates the above combinatorial characterization of monotonicity.

Of course, one should not expect complex biological systems to have an associated signed graph that is consistent. However, if the number of inconsistent pairs of path in the undirected versions of a sign graph is *small*, it may well be the case that the network is in fact consistent in a practical sense. For example, a gene regulatory network represents all *potential* effects among genes. These effects are mediated

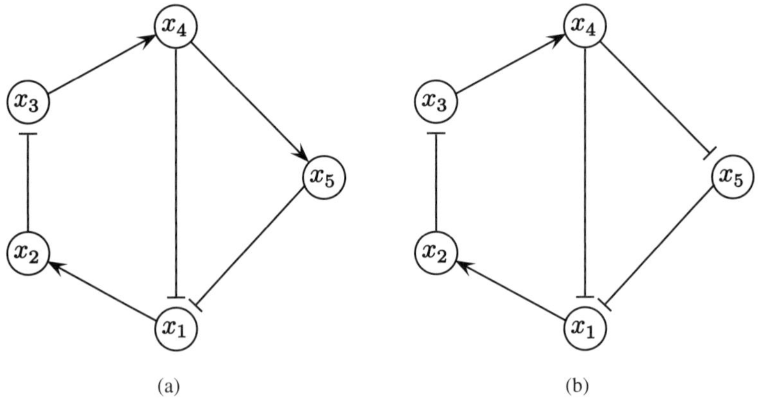

FIGURE 6.9 Two signed graphs. The graph in **(a)** is sign-consistent, but the graph in **(b)**, which differs in just one edge from **(a)**, is not sign-consistent since it has two paths in its undirected version with different parity between nodes x_1 x_4, namely a direct path of odd parity and a path of even parity transversing node x_5. Self-loops, which in biochemical systems often represent degradation terms, are ignored in the definition.

Problem name: sign consistency (SGN-CONST)

Instance: a directed graph $G = (V, E)$ with an arc labeling function $\mathcal{L}\colon E \mapsto \{-1, 1\}$.

Valid Solutions: a node labeling function $\ell\colon V \to \{-1, 1\}$.

Objective: maximize $|F_\ell|$, where $F_\ell = \{(u,v) \in E \mid \mathcal{L}(u,v) = \ell(u)\ell(v)\}$ is the set of consistent arcs for the node labeling function ℓ.

FIGURE 6.10 Definition of the sign consistency problem.

by proteins which themselves may need to be activated in order to perform their function, and this activation may, in turn, depend on certain extracellular ligands being present. Thus, depending on the particular combination of external signals present, different subgraphs of the original signed graph describe the system under those conditions, and these graphs may individually be consistent. For example, in Fig. 6.9b, the edges $x_4 \dashv x_1$ and $x_4 \dashv x_5$ may be present under completely different environmental conditions *A* and *B*. Thus, under either of the conditions *A* or *B*, the signed graph would be consistent, even though the entire signed graph is not consistent. Evidence that this is indeed the case is provided by reference 69, where the authors compare certain biological networks and appropriately randomized versions of them and show that the original networks are closer to being consistent. Thus, we are led to the computational problem of computing the smallest number of arcs that have to be removed so that there remains a consistent graph. We formalize this computational question as the *sign consistency* (SGN-CONST) problem (see Fig. 6.10) to determine how close to being monotone a system is [27,57]. For example, for the particular signed graph shown in Fig. 6.11, removal of just one arc (x_2, x_4) suffices (in this case, the solution is unique: no single other arc would suffice; for other graphs, there may be many minimal solutions).

We remind the reader that the above combinatorial characterization of monotonicity is via the absence of *undirected* closed chains of parity 1. Thus, in particular, any monotone system has **(i)** *no* negative feedback loops and **(ii)** *no* incoherent

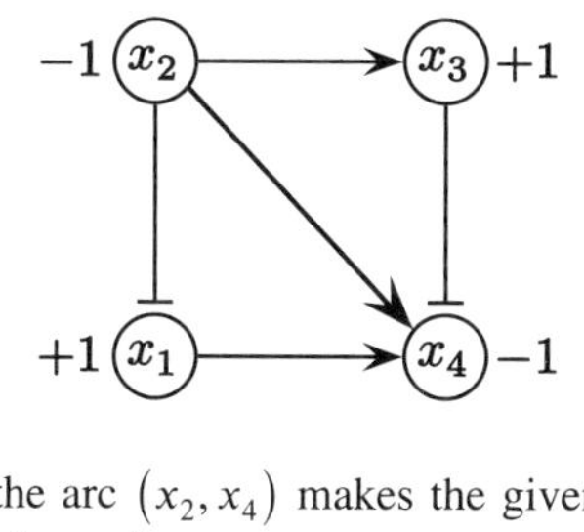

FIGURE 6.11 Deletion of the arc (x_2, x_4) makes the given signed graph consistent. The node labels are shown besides the nodes.

feed-forward-loops. However, some systems may not be monotone *even if* **(i)** *and* **(ii)** *hold*; the following example was shown in reference 1:

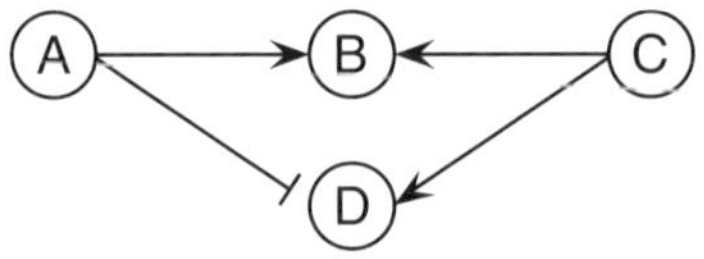

Based on the formulation of the SGN-CONST problem, Albert *et al.* [1] defined the *degree* of monotonicity of a signed graph to be

$$\mathsf{M} = \frac{|F_{\ell_{\mathrm{opt}}}|}{|E|}; \tag{6.10}$$

where $F_{\ell_{\mathrm{opt}}} \subseteq E$ is the set of consistent arcs in an optimal node labeling ℓ_{opt}, and the $|E|$ term in the denominator in Eq. (6.10) is a min–max normalization to ensure that $0 < \mathsf{M} \leq 1$. Note that the higher the value of M, is the more monotone the network.

An interesting interpretation of the SGN-CONST problem in statistical mechanics terms was discussed in reference 27. We briefly recount the interpretation here. Think of the arc labels as "interaction energies," and think of the node labels in the SGN-CONST problem as the (magnetic) "spin configurations." Note that an arc $\{u, v\} \in E$ is consistent in a node labeling ℓ provided $\mathcal{L}(u, v)\ell_u \ell_v = 1$. A graph with ± 1 arc labels is called an Ising spin-glass model in statistical physics. A "non-frustrated" spin-glass model is one for which there is a spin configuration for which *every* arc is consistent [11,24,36,58]. This is the same as a consistent graph in our previous discussion. Moreover, a spin configuration that maximizes the number of consistent edges is a "ground state," namely one for which the "free energy" with no exterior magnetic field $\left(- \sum_{(u,v)\in E} \mathcal{L}(u, v)\ell_u \ell_v \right)$ is minimized. Thus, solving the SGN-CONST problem amounts to finding ground states.

Incorporating Control Inputs. As we illustrated in Fig. 6.8, a useful approach to the analysis of biological systems consists of decomposing a given system into an interconnection of monotone subsystems. The formulation of the notion of interconnection requires subsystems to be endowed with "input and output channels" through which information is to be exchanged. In order to address this we may use the "controlled" dynamical systems defined by Eq. (5.1) in Section 5.1.2, namely

$$\dot{\mathbf{x}} = \mathbf{f}\left(\mathbf{x}, \mathbf{u}\right).$$

Specifying the time-dependency of the input vector $\mathbf{u}$ by a function $\mathbf{u}(t) \in \mathbb{R}^m$ for $t \geq 0$, it follows that each input defines a time-dependent dynamical system in the usual sense. We associate a *feedback function* $\mathfrak{h}: \mathbb{R}^n \mapsto \mathbb{R}^m$ with system (5.1) to create the closed-loop system $\dot{\mathbf{x}} = \mathbf{f}\left(\mathbf{x}, \mathfrak{h}(\mathbf{x})\right)$. Finally, if $\mathbf{x} \in \mathbb{R}^n$ and $\mathbf{u} \in \mathbb{R}^m$ are ordered by the orthant orders $\preceq_{(g_1,\ldots,g_n)}$ and $\preceq_{(q_1,\ldots,q_m)}$ respectively, then we say that the system is monotone if it satisfies Kamke's condition for every $\mathbf{u}$, and also

$$\forall\, k, j:\; q_k g_j \frac{\partial f_j}{\partial u_k} \geq 0.$$

See reference 9 for further discussions on this.

6.5.3 Algorithmic Issues in Computing the Degree of Monotonicity M

To calculate the degree of monotonicity M via Eq. (6.10), we obviously need to develop an efficient algorithm to compute an exact or approximate solution to the SGN-CONST problem. A special case of SGN-CONST , namely when $\mathcal{L}(u, v) = -1$ for every edge $(u, v) \in E$, is the well-known MAX-CUT problem (e.g. , see reference 99) which is NP-hard. Thus, we *cannot* hope for an efficient algorithm to find an *exact* solution for SGN-CONST when the size of the signed graph is *large*.

SGN-CONST can be posed as a special type of ''constraint satisfaction problem'' in the following manner. Let $\tau\colon \{-1, 1\} \mapsto \{0, 1\}$ be the linear transformation defined by $\tau(x) = \frac{1-x}{2}$. Then, $\tau(1) = 0$, $\tau(-1) = 1$ and the constraint $\mathcal{L}(u, v) = \ell(u)\ell(v)$ is equivalent to $\mathcal{L}(u, v) = \tau\big(\ell(u)\big) + \tau\big(\ell(v)\big) \pmod 2$. Thus, SGN-CONST can be posed as an optimization problem in which we have $|E|$ linear equations over GF(2) involving $|V|$ Boolean variables, with one equation per arc and each equation involving *exactly* two variables, and the goal is to assign truth values to the Boolean variables to satisfy a *maximum* number of equations. For algorithms and lower-bound results for general cases of these types of problems, such as when the equations are over GF(p) for an arbitrary prime p or when there are an arbitrary number of variables per equation or when the goal is to minimize the number of unsatisfied equations, see references such as 25 and 51. Below we outline an approximation algorithm for SGN-CONST used by DasGupta *et al.* [27] based on the semidefinite programming (SDP) technique used by Goemans and Williamson [44] for MAX-CUT; readers not very familiar with the SDP technique are also referred to an excellent treatment of this technique in the book by Vazirani [99].

The starting point in applying the SDP method is to observe that SGN-CONST can be written down as the following quadratic integer programming (QIP) problem:

$$\begin{array}{ll} \text{maximize} & \dfrac{1}{2}\left(\displaystyle\sum_{\substack{(u,v)\in E \\ \mathcal{L}(u,v)=-1}} \left(1 - \ell_u \ell_v\right) + \sum_{\substack{(u,v)\in E \\ \mathcal{L}(u,v)=1}} \left(1 + \ell_u \ell_v\right) \right) \\ \text{subject to} & \forall\, v \in V\colon\; \left(\ell_v\right)^2 = 1. \end{array} \tag{6.11}$$

Since this QIP formulation is also NP-hard, we cannot solve it directly. Thus, we ''relax'' this to a ''vector program'' (VP) by allowing each variable ℓ_v to be a real vector $\overrightarrow{\ell_v}$ over an n-dimensional unit-norm hypersphere in the following manner (where "·" denotes the vector inner product):

$$
\begin{array}{ll}
\text{maximize} & \dfrac{1}{2}\left(\displaystyle\sum_{\substack{(u,v)\in E\\ \mathcal{L}(u,v)=-1}} \left(1 - \overrightarrow{\ell_u}\cdot\overrightarrow{\ell_v}\right) + \sum_{\substack{(u,v)\in E\\ \mathcal{L}(u,v)=1}} \left(1 + \overrightarrow{\ell_u}\cdot\overrightarrow{\ell_v}\right)\right) \\
\text{subject to} & \forall\, v\in V:\ \overrightarrow{\ell_v}\cdot\overrightarrow{\ell_v} = 1. \\
 & \forall\, v\in V:\ \overrightarrow{\ell_v}\in\mathbb{R}^{|V|}
\end{array}
\tag{6.12}
$$

The above vector program can in fact be solved *exactly* in polynomial time. For this purpose we first formulate the following semi-definite programming (SDP) optimization problem; this SDP optimization problem can be solved exactly in polynomial time using any existing algorithm for SDP using ellipsoid, interior-point, or convex-programming methods [3,48,75,76,97], or by using any existing software for solving SDP optimization problems (e.g. , SDP optimization solving module in MATLAB):

$$
\begin{array}{l}
\text{maximize } \dfrac{1}{2}\left(\displaystyle\sum_{\substack{(u,v)\in E\\ \mathcal{L}(u,v)=-1}} \left(1 - y_{u,v}\right) + \sum_{\substack{(u,v)\in E\\ \mathcal{L}(u,v)=1}} \left(1 + y_{u,v}\right)\right) \\
\text{subject to: } \forall v\in V:\ y_{v,v} = 1 \\
\qquad\qquad Y = \left[y_{u,v}\right] \text{ is a symmetric positive semidefinite matrix.}
\end{array}
\tag{6.13}
$$

Since any solution matrix Y to (6.13) is positive semidefinite, such a solution matrix Y can be written as $Y = B^{\mathrm{T}}B$ for some real matrix $B\in\mathbb{R}^{|V|\times|V|}$. Such a decomposition of Y can be found by the well-known Cholesky decomposition algorithm [47] (MATLAB provides an implementation of this algorithm.) Once the matrix B is found, the solution vectors $\left\{\overrightarrow{\ell_v} \mid v\in V\right\}$ to (6.12) can be found by setting $\overrightarrow{\ell_v}$ to the column of B that consists of the entries $\left\{y_{u,v} \mid u\in V\right\}$.

What remains now is to compute mappings $P_v\colon \overrightarrow{\ell_v}\mapsto\{-1,1\}$ for every $v\in V$ such that the set of values $\left\{\ell_v = P_v\left(\overrightarrow{\ell_v}\right) \mid v\in V\right\}$ provide a solution of (6.11) of desired quality. Mappings P_v is computed by the following *randomized* algorithm.

(a) Select a vector $\vec{r} = \left(r_1, r_2, \ldots, r_{|V|}\right)$ over the $|V|$-dimensional unit-norm hyper-sphere *uniformly at random*. For example, this can be done by sampling r'_j, for each $1\le j\le|V|$, from a normal distribution of zero mean and unit standard deviation (i.e., $\Pr\left[a\le r'_j\le b\right] = \int_a^b (1/\sqrt{2\pi})\, e^{-x^2/2}\,\mathrm{d}x$ for any $\emptyset\subset\left[a,b\right]\subset\mathbb{R}$) and then normalizing the coordinates so that the 2-norm of the resulting vector is 1, that is, setting $r_j = \dfrac{r'_j}{\sqrt{\sum_{i=1}^{n}(r'_i)^2}}$ for $j = 1, 2, \ldots, |V|$.

(b) For each $v \in V$, set $\ell_v = \begin{cases} 1 & \text{if } \vec{r} \cdot \overrightarrow{\ell_v} \geq 0 \\ -1 & \text{otherwise} \end{cases}$ where "·" denotes the vector inner product.

Let ℓ_{opt} be an optimum node labeling function for SGN-CONST with $|F_{\ell_{\text{opt}}}|$ consistent edges. It can be shown that the above randomized algorithm provides a node labeling function ℓ_{approx} to SGN-CONST with $|F_{\ell_{\text{approx}}}|$ consistent edges such that

$$\mathbb{E}\left[|F_{\ell_{\text{approx}}}| \right] \geq \kappa \, |F_{\ell_{\text{opt}}}|,$$

where $\kappa = \min \left\{ \min_{0 \leq \theta \leq \pi} \frac{2\theta/\pi}{1 - \cos\theta}, \min_{0 \leq \theta \leq \pi} \frac{2 - (2\theta/\pi)}{1 + \cos\theta} \right\}$. It can be shown using elementary calculus that $\kappa > 0.87856$; thus on an average the approximate solution retains *at least* 87.85% of the number of consistent edges in an optimal solution. The above randomized approach to compute the mappings P_v can be made deterministic (i.e., can be "de-randomized") [70], but the derandomization procedure is complicated. Instead, one usually runs the randomized algorithm for computing the P_vs many times and accepts the best of these solutions; it can be shown that such an approach retains at least κ fraction of the optimal number of consistent edges with *very high probability* [84].

REFERENCES

1. R. Albert, B. DasGupta, A. Gitter, G. Gürsoy, R. Hegde, P. Pal, G. S. Sivanathan, and E. Sontag. A new computationally efficient measure of topological redundancy of biological and social networks. *Phys. Rev. E*, **84**(3):036117, 2011.
2. R. Albert and H. Othmer. The topology of the regulatory interactions predicts the expression pattern of the segment polarity genes in *Drosophila melanogaster*. *J. Theoret. Biol.*, **223**:1–18, 2003.
3. F. Alizadeh. Interior point methods in semidefinite programming with applications to combinatorial optimization. *SIAM J. Opti.*, **5**:13–51, 1995.
4. R. Alur, C. Belta, F. Ivanc, V. Kumar, M. Mintz, G.J. Pappas, H. Rubin, and J. Schlug. Hybrid modeling and simulation of biomolecular networks. In M. D. Di Benedetto and A. Sangiovanni-Vincentelli, editors. *Hybrid Systems: Computation and Control*, Lecture Notes in Computer Science, 2034, pp. 19–32, Springer Verlag, New York, 2001.
5. R. Alur, T. Henzinger, G. Lafferriere, and G. J. Pappas. Discrete abstractions of hybrid systems. *Proc. IEEE*, **88**(7):971–984, 2000.
6. D. Angeli, P. De Leenheer, and E. D. Sontag. A small-gain theorem for almost global convergence of monotone systems. *Syst. Control Lett.*, **51**:185–202, 2004.
7. D. Angeli, J. E. Ferrell, Jr., and E. D. Sontag. Detection of multi-stability, bifurcations, and hysteresis in a large class of biological positive-feedback systems, *Proc. Nat. Acad. Sci. USA*, **101**:1822–1827, 2004.
8. D. Angeli and E.D. Sontag. Monotone control systems. *IEEE Trans. Autom. Control*, **48**:1684–1698, 2003.

9. D. Angeli and E. D. Sontag. Multistability in monotone I/O systems, *Syst. Control Lett.*, **51**:185–202, 2004.
10. D. Angeli and E. D. Sontag. An analysis of a circadian model using the small-gain approach to monotone systems. In *IEEE Conference on Decision and Control*, pp. 575–578, 2004.
11. F. Barahona. On the computational complexity of Ising spin glass models. *J. Phys. A: Math. General*, **15**(10):3241–3253, 1982.
12. E. B. Baum and D. Haussler. What size net gives valid generalization? *Neural Comput.*, **1**:151–160, 1989.
13. C. Belta, P. Finin, L. C. G. J. M. Habets, A. Halasz, M. Imielinksi, V. Kumar and H. Rubin. Understanding the bacterial stringent response using reachability analysis of hybrid systems. In R. Alur and G. Pappas, editors, *Hybrid Systems: Computation and Control, Hybrid Systems: Computation and Control, Lecture Notes in Computer Science*, 2993, pp. 111–125, Springer Verlag, New York, 2004.
14. A. Bemporad, G. Ferrari-Trecate, and M. Morari. Observability and controllability of piecewise affine and hybrid systems. *IEEE Trans. Autom. Control*, **45**(10):1864–1876, 2000.
15. A. Bemporad and M. Morari. Control of systems integrating logic, dynamics, and constraints. *Automatica*, **35**:407–428, 1999.
16. D. P. Bertsekas and J. N. Tsitsiklis. *Parallel and Distributed Computation, Numerical Method*. Prentice Hall, Englewood Cliffs, NJ, 1989.
17. A. Blum and R. L. Rivest. Training a 3-node neural network is NP-complete. *Neural Networks*, **5**:117–127, 1992.
18. A. Blumer, A. Ehrenfeucht, D. Haussler, and M. Warmuth. Learnability and the Vapnik–Chervonenkis dimension. *J. ACM*, **36**:929–965, 1989.
19. J. W. Bodnar. Programming the drosophila embryo. *J. Theoret. Biol.*, **188**:391–445, 1997.
20. R. L. Casey, H. L. Jong, and J. L. L. Gouzé. Piecewise-linear models of genetic regulatory networks: Equilibria and their stability, *J. Math. Biol.*, **52**:27–56, 2006.
21. M. Chaves, R. Albert and E. D. Sontag. Robustness and fragility of boolean models for genetic regulatory networks, *J. Theoret. Biol.*, **235**:431–449, 2005.
22. M. Chaves, E.D. Sontag, and R. Albert. Methods of robustness analysis for boolean models of gene control networks, *IEE Proc. Syst. Biol.*, **153**(4):154–167, 2006.
23. O. Cinquin and J. Demongeot. Positive and negative feedback: Striking a balance between necessary antagonists, *J. Theoret. Biol.*, **216**:229–241, 2002.
24. B. A. Cipra. The Ising model is NP-complete, *SIAM News*, **33**(6):2000.
25. N. Creignou, S. Khanna and M. Sudan. Complexity classifications of Boolean constraint satisfaction problems. *SIGACT News*, **32**(4):24–33, 2001.
26. G. Cybenko. Approximation by superposition of a sigmoidal function, *Math. Control Signals Syst.*, **2**:303–314, 1989.
27. B. DasGupta, G. A. Enciso, E. Sontag, and Y. Zhang. Algorithmic and complexity results for decompositions of biological networks into monotone subsystems, *Biosystems*, **90**(1):161–178, 2007.
28. B. DasGupta and G. Schnitger. The power of approximating: A comparison of activation functions. In C. L. Giles. S. J. Hanson, and J. D. Cowan, editors, *Advances in*

Neural Information Processing Systems, 5, pp. 615–622. Morgan Kaufmann Publishers, Burlington, MA, 1993.

29. B. DasGupta and G. Schnitger. Analog versus discrete neural networks. *Neural Comput.*, **8**(4):805–818, 1996.
30. B. DasGupta, H. T. Siegelmann, and E. Sontag. On the intractability of loading neural networks. In V. P. Roychowdhury, K. Y. Siu and A. Orlitsky, editors, *Theoretical Advances in Neural Computation and Learning*, pp. 357–389. Kluwer Academic Publishers, Dordrecht, 1994.
31. B. DasGupta, H. T. Siegelmann, and E. Sontag. On the complexity of training neural networks with continuous activation functions. *IEEE Trans. Neural Networks*, **6**(6):1490–1504, 1995.
32. B. DasGupta and E. D. Sontag. A polynomial-time algorithm for checking equivalence under certain semiring congruences motivated by the state-space isomorphism problem for hybrid systems. *Theoreti. Comput. Sci.*, **262**:161–189, 2001.
33. D. L. DeAngelis, W. M. Post, and C. C. Travis. *Positive Feedback in Natural Systems.* Springer-Verlag, New York, 1986.
34. H. de Jong. Modeling and simulation of genetic regulatory systems: A literature review. *J. Comput. Biol.*, **9**(1):67–103, 2002.
35. P. De Leenheer, D. Angeli and E.D. Sontag. On predator-prey systems and small-gain theorems. *J. Math. Biosci. Engi.*, **2**:25–42, 2005.
36. C. De Simone, M. Diehl, M. Junger, P. Mutzel, G. Reinelt, and G. Rinaldi. Exact ground states of Ising spin glasses: New experimental results with a branch and cut algorithm. *J. Stat. Phys.*, **80**:487–496, 1995.
37. R. Edwards, H. T. Siegelmann, K. Aziza and L. Glass. Symbolic dynamics and computation in model gene networks. *Chaos*, **11**(1):160–169, 2001.
38. G. Enciso and E. Sontag. Global attractivity, I/O monotone small-gain theorems, and biological delay systems. *Discrete Contin. Dynam. Syst.*, **14**:549–578, 2006.
39. T. Gedeon and E. D. Sontag. Oscillation in multi-stable monotone system with slowly varying positive feedback. *J. Differential Equations*, **239**:273–295, 2007.
40. R. Ghosh and C. J. Tomlin. Symbolic reachable set computation of piecewise affine hybrid automata and its application to biological modeling: Delta-notch protein signaling. *IEE Proc. Syst. Biol.*, **1**:170–183, 2004.
41. R. Ghosh and C. J. Tomlin. Lateral inhibition through Delta-notch signaling: A piecewise affine hybrid model. In M. D. D. Benedetto and A. Sangiovanni-Vincentelli, editors, *Hybrid Systems: Computation and Control*, pp. 232–246. LNCS 2034, Springer Verlag, New York, 2001.
42. A. Ghysen and R. Thomas. The formation of sense organs in drosophila: A logical approach, *BioEssays*, **25**:802–807, 2003.
43. L. Glass and S. A. Kauffman. The logical analysis of continuous, non-linear biochemical control networks, *Journal of Theoretical Biology*, **39**(1):103–129, 1073.
44. M. Goemans and D. Williamson. Improved approximation algorithms for maximum cut and satisfiability problems using semidefinite programming. *J. ACM*, **42**(6):1115–1145, 1995.
45. P. Goldberg and M. Jerrum. Bounding the Vapnik–Chervonenkis dimension of concept classes parameterized by real numbers. *Machine Learning*, **18**:131–148, 1995.

46. M. Goldmann and J. Hastad. On the power of small-depth threshold circuits, *Comput. Complexity*, **1**(2):113–129, 1991.
47. G. H. Golub and C. F. Van Loan. *Matrix Computations*, 3rd edition. Johns Hopkins University Press, Baltimore, 1996.
48. M. Grötschel, L. Lovász and A. Schrijver. *Geometric Algorithms and Combinatorial Optimization*. Springer-Verlag, New York, 1988.
49. P. Gupta, R. Janardan, M. Smid, and B. DasGupta. The rectangle enclosure and point-dominance problems revisited. *Int. J. Comput. Geom. Appli.*, **7**(5):437–455, 1997.
50. V. V. Gursky, J. Reinitz, and A. M. Samsonov. How gap genes make their domains: An analytical study based on data driven approximations, *Chaos*, **11**:132–141, 2001.
51. J. Håstad and S. Venkatesh. On the advantage over a random assignment. *Random Struct. Algorithms*, **25**(2):117–149, 2004.
52. A. Hajnal, W. Maass, P. Pudlak, M. Szegedy and G. Turan. Threshold circuits of bounded depth, *J. Comput. Syst. Sci.*, **46**:129–154, 1993.
53. D. Haussler. Decision theoretic generalizations of the PAC model for neural nets and other learning applications, *Inform. Comput.*, **100**:78–150, 1992.
54. M. Hirsch. Systems of differential equations that are competitive or cooperative II: Convergence almost everywhere, *SIAM J. Math. Anal.*, **16**:423–439, 1985.
55. M. Hirsch. Differential equations and convergence almost everywhere in strongly monotone flows. *Contemp. Math.*, **17**:267–285, 1983.
56. J. E. Hopcroft, R. Motwani and J. D. Ullman. *Introduction to Automata Theory, Languages, and Computation*, 3rd edition. Addison Wesley; Boston, 2006.
57. F. Hüffner, N. Betzler and R. Niedermeier. Optimal edge deletions for signed graph balancing. In *6th Workshop on Experimental Algorithms*, LNCS 4525, pp. 297–310, Springer-Verlag, New York, 2007.
58. S. Istrail. Statistical mechanics, three-dimensionality and NP-completeness: I. Universality of intractability of the partition functions of the ising model across non-planar lattices, Thirty-Second Annual ACM Symposium on Theory of Computing, pp. 87–96, 2000.
59. M. Karpinski and A. Macintyre. Polynomial bounds for VC dimension of sigmoidal and general Pfaffian neural networks, *J. Comput. Syst. Sci.*, **54**:169–176, 1997.
60. S. Kauffman, C. Peterson, B. Samuelsson, and C. Troein. Random Boolean network models and the yeast transcriptional network. *Proc. Nat. Acad. Sci. USA*, **100**:14796–14799, 2003.
61. A. N. Kolmogorov. On the representation of continuous functions of several variables by superposition of continuous functions of one variable and addition. *Doklady Akad. Nauk (Proc. Russian Acad. Sci.)*, **114**:953–956, 1957.
62. K. M. Kyoda, M. Morohashi, S. Onami and H. Kitano. A gene network inference method from continuous-value gene expression data of wild-type and mutants. *Genome Inform.*, **11**:196–204, 2000.
63. J. Lewis, J. M. Slack and L. Wolpert. Thresholds in development. *J. Theoret. Biol.*, **65**:579–590, 1977.
64. G. Marnellos, G. A. Deblandre, E. Mjolsness, and C. Kintner. Delta-notch lateral inhibitory patterning in the emergence of ciliated cells in *Xenopus*: Experimental observations and a gene network model. *Pacific Symp. Biocomput.*, 326–337, 2000.

65. W. Maass. Perspectives of current research about the complexity of learning in neural nets. In V. P. Roychowdhury, K. Y. Siu, and A. Orlitsky, editors, *Theoretical Advances in Neural Computation and Learning*, pp. 295–336, Kluwer Acedemic Publishers, Dordrecht, 1994.

66. W. Maass. Bounds for the computational power and learning complexity of analog neural nets. *SIAM J. Comput.*, **26**(3):708–732, 1997.

67. W. Maass, G. Schnitger, and E. D. Sontag. On the computational power of sigmoid versus Boolean threshold circuits. In *Proceedings of the* 32nd *Annual IEEE Symposium on Foundations of Computer Science*, pp. 767–776, 1991.

68. W. Maass and E. D. Sontag. Analog neural nets with gaussian or other common noise distributions cannot recognize arbitrary regular languages. *Neural Comput.*, **11**(3):771–782, 1999.

69. A. Maayan, R. Iyengar and E.D. Sontag. Intracellular regulatory networks are close to monotone systems. *IET Systems Biol.*, **2**:103–112, 2008.

70. S. Mahajan and H. Ramesh. Derandomizing semidefinite programming based approximation algorithms. *SIAM J. Comput.*, **28**(5):1641–1663, 1999.

71. H. Meinhardt. Space-dependent cell determination under the control of morphogen gradient. *J. Theoret. Biol.*, **74**:307–321, 1978.

72. L. Mendoza, D. Thieffry, and E. R. Alvarez-Buylla. Genetic control of flower morphogenesis in *Arabidopsis thaliana*: A logical analysis. *Bioinformatics*, **15**:593–606, 1999.

73. M. Minsky and S. Papert. *Perceptrons*, The MIT Press, Cambridge, MA, 1988.

74. J. Monod and F. Jacob. General conclusions: Telenomic mechanisms in cellular metabolism, growth, and differentiation. *Cold Spring Harbor Symp. Quan. Biol.*, **26**:389–401, 1961.

75. Y. Nesterov and A. Nemirovskii. *Self-Concordant Functions and Polynomial Time Methods in Convex Programming*. Central Economic and Mathematical Institute, USSR Academy of Science, 1989.

76. Y. Nesterov and A. Nemirovskii. *Interior Point Polynomial Methods in Convex Programming*, Society of Industrial and Applied Mathematics, Philadelphia, 1994.

77. I. Parberry. A primer on the complexity theory of neural networks. In R. B. Banerji, editor, *Formal Techniques in Artificial Intelligence: A Sourcebook*, pp. 217–268, Elsevier Science Publishers B. V. (North-Holland), Amsterdam, 1990.

78. E. Plathe, T. Mestl, and S.W. Omholt. Feedback loops, stability and multistationarity in dynamical systems. *J. Biol. Syst.*, **3**:409–413, 1995.

79. C. V. Rao, D. M. Wolf, and A. P. Arkin. Control, exploitation and tolerance of intracellular noise. *Nature*, **420**:231–237, 2002.

80. J. H. Reif. On threshold circuits and polynomial computation. *SIAM J. Comput.*, **21**(5):896–908, 1992.

81. J. Reinitz and D. H. Sharp. Mechanism of eve stripe formation. *Mech. Dev.*, **49**(1–2):133–158, 1995.

82. E. Remy, B. Mosse, C. Chaouiya, and D. Thieffry. A description of dynamical graphs associated to elementary regulatory circuits. *Bioinformatics*, **19**(Suppl. 2):i172–i178, 2003.

83. D. E. Rumelhart and J. L. McClelland. *Parallel Distributed Processing: Explorations in the Microstructure of Cognition*, The MIT Press, Cambridge, MA, 1986.

84. L. Sánchez and D. Thieffry. A logical analysis of the drosophila gap-gene system. *J. Theoret. Biol.*, **211**:115–141, 2001.
85. H. T. Siegelmann. *Neural Networks and Analog Computation: Beyond the Turing Limit*, Birkhäuser Publishers, 1998.
86. H. T. Siegelmann and E. D. Sontag. Analog computation, neural networks, and circuits. *Theoret. Comput. Sci.*, **131**:331–360, 1994.
87. H. T. Siegelmann and E. D. Sontag. On the computational power of neural nets. *J. Comput. Syst. Sci.*, **50**:132–150, 1995.
88. H. L. Smith. Systems of ordinary differential equations which generate an order-preserving flow: A survey of results. *SIAM Rev.*, **30**:87–111, 1988.
89. H. L. Smith. *Monotone Dynamical Systems*, AMS, Providence, R.I., 1995.
90. E. H. Snoussi. Necessary conditions for multistationarity and stable periodicity. *J. Biol. Syst.*, **6**:3–9, 1998.
91. E. D. Sontag. Nonlinear regulation: The piecewise linear approach. *IEEE Trans. Autom. Control*, **26**(2):346–358, 1981.
92. E. D. Sontag. Remarks on piecewise-linear algebra. *Pacific J. Math.*, **98**:183–201, 1982.
93. E. D. Sontag. Molecular systems biology and control. *Eur. J. Control*, **11**:396–435, 2005.
94. E. D. Sontag. Some new directions in control theory inspired by systems biology. *Syst. Biol.*, **1**:9–18, 2004.
95. R. Thomas. Boolean formalization of genetic control circuits. *J. Theoret. Biol.*, **42**:563–585, 1973.
96. R. Thomas. Logical analysis of systems comprising feedback loops. *J. Theoret. Biol.*, **73**:631–656, 1978.
97. P. Vaidya. A new algorithm for minimizing convex functions over convex sets. *Math. Programming*, **73**(3):291–341, 1996.
98. V. N. Vapnik and A. Chervonenkis. *Theory of Pattern Recognition* (in Russian). Nauka, Moscow, 1974.
99. V. V. Vazirani. *Approximation Algorithms*. Springer-Verlag, Berlin, 2001.
100. G. von Dassow, E. Meir, E. M. Munro, and G. M. Odell. The segment polarity network is a robust developmental module. *Nature*, **406**:188–192, 2000.
101. S. M. Welch, J. L. Roe and Z. Dong. A genetic neural network model of flowering time control in *arabidopsis thaliana*. *Agronomy J.*, **95**:71–81, 2003.
102. C. H. Yuh, H. Bolouri, J. M. Bower, and E. H Davidson. A logical model of cis-regulatory control in a eukaryotic system. In J. M. Bower and H. Bolouri, editors, *Computational Modeling of Genetic and Biochemical Networks*, pp. 73–100, The MIT Press, Cambridge, MA, 2001.

EXERCISES

6.1 Convince yourself that delays may affect the dynamic behavior in a nontrivial manner by considering the following two discrete-time synchronous-update two-species interactions:

$x_1(t) = \frac{1}{4}\, x_1(t-1) + \frac{1}{4}\, x_2(t-2)$	$x_1(t) = \frac{1}{4}\, x_1(t-2) + \frac{1}{4}\, x_2(t-2)$
$x_2(t) = \frac{1}{4}\, x_2(t-1) + \frac{1}{4}\, x_1(t-2)$	$x_2(t) = \frac{1}{4}\, x_1(t-2) + \frac{1}{4}\, x_2(t-2)$

with $x_1(1) = x_2(1) = 0$ and $x_1(2) = x_2(2) = 1$, and observing the asymptotic behaviors of $x_1(t)$ and $x_2(t)$. Can you give an estimate of the asymptotic growth of the two variables (as a function of t) in the two systems?

6.2 Determine if the following system is controllable by constructing the controllability matrix:

$$\begin{bmatrix} x_1(t+1) \\ x_2(t+1) \end{bmatrix} = \begin{bmatrix} 2 & 5 \\ -1 & 7 \end{bmatrix} \begin{bmatrix} x_1(t) \\ x_2(t) \end{bmatrix} + \begin{bmatrix} 3 & 5 \\ 9 & -1 \end{bmatrix} \begin{bmatrix} u_1(t) \\ u_2(t) \end{bmatrix}.$$

6.3 Show that a Boolean network can be simulated by an ANN with threshold gate function by showing that each of the logical gates AND, OR, and NOT can be simulated by a node with threshold gate function by appropriately choosing the parameters of the threshold function.

6.4 Consider the network model shown in Fig. 6.4.

(**a**) Starting with the initial states $x_1(0) = x_2(0) = x_3(0) = x_4(0) = 1/2$, generate a time-series data by running the network for 4 time steps. Now, use the method of Jarrah et al. (discussed in Section 5.4.2.2) to reverse engineer a causal network and its corresponding Boolean counterpart.

(**b**) Repeat Exercise 6.4 by running the network for 10 time steps. Is there any improvement in the quality of the reconstructed network?

6.5 The goal of this exercise is to convince the reader that PL systems can simulate Boolean circuits.

(**a**) Consider the Boolean AND function on n Boolean inputs: $x_{n+1}^{+} = x_1 \wedge x_2 \wedge \ldots \wedge x_n$. Assume that each Boolean variable $x_i \in \{0, 1\}$ is obtained from a corresponding real-valued variable $0 \leq y_i \leq 1$ by the thresholding rule $y_i = \mathcal{H}\left(x_i - 1/2\right)$, where $\mathcal{H}$ is the threshold function described in Fig. 6.2. Write down a PL system of the form as shown in Eq. (6.6) that produced the output of the AND gate on $y_1, y_2, \ldots, y_n$.

(**b**) Repeat Exercise 6.5a for a Boolean OR gate: $x_{n+1}^{+} = x_1 \vee x_2 \vee \ldots \vee x_n$.

6.6 Show that NFAS and the PL systems (6.6) are subclasses of the PLHAS.

6.7 Prove that transformations of the form outlined in Eq. (6.5) can be used to simulate an arbitrary interconnection of linear systems and finite automata by a PL system.

6.8 Suppose that Definition 2 is used for the definition of observability of the system (6.2). Show that if an initial state $\mathbf{x}(0)$ is unobservable for the time duration $\left[0, n\right]$ then it is unobservable over any time duration.

6.9 Write a computer program in your favorite programming language to simulate the dynamics of the PLHA model for Delta-Notch signaling for one cell (thus, $x_{i,\text{Delta}} = 0$ for $i = 1, 2, \ldots, 6$). Test your program with various initial conditions.

6.10 Show that the system (6.9) is not monotone with respect to $\preceq_s$ for any s.

7

CASE STUDY OF BIOLOGICAL MODELS

In the preceding two chapters, we have seen the underlying principles behind synthesizing and analyzing several types of biological models. In this chapter, we discuss dynamical and topological properties of a few *specific* biological models. Our goal in this chapter is not to provide every possible details of these models, but rather to point out salient features of these models that have made them attractive in their applications.

7.1 SEGMENT POLARITY NETWORK MODELS

An important part of the development of the early *Drosophila* (fruit fly) embryo is the differentiation of cells into several stripes (''segments''), each of which eventually gives rise to an identifiable part of the body such as the *head*, the *wings* and the *abdomen*. Each segment then *differentiates* into a posterior and an anterior part, in which case the segment is said to be *polarized*; this differentiation process continues up to the point when all identifiable tissues of the fruit fly have developed. Differentiation at this level starts with differing concentrations of certain key proteins in the cells; these proteins form striped patterns by reacting with each other and by diffusion through the cell membranes (see Fig. 7.1 for an illustration). The genes involved in the process include *engrailed* (en), *wingless* (wg), *hedgehog* (hh), *patched* (ptc), *cubitus interruptus* (ci), and *sloppy paired* (slp), coding for the proteins (denoted by corresponding capitalized names) EN, WG, HH, PTC, CI, and SLP, respectively.

Models and Algorithms for Biomolecules and Molecular Networks, First Edition. Bhaskar DasGupta and Jie Liang.

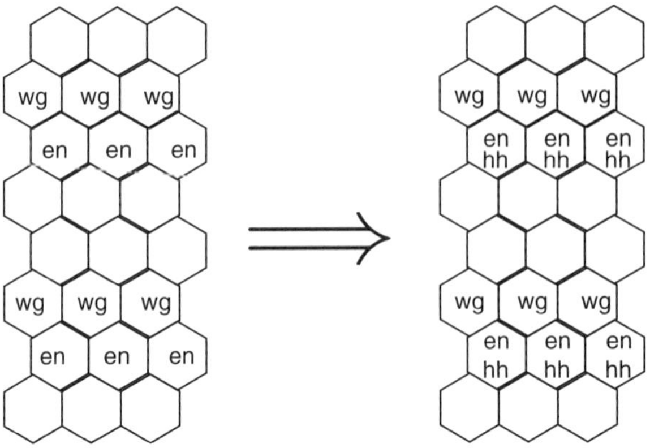

FIGURE 7.1 A schematic diagram of the early development of a *Drosophila* embryo. Each hexagon represents a cell, and neighboring cells interact to form a collective behavior. In this figure, an initial striped pattern of the genes en and wg induces the production of the gene hh, but only in those cells that are producing en.

Two additional proteins resulting from transformations of the protein CI also play important roles: CI may be converted into a *transcriptional activator* CIA, or it may be cleaved to form a *transcriptional repressor* CIR.

7.1.1 Boolean Network Model

Albert and Othmer [4] developed and analyzed a Boolean model based on the binary ON/OFF representation of mRNA and protein levels of five segment polarity genes. This model was constructed based on the known topology and was validated using published gene and expression data. The expressions of the segment polarity genes occur in stripes encircling the embryo. The key features of these patterns can be represented in one dimension by a line of 12 interconnected cells, grouped into 3 parasegment primordia, in which the genes are expressed in every fourth cell. In Albert and Othmer's model, parasegments are assumed to be identical, and thus only one parasegment of four cells is considered. Therefore, in their model the Boolean variables are the expression levels of the segment polarity genes and proteins in each of the four cells. For further details, the reader is referred to reference 4.

7.1.2 Signal Transduction Network Model

In the gene regulatory network model for segment polarity, the interactions incorporated include translation (protein production from mRNA), transcriptional regulation, and protein–protein interactions. Two of the interactions are *intercellular*; that is, the proteins WG and HH may leave the cell in which they are produced and interact with receptor proteins in the membranes of *neighboring* cells. The network for a single

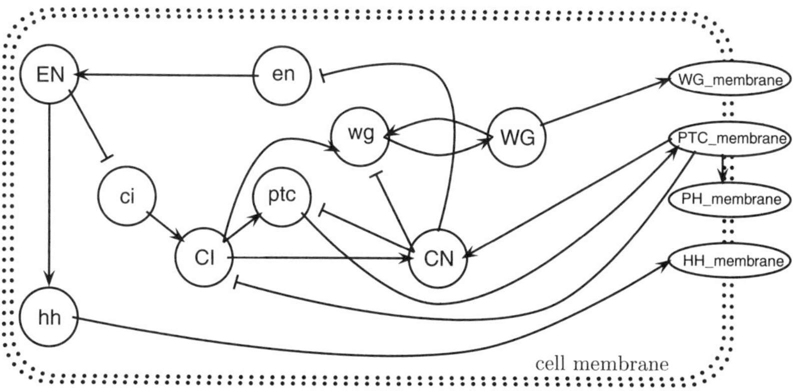

FIGURE 7.2 The *Drosophila* segment polarity regulatory network for one cell with the interpretation of the regulatory role of PTC on the reaction CI → CN as PTC → CN and PTC ⊣ CI [8].

cell was first published in reference 8 and later appeared in slightly modified form in references 4 and 12. Figure 7.2 shows the network.

As illustrated in papers such as references 2 and 7, one can build a one-dimensional *multicellular* version of this network by considering a row of *n* cells, each of which has *separate* variables for each of the compounds, and letting the cell-to-cell interactions be as shown in Fig. 7.3 using cyclic boundary conditions. One can calculate the degrees of monotonicity and redundancy for this network as follows:

- Albert et al. [2] show that, for $n > 1$, $\mathfrak{R}_{\mathcal{T}_{\mathrm{R}}} > 3/11$.
- DasGupta *et al.* [7] show that the one-dimensional network follows a monotone dynamics if we remove 3 edges from the network for every cell (and, similarly, remove $3n$ edges from the n-node network).

7.2 ABA-INDUCED STOMATAL CLOSURE NETWORK

Microscopic stomatal pores of plants have a surrounding pair of *guard cells*. It is known that plants take in carbon dioxide as well as lose water through these pores, and this process is controlled by the surrounding guard cells [11]. During drought, the plant hormone *abscisic acid* (ABA), inhibits stomatal opening and promotes stomatal closure, thereby promoting water conservation.

Dozens of cellular components have been identified to function in ABA regulation of guard cell volume and thus of stomatal aperture [6,10,16]. Based on these and other known interactions, Li *et al.* [15, Table S1] compiled a list of about 140 direct interactions and double-causal inferences, both of type "A promotes B" and "C promotes process (A promotes B)," and *manually* synthesized a network of 54

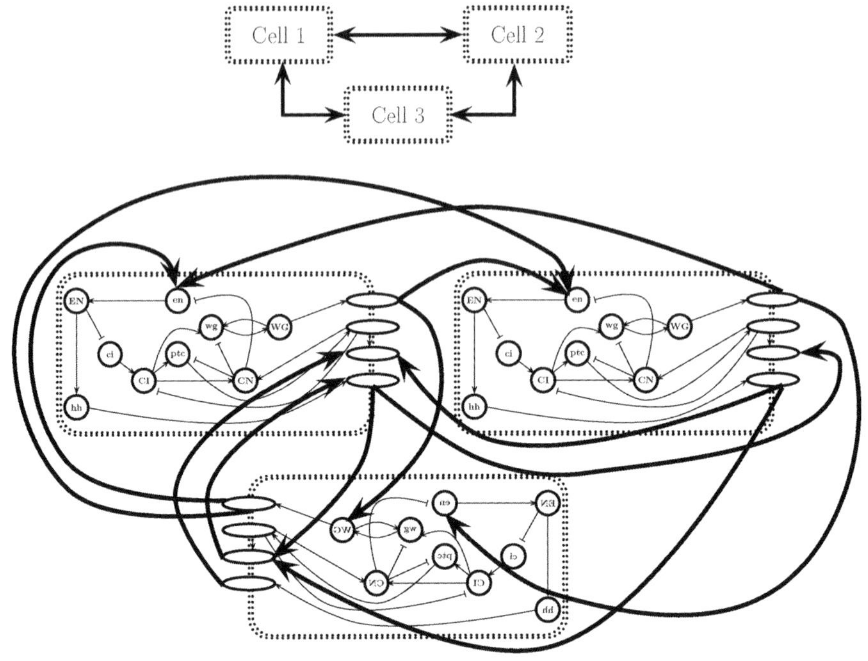

FIGURE 7.3 A one-dimensional three-cell *Drosophila* segment polarity regulatory network with cyclic boundary conditions.

nodes and 92 (activation or inhibitory) arcs that accurately portrays the dynamics of the regulation (see Fig. 7.4). Because the interaction data contain both direct and double-causal interactions, with additional information about interactions that are known to be direct with absolute certainty (''mandatory'' edges in the terminology in Chapter 5.3.1), Albert *et al.* [1,3] were able to exploit the algorithmic framework shown in Fig. 5.4 and use this framework in their NET-SYNTHESIS software [14] to automatically generate a slightly more minimal network of 57 nodes and 84 edges.

7.3 EPIDERMAL GROWTH FACTOR RECEPTOR SIGNALING NETWORK

Epidermal Growth Factor (EGF) is a protein that is frequently stored in skin and other epithelial tissues, and it is released when rapid cell division is needed (e.g., after an injury). The function of EGF is to bind to a receptor, the *Epidermal Growth Factor Receptor* (EGFR), on the membrane of the cells. The EGFR, on the inner side of the membrane, has the appearance of a scaffold with dozens of docks to bind with numerous agents, and it starts a large number of reactions at the cell level that ultimately induces cell division. In 2005, Oda *et al.* [17] integrated the information about this process from multiple sources to define a network with 330 known molecules

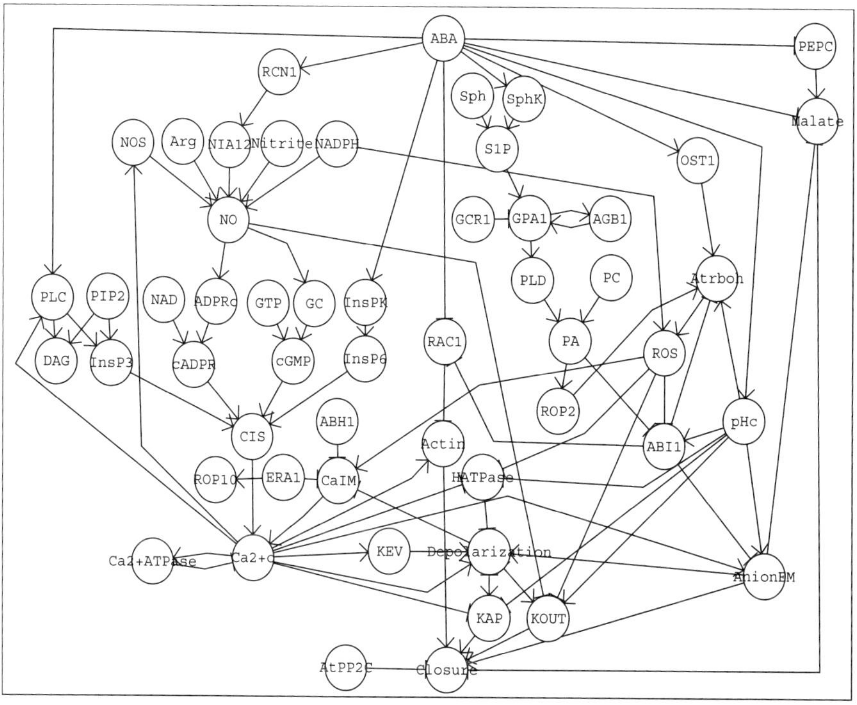

FIGURE 7.4 A pictorial representation of the network manually synthesized by Li *et al.* [15]. Nodes in the network correspond to proteins, genes, and other small molecules (e.g., RAC1 is a small GTPase protein).

under 211 chemical reactions. Each reaction in the network classifies the molecules as reactants, products, and/or modifiers (enzymes). The network was made available in the supplementary material of [17] in SBML[1] format.

For the purpose of analyzing the monotonicity of dynamics of the above biochemical network, DasGupta *et al.* [7] imported the information about the classification of the molecules as reactants, products or enzymes to MATLAB using the Systems Biology Toolbox, and defined a network G in the following manner:

(a) There are 330 nodes $v_1, \ldots, v_{330}$ corresponding to the 330 molecules $m_1, \ldots, m_{330}$.

(b) The network has 855 arcs which were added in the following manner:

(b1) If there exists a reaction in which m_j is a product and m_i is either a reactant or a modifier, then we add the arc $v_i \to v_j$ to the network.

[1] Systems Biology Markup Language, see http://www.sbml.org.

(ℓ2) If there exists a reaction in which m_j is a reactant and m_i is either a reactant or a modifier, then we add the arc $v_i \dashv v_j$ to the network.

(ℓ3) If both the arcs $v_i \rightarrow v_j$ and $v_i \dashv v_j$ exist by (ℓ1) and (ℓ2) for any pair of nodes v_i and v_j, then we remove both the arcs $v_i \rightarrow v_j$ and $v_i \dashv v_j$ from the network (in the terminology of reference 7, both the arcs are removed and an arc from v_i to v_j is added with the label ‘‘undefined’’ (NaN)). There are exactly 7 such pairs of nodes in the network.

(ℓ4) In a few reactions there was a modifier or a reactant involved which had an *inhibitory* effect in the reaction, but the effect of this compound on the remaining participants of the reaction was the opposite from that described above. In such cases, the network was corrected manually by looking at the *annotations* given for each reaction.

G as constructed above has more arcs than the digraph displayed in reference 17. The reason for this is as follows: If molecules m_i and m_j are both reactants in the same reaction, then the presence of m_i will have an *indirect* inhibiting effect on the concentration of m_j, since it would accelerate the reaction which consumes m_j (assuming m_j is not also a product). Therefore an inhibitory arc must also appear from m_i to m_j, and vice versa. Similarly, modifiers have an inhibiting effect on reactants.

DasGupta *et al.* [7] applied the algorithmic procedures discussed in Section 6.5.3 and found that the network can be made monotone by deleting 219 arcs, and thus the entire network can be decomposed as the feedback loop of a controlled monotone system using 219 inputs. To check whether removing significantly fewer than 219 arcs may also provide a monotone network, DasGupta *et al.* [7] suggested the following two *heuristic* approaches:

- For a suitable positive integer k, consider the k nodes with the highest out-degrees (‘‘hubs’’), and eliminate all the outgoing arcs associated to these hubs from the reaction network G to form a new network G'. Then, use the algorithm for the combinatorial characterization via the sign-consistency (SGN-CONST) problem as discussed in Section 6.5.2 on G' to find a node labeling function ℓ_V of the nodes and a set of m arcs that can be removed to eliminate all remaining undirected closed chains of parity -1. Finally, put back to the reduced network among those arcs that were taken out the ones which are consistent with respect to the node labels induced by ℓ_V.
- Make suitable changes of variables in the original system using the mass conservation laws. Such changes of variables are discussed in many places; for example, see references 5 and 19. In terms of the associated signed graph, the result of the change of variables is often the elimination of one of the cycles. The simplest target for a suitable change of variables is a set of three nodes that form part of the same chemical reaction—for instance, two reactants and one product, or one reactant, one product, and one modifier. It is easy to see that such nodes are connected in the associated signed graph by a cycle of three arcs. To the

extent to which most of these triangles can be eliminated by suitable changes of variables, this can yield a much lower number of arcs to be removed.

7.4 *C. ELEGANS* METABOLIC NETWORK

This network was constructed and studied by Jeong *et al.* [13] and was also subsequently investigated by other researchers such as references 2 and 9.

Data sources for the *C. elegans* metabolic network include two types of nodes, namely the *metabolites* and *reaction* nodes, and the arcs are directed either from those metabolites that are the reactants of a reaction to the reaction node or from the reaction node to the products of the reaction. The network constructed in reference 13 had 651 nodes and about 2040 arcs (after removal of duplicate arcs). Thus, the network is dense with an average degree of 3.13. Albert *et al.* [2] found by empirically evaluation that for this network the degree of redundancy $\mathfrak{R}_{T_R}$ is 0.669 and the degree of monotonicity M is 0.444.

The degree of redundancy of the metabolic network is surprisingly high among similar biological networks. A biological basis for this could be due to the existence of *currency metabolites*. Currency metabolites (also called *carrier* or *current* metabolites) are abundant in normal cells and occur in widely different exchange processes. For example, ATP can be seen as the energy currency of the cell. Because of their wide participation in diverse reactions, currency metabolites tend to be the highest degree nodes of metabolic networks. For the metabolic network, redundant arcs appear if both (one of) the reactant(s) and (one of) the product(s) of a reaction appear as reactants of a different reaction, or conversely, both (one of) the reactant(s) and (one of) the product(s) of a reaction appear as products of a different reaction. Thus, intuitively, metabolites that participate in a large number of reactions will have a higher chance to be the reactant or product of redundant arcs. Based on their empirical analysis, Albert *et al.* [2] concluded that the high redundancy of the *C. elegans* metabolic network is in fact *mostly* due to inclusion of currency metabolites.

7.5 NETWORK FOR T-CELL SURVIVAL AND DEATH IN LARGE GRANULAR LYMPHOCYTE LEUKEMIA

Large granular lymphocytes (LGLs) are medium to large size cells with eccentric nuclei and abundant cytoplasm. In normal adults, LGLs comprise about 10% to 15% of the total peripheral blood mononuclear cells. LGLs can be further divided into two major lineages: CD3-natural-killer (NK) cell lineage, which comprises 85% of LGL cells and mediates non-major histocompatibility complex-restricted cytotoxicity, and CD3+ lineage, which comprises 15% of LGLs and represents activated cytotoxic T cells. LGL leukemia is a type of disordered clonal expansion of LGL and their invasions in the marrow, spleen, and liver. LGL leukemia was further divided into T-cell LGL leukemia and NK-cell LGL leukemia. Ras is a small GTPase which is essential for controlling multiple essential signaling pathways, and its deregulation is frequently seen in human cancers. Activation of H-Ras requires its farnesylation, which can be blocked by farnesyltransferase inhibitiors (FTIs). This envisions FTIs as future drug target for anti-cancer therapies. One such FTI is tipifarnib, which shows apoptosis induction effect to leukemic LGL *in vitro*. This observation, together with the finding that Ras is constitutively activated in leukemic LGL cells, leads to the hypothesis that Ras plays an important role in LGL leukemia and may function through influencing Fas/FasL pathway.

Kachalo *et al.* [14] synthesized a cell-survival/cell-death regulation-related signaling network for T-LGL leukemia from the TRANSPATH 6.0 database with additional information manually curated from literature search. The 359 nodes of this network represented proteins/protein families and mRNAs participating in pro-survival and Fas-induced apoptosis pathways, and the 1295 arcs in the network represented regulatory relationships between these nodes, including protein interactions, catalytic reactions, transcriptional regulation (for a total of 766 direct interactions), and known indirect causal regulation. The approach used by Kachalo *et al.* [14] was to focus special interest on the effect of Ras on apoptosis response through Fas/FasL pathway by designating nodes that correspond to proteins with no evidence of being changed during this effect as pseudo-nodes and simplifying the network via iterations of the PNC and TR operations described in Sections 5.3.3.1 and 5.3.3.2 to simplify the network to contain 267 nodes and 751 arcs.

Saadatpour *et al.* [18] further investigated the T-LGL network and found that 14 nodes of the network have high importance in the sense that blocking any of these nodes disrupts (almost) all signaling paths from the complementary node to apoptosis, thus providing these nodes as possible candidate therapeutic targets. All of these nodes are also found to be essential for the T-LGL survival state according to a dynamic model; that is, reversing their states causes apoptosis to be the *only possible* outcome of the system. Moreover, experimental verification of the importance of these nodes exists for 10 of the 14 nodes [18].

REFERENCES

1. R. Albert, B. DasGupta, R. Dondi, S. Kachalo, E. Sontag, A. Zelikovsky, and K. Westbrooks. A novel method for signal transduction network inference from indirect experimental evidence. *J. Comput. Biol.*, **14**(7):927–949, 2007.
2. R. Albert, B. DasGupta, A. Gitter, G. Gürsoy, R. Hegde, P. Pal, G. S. Sivanathan, and E. Sontag. A new computationally efficient measure of topological redundancy of biological and social networks. *Phys. Rev. E*, **84**(3):036117, 2011.
3. R. Albert, B. DasGupta and E. Sontag. Inference of signal transduction networks from double causal evidence. In D. Fenyo, editor, *Methods in Molecular Biology: Topics in Computational Biology*, p. 673, Chapter 16. Springer, New York, 2010.
4. R. Albert and H. Othmer. The topology of the regulatory interactions predicts the expression pattern of the segment polarity genes in *Drosophila melanogaster*. *J. Theoret. Biol.*, **223**:1–18, 2003.

5. D. Angeli and E.D. Sontag. Monotone control systems, *IEEE Trans. Autom. Control*, **48**:1684–1698, 2003.
6. M. R. Blatt and A. Grabov. Signal redundancy, gates and integration in the control of ion channels for stomatal movement. *J. Exp. Botany*, **48**:529–537, 1997.
7. B. DasGupta, G. A. Enciso, E. Sontag, and Y. Zhang. Algorithmic and complexity results for decompositions of biological networks into monotone subsystems. *Biosyst.*, **90**(1):161–178, 2007.
8. G. von Dassow, E. Meir, E. M. Munro, and G. M. Odell. The segment polarity network is a robust developmental module. *Nature*, **406**:188–192, 2000.
9. J. Duch and A. Arenas. Community identification using extremal optimization. *Phys. Rev. E*, **72**:027104, 2005.
10. L. M. Fan, Z. Zhao and S. M. Assmann. Guard cells: A dynamic signaling model. *Curr. Opin. Plant Biol.*, **7**:537–546, 2004.
11. A. M. Hetherington and F. I. Woodward. The role of stomata in sensing and driving environmental change. *Nature*, **424**:901–908, 2003.
12. N. T. Ingolia. Topology and robustness in the *Drosophila* segment polarity network. *PLoS Biol.*, **2**(6):e123, 2004.
13. H. Jeong, B. Tombor, R. Albert, Z. N. Oltvai, and A.-L. Barabasi. The large-scale organization of metabolic networks. *Nature*, **407**:651–654, 2000.
14. S. Kachalo, R. Zhang, E. Sontag, R. Albert, and B. DasGupta. NET-SYNTHESIS: A software for synthesis, inference and simplification of signal transduction networks. *Bioinformatics*, **24**(2):293–295, 2008.
15. S. Li, S. M. Assmann and R. Albert. Predicting essential components of signal transduction networks: A dynamic model of guard cell abscisic acid signaling. *PLoS Biol.*, **4**(10):e312, 2006.
16. E. A. MacRobbie. Signal transduction and ion channels in guard cells. *Philos. Trans. Royal Soc. B: Biol. Sci.*, **353**:1475–1488, 1998.
17. K. Oda, Y. Matsuoka, A. Funahashi, and H. Kitano. A comprehensive pathway map of epidermal growth factor receptor signaling. *Mol. Syst. Biol.*, **1**(1):1–17, 2005.
18. A. Saadatpour, R. S. Wang, A. Liao, X. Liu, T. P. Loughran, I. Albert and R. Albert R. Dynamical and structural analysis of a T cell survival network identifies novel candidate therapeutic targets for large granular lymphocyte leukemia. *PLoS Comput. Biol.*, **7**:e1002267, 2011.
19. A. I. Volpert, V. A. Volpert, and V. A. Volpert. *Traveling Wave Solutions of Parabolic Systems,* Vol. 140 of Translations of Mathematical Monographs, American Mathematical Society, Providence, RT, 2000.

EXERCISES

7.1 The purpose of this exercise is to get the reader familiar with the algorithmic framework shown in Fig. 5.4 and the NET-SYNTHESIS software in reference 14 that uses this algorithmic framework. Consider the following small subset of the interactions reported in reference 15:

ABA ⊣ NO
ABA → PLD
ABA → GPA1
ABA → PLC
GPA1 → PLD
PLD → PA
NO ⊣ KOUT
KOUT → Closure
PA → Closure
PLC → (ABA → KOUT)

For each of the following tasks, report the network generated and verify that it is correct.

(**a**) Generate the network using only the direct interactions and perform transitive reduction on the graph (e.g., in NET-SYNTHESIS software, select ''Reduction (slower)'' from the Action menu).

(**b**) Add double-causal inferences to the network (e.g., in NET-SYNTHESIS software, select ''Add pseudonodes'' from the Action menu).

(**c**) Perform pseudo-node collapse (e.g., in NET-SYNTHESIS software, select ''Collapse pseudonodes'' from the Action menu).

(**d**) Perform a follow-up round of binary transitive reduction and pseudo-node collapse until the graph cannot be reduced further.

7.2 Collect five biological networks from existing bioinformatics literature. For each network, do the following.

(**a**) Explain in detail the biological process that is modeled by the network.

(**b**) Investigate topological and dynamical properties such as

- degree distribution and connectivity,
- degree of redundancy $\mathfrak{R}_{T_R}$,
- degree of monotonicity M.

7.3 Show that $\mathfrak{R}_{T_R} > 3/11$ for the one-dimensional n-cell *Drosophila* segment polarity regulatory network with cyclic boundary conditions.

GLOSSARY

GF(2) Galois field of two elements 0 and 1. The addition rule for this field is $c = a + b \pmod 2$ and the multiplication rule is $c = a\,b$.

$\mathbb{R}$ Set of real numbers.

$\mathbb{R}^n$ n-dimensional space whose each component is a real number.

e Base of natural logarithm.

$g_1 \circ g_2$ The composition of two functions g_1 and g_2, *i.e.*, $g_1 \circ g_2(x) = g_1\big(g_2(x)\big)$.

Apoptosis Programmed cell death.

Chemical reactions, reactants and products

A chemical reaction is a chemical transformation in which a set of substances (called reactants) together produce another set of substances (called products).

Currency metabolites

Currency metabolites (sometimes also referred to as *carrier* or *current* metabolites) are ubiquitous substrates having a high turnover and occurring in widely different exchange processes. For example, ATP can be seen as the energy currency of the cell. There is some discussion in the literature on how large the group of currency metabolites is, but the consensus list includes H_2O, ATP, ADP, and NAD and its variants, NH_4^+, and $PO4^{3-}$ (phosphate).

Equivalence relation
A relation R on a set X is a set of ordered pairs of elements of X. R is an equivalence relation if it is reflexive (*i.e.*, $(x, x) \in R$ for every $x \in X$), symmetric (*i.e.*, $(x, y) \in R$ implies $(y.x) \in R$) and transitive (*i.e.*, $(x, y) \in R$ and $(y, z) \in R$ imply $(x, z) \in R$). Any equivalence relation produced a partition of X where elements in the same partition are mutually related to each other.

Hyper-rectangle
A generalization of two-dimensional rectangles to more than two dimensions. A n-dimensional hyper-rectangle is a Cartesian product of n intervals.

Occam's razor
The general principle that recommends selection among competing hypotheses the one that makes the fewest new assumptions.

Pocket
A pocket on a protein is a concave unfilled region connected to the outside with a constriction, namely, it has an opening that is narrower than one interior cross section.

Power set of a set S
The set of all subsets of S.

Rational number
A number of the form p/q where p and q are integers and $q \neq 0$.

T-LGL
T-cell large granular lymphocyte leukemia represents a spectrum of lymphoproliferative diseases.

Void
A void inside a protein is a connected interior empty space fully buried from the outside of the protein.

INDEX

Models and Algorithms for Biomolecules and Molecular Networks, First Edition. Bhaskar DasGupta and Jie Liang.

IEEE Press Series in Biomedical Engineering

The focus of our series is to introduce current and emerging technologies to biomedical and electrical engineering practitioners, researchers, and students. This series seeks to foster interdisciplinary biomedical engineering education to satisfy the needs of the industrial and academic areas. This requires an innovative approach that overcomes the difficulties associated with the traditional textbooks and edited collections.

Series Editor: Metin Akay, University of Houston, Houston, Texas

1. *Time Frequency and Wavelets in Biomedical Signal Processing*
Metin Akay

2. *Neural Networks and Artificial Intelligence for Biomedical Engineering*
Donna L. Hudson, Maurice E. Cohen

3. *Physiological Control Systems: Analysis, Simulation, and Estimation*
Michael C. K. Khoo

4. *Principles of Magnetic Resonance Imaging: A Signal Processing Perspective*
Zhi-Pei Liang, Paul C. Lauterbur

5. *Nonlinear Biomedical Signal Processing, Volume 1, Fuzzy Logic, Neural Networks, and New Algorithms*
Metin Akay

6. *Fuzzy Control and Modeling: Analytical Foundations and Applications*
Hao Ying

7. *Nonlinear Biomedical Signal Processing, Volume 2, Dynamic Analysis and Modeling*
Metin Akay

8. *Biomedical Signal Analysis: A Case-Study Approach*
Rangaraj M. Rangayyan

9. *System Theory and Practical Applications of Biomedical Signals*
Gail D. Baura

10. *Introduction to Biomedical Imaging*
Andrew G. Webb

11. *Medical Image Analysis*
Atam P. Dhawan

12. *Identification of Nonlinear Physiological Systems*
David T. Westwick, Robert E. Kearney

13. *Electromyography: Physiology, Engineering, and Non-Invasive Applications*
Roberto Merletti, Philip Parker

14. *Nonlinear Dynamic Modeling of Physiological Systems*
Vasilis Z. Marmarelis

15. *Genomics and Proteomics Engineering in Medicine and Biology*
Metin Akay

16. *Handbook of Neural Engineering*
Edited by Metin Akay

17. *Medical Image Analysis, Second Edition*
Atam P. Dhawan

18. *Advanced Methods of Biomedical Signal Processing*
Edited by Sergio Cerutti, Carlo Marchesi

19. *Epistemology of the Cell: A Systems Perspective on Biological Knowledge*
Edward R. Dougherty, Michael L. Bittner

20. *Micro and Nanotechnologies for Engineering Stem Cells and Tissues*
Murugan Ramalingam, Esmaiel Jabbari, Seeram Ramakrishna, Ali Khademhosseini

21. *Introduction to Neural Engineering for Motor Rehabilitation*
Dario Farina, Winnie Jensen, Metin Akay

22. *Introduction to Tissue Engineering: Applications and Challenges*
Ravi Birla

23. *Handbook of Biomedical Telemetry*
Edited by Konstantina S. Nikita

24. *Biomedical Signal Analysis, Second Edition*
Rangaraj M. Rangayyan

25. *Models and Algorithms for Biomolecules and Molecular Networks*
Bhaskar DasGupta and Jie Liang